# Advances in Sustainability Science and Technology

The book series aims at bringing together valuable and novel scientific contributions that address the critical issues of renewable energy, sustainable building, sustainable manufacturing, and other sustainability science and technology topics that have an impact in this diverse and fast-changing research community in academia and industry.

The areas to be covered are

- Climate change and mitigation, atmospheric carbon reduction, global warming
- Sustainability science, sustainability technologies
- Sustainable building technologies
- Intelligent buildings
- Sustainable energy generation
- Combined heat and power and district heating systems
- Control and optimization of renewable energy systems
- Smart grids and micro grids, local energy markets
- Smart cities, smart buildings, smart districts, smart countryside
- Energy and environmental assessment in buildings and cities
- Sustainable design, innovation and services
- Sustainable manufacturing processes and technology
- Sustainable manufacturing systems and enterprises
- Decision support for sustainability
- Micro/nanomachining, microelectromechanical machines (MEMS)
- Sustainable transport, smart vehicles and smart roads
- Information technology and artificial intelligence applied to sustainability
- Big data and data analytics applied to sustainability
- Sustainable food production, sustainable horticulture and agriculture
- Sustainability of air, water and other natural resources
- Sustainability policy, shaping the future, the triple bottom line, the circular economy

High quality content is an essential feature for all book proposals accepted for the series. It is expected that editors of all accepted volumes will ensure that contributions are subjected to an appropriate level of reviewing process and adhere to KES quality principles.

The series will include monographs, edited volumes, and selected proceedings.

More information about this series at https://link.springer.com/bookseries/16477

Srikanta Patnaik · Roumen Kountchev · Vipul Jain
Editors

# Smart and Sustainable Technologies: Rural and Tribal Development Using IoT and Cloud Computing

Proceedings of ICSST 2021

Springer

*Editors*
Srikanta Patnaik
Department of Computer Science
and Engineering
SOA University
Bhubaneswar, Odisha, India

Roumen Kountchev
Department of Radio Communications
and Video Technology
Technical University of Sofia
Sofia, Bulgaria

Vipul Jain
School of Management
Victoria University of Wellington
Wellington, Wellington, New Zealand

ISSN 2662-6829 ISSN 2662-6837 (electronic)
Advances in Sustainability Science and Technology
ISBN 978-981-19-2276-3 ISBN 978-981-19-2277-0 (eBook)
https://doi.org/10.1007/978-981-19-2277-0



This Springer imprint is published by the registered company Springer Nature Singapore Pte Ltd.
The registered company address is: 152 Beach Road, #21-01/04 Gateway East, Singapore 189721, Singapore

# Committee Members

## Chief Patron

Dr. Satya Prakash Panda, President, GIET University
Dr. Chandra Dhwaj Panda, Vice President, GIET University
Dr. Jagadish Panda, Director-General, GIET University

## Patron

Dr. Goutam Ghosh, Vice Chancellor, GIET University
Dr. N. V. Jagannadha Rao, Registrar, GIET University

## Honorary Chair

Lakhmi C. Jain, University of Technology Sydney, Australia

## Finance Chair

Dr. Ajit Patro, GIET University

## General Chair

Dr. Srikanta Patnaik, SOA University, Bhubaneswar, Odisha
Dr. Roumen Kountchev, Technical University of Sofia, Bulgaria

## International Advisory Committee

Dr. Ing. Mario José Diván, National University of La Pampa, Argentina
Dr. Saraju P. Mohanty, University of North Texas, Denton, USA
Dr. Parthasarathi Roop, The University of Auckland, Auckland, New Zealand
Dr. Sanjeevikumar Padmanaban, Aalborg University, Esbjerg, Denmark
Dr. Yara C. Almanza Arjona, ICAT, National Autonomous University of Mexico, Mexico
Dr. Kingsley Arinze Okoye, Tecnologico de Monterrey, Mexico
Dr. Salahddine Krit, Ibn Zohr University, Agadir, Morocco
Dr. Nada Al Hakkak, Baghdad College of Economic Sciences University, Iraq
Dr. Biswa Nath Chatterji, IIT Kharagpur, West Bengal
Dr. Ganapati Panda, IIT Bhubaneswar, Odisha
Dr. Sarat Kumar Patra, IIIT Vadodara, Gujarat
Dr. Pradyut Kumar Biswal, IIIT Bhubaneswar, Odisha
Dr. Bijayananda Patnaik, IIIT Bhubaneswar, Odisha
Dr. Bidyadhar Subudhi, IIT, Goa
Dr. Saradindu Ghosh, NIT Durgapur, West Bengal
Dr. Subrata Banerjee, NIT Durgapur, West Bengal
Dr. Siba Sankar Mahapatra, NIT, Rourkela, Odisha
Dr. Ch. Sai babu, JNTU, Kakinada, Andhra Pradesh
Dr. Jamuna Kanta Sing, Jadavpur University, Kolkata, West Bengal
Dr. Gadi Venkata Siva Krishna Rao, Andhra University, Visakhapatnam, Andhra Pradesh
Dr. Prakash Kumar Hota, VSSUT, Burla, Odisha
Dr. Bibhuti Bhusan Pati, VSSUT, Burla, Odisha
Dr. Sivkumar Mishra, BPUT Rourkela, Odisha
Dr. Prasanta Kumar Satpathy, CET, Bhubaneswar, Odisha
Dr. Tapas Kumar Panigrahi, PMEC, Berhampur, Odisha
Dr. Mrutyunjaya Panda, Utkal University, Bhubaneswar, Odisha
Dr. Prasant Kumar Sahu, IIT Bhubaneswar, Odisha
Dr. Santos Kumar Das, NIT, Rourkela, Odisha
Dr. Debalina Ghosh, IIT Bhubaneswar, Odisha

## Program Chair

Dr. Subhrajit Pradhan, GIET University
Prof. Dhananjaya Sarangi, IIMT, Bhubaneswar

## Program Committee

Dr. A. V. N. L. Sharma, Dean Academic, GIET University
Dr. P. Vijayakumar, CoE, GIET University
Dr. Santosh Kumar Panda, Dean SoAg, GIET University
Dr. Tarini Charana Behera, HoD SoS, GIET University
Dr. Sanjay Kumar Kuanar, HoD CSE and IT, SoET, GIET University
Dr. Manoja Das, HoD BT, SoET, GIET University
Mr. G. R. K. D. Satya Prasad, HoD EE and EEE, SoET, GIET University
Mr. Ashis Kumar Samal, HoD CE, SoET, GIET University
Dr. Ajit Kumar Senapati, HoD ME, SoET, GIET University
Dr. Radha Krushna Padhi, HoD ChE, SoET, GIET University
Dr. Dillip Kumar Pattanayak, HoD BSH, SoET, GIET University

## Organizing Chair

Dr. Priyadarsan Parida, Department of Electronics and Communication Engineering, GIET University

## Organizing Co-Chair

Dr. Ajit Patro, Department of Electronics and Communication Engineering, GIET University

## Conference Coordinator

Dr. Priyadarsan Parida, Department of Electronics and Communication Engineering, GIET University

## Convener

Dr. Subhrajit Pradhan, Department of Electronics and Communication Engineering, GIET University

## Co-Convenor

Dr. Ajit Patro, Department of Electronics and Communication Engineering, GIET University

## Technical Program Committee Chairs

Dr. Rutuparna Panda, VSSUT, Burla, Odisha
Dr. Tusarakanta Panda, GIET University
Dr. Shasanka Sekhar Rout, GIET University
Dr. Shirshendu Roy, GIET University
Dr. Jagan Bihari Padhy, GIET University
Dr. Bibhu Prasad, GIET University

## Organizing Committee

Mrs. Ranjita Rout, GIETU, Gunupur, Odisha
Mrs. Bandana Mallick, GIETU, Gunupur, Odisha
Mrs. Padmini Mishra, GIETU, Gunupur, Odisha
Mrs. Ashima Sindhu Mohanty, GIETU, Gunupur, Odisha
Mrs. Priyambada Parida, GIETU, Gunupur, Odisha
Mrs. N. Sowmya, GIETU, Gunupur, Odisha
Mr. Tolada Appa Rao, GIETU, Gunupur, Odisha
Mr. Radhanath Patra, GIETU, Gunupur, Odisha
Dr. Ami Kumar Parida, GIETU, Gunupur, Odisha
Mr. Ashish Tiwary, GIETU, Gunupur, Odisha
Mr. Ribhu Abhusan Panda, GIETU, Gunupur, Odisha

# Preface

The 1st International Conference on Smart & Sustainable Technologies (ICSST 2021) was organized by the Department of Electronics and Communication Engineering, School of Engineering and Technology, GIET University, during December 16–18, 2021. ICSST 2021 has provided a forum that brought together researchers, academia, and practitioners from industry to meet and exchange their ideas and recent research achievements in all aspects of smart and sustainable technologies, together with their applications in the contemporary world. The conference aims at attracting contributions of the system and network design that can support existing and future applications and services. In recent years, sustainable technologies attracted significant attention in both research and industry. This approach corresponds to natural human vision and is the best way to represent, generate, and implement various contemporary achievements. The conference has received a good response with a large number of submissions. The total number of relevant papers accepted for publication has been broadly divided into five major parts: (i) IoT, big data, and cloud computing, (ii) smart communication, (iii) smart energy systems, (iv) machine learning techniques and applications, and (v) smart technology for smart living.

ICSST 2021 has not only encouraged the submission of unpublished original articles in the field of smart and sustainable technology issues but also considered the several cutting-edge applications across organizations and firms while scrutinizing the relevant papers.

Bhubaneswar, India — Srikanta Patnaik
Sofia, Bulgaria — Roumen Kountchev
Gunupur, Odisha, India — Subhrajit Pradhan
Gunupur, Odisha, India — Priyadarsan Parida

# Contents

# About the Editors

**Prof. Srikanta Patnaik** is Professor in the Department of Computer Science and Engineering, Faculty of Engineering and Technology, SOA University, Bhubaneswar, India. He has received his Ph.D. in Engineering on Computational Intelligence from Jadavpur University, India, in 1999 and supervised 25 Ph.D. theses and more than 60 master theses in the area of computational intelligence, soft computing applications and re-engineering. Dr. Patnaik has published around 100 research papers in international journals and conference proceedings. He is author of two textbooks and 52 edited volumes and few invited chapters, published by leading international publisher like Springer-Verlag, Kluwer Academic, etc. Dr. Patnaik is Editor-in-Chief of *International Journal of Information and Communication Technology* and *International Journal of Computational Vision and Robotics* published by Inderscience Publishing House, England, and also Editor-in-chief of Book Series on *Modeling and Optimization in Science and Technology* published by Springer, Germany, Book Series on *Advances in Computer and Electrical Engineering* (ACEE), and Book Series on Advances in Medical Technologies and Clinical Practices (AMTCP), published by IGI-Global, USA. He is Editor of *Journal of Information and Communication Convergence Engineering*, published by Korean Institute of Information and Communication Convergence Engineering. He is also Associate Editor of *International Journal of Granular Computing, Rough Sets and Intelligent Systems* (IJGCRSIS) published by Inderscience Publishing House, England.

**Prof. Roumen Kountchev** Ph.D., D.Sc. is Professor at the Faculty of Telecommunications, Department of Radio Communications and Video Technologies at the Technical University of Sofia, Bulgaria. His scientific areas of interest are digital signal and image processing, image compression, multimedia watermarking, video communications, pattern recognition and neural networks. Professor Kountchev has 341 papers published in magazines and conference proceedings (71 international), 15 books, 46 chapters, and 20 patents (three international). He had been Principle Investigator of 38 research projects (six international). At present, he is Member of Euro Mediterranean Academy of Arts and Sciences (EMAAS), President of Bulgarian Association for Pattern Recognition (member of Intern. Association for Pattern Recognition), and

Editorial Board Member of: *International Journal of Reasoning-based Intelligent Systems*; *International Journal Broad Research in Artificial Intelligence* and *Neuroscience*; *KES Focus Group on Intelligent Decision Technologies*; *Egyptian Computer Science Journal*; *International Journal of Bio-Medical Informatics and e-Health*, and *International Journal Intelligent Decision Technologies*. He has been Plenary Speaker at WSEAS International Conference on Signal Processing 2009, Istanbul; WSEAS International Conference on Signal Processing, Robotics and Automation, University of Cambridge 2010, UK; WSEAS Intern. Conf. on Signal Processing, Computational Geometry and Artificial Vision 2012, Istanbul, Turkey; International Workshop on Bioinformatics, Medical Informatics and e-Health 2013, Ain Shams University, Cairo, Egypt; Workshop SCCIBOV 2015, Djillali Liabès University, Sidi Bel Abbès, Algérie; International Conference on Information Technology 2015 and 2017, Al Zayatoonah University, Amman, Jordan; WSEAS European Conference of Computer Science 2016, Rome, Italy.

**Dr. Vipul Jain** is working in the area of operations and supply chain management at Victoria Business School, Victoria University of Wellington, New Zealand. He has also worked as French Government Researcher for the French National Institute for Research in Computer Science and Control at Nancy, France, and was also involved in the European project I*PROMS (Innovative Production Machines and Systems) from the UK. Vipul has more than 85 archival publications to his credit in high impact factor journals, as well as conference papers and chapters, is Editor-in-Chief of *International Journal of Intelligent Enterprise*, and Editorial Board Member for seven international journals. Vipul is ranked 7th in India in a list of top 20 Indian leading academicians publishing in the area of logistics and supply chain management in an exhaustive study entitled "*Analysis of the Logistics Research in India—White Paper* (2012), pp. 1–11" conducted by researchers from Institute of Transport Logistics, Dortmund University, Institute for Supply Chain and Network Management, Technische Universität Darmstadt, and University of Münster, Germany. Vipul is involved as Scientific and Advisory Committee Member for various international conferences. He was Program Coordinator for the Masters in Industrial Engineering Program at IIT Delhi and has been actively involved in conducting Management Development Programs/Training Programs for Senior Managers and Executives of Public and Private Organizations in India on various issues and challenges encountered in operations, logistics, and supply chain management.

# IOT, Big Data and Cloud Computing

# Big Data: A Boon to Fight Against Cancer Using Map Reduce Framework

**Sudipta Priyadarshinee and Madhumita Panda**

**Abstract** Healthcare data consist of an enormous amount of information, which is challenging to be maintained by manual methods. The use of big data in health care is giving solutions for improving patient care and generating value. Cancer is the second main reason for death in the world. The battle against cancer has made a big task. The aim of big data analytics is to connect present people databases and collect complete data of every single patient suffering from cancer for further detailed analysis. Integration of these sources is key and will be advantageous for improvement in cancer research, patient care, and monitoring quality of care. The present paper discusses Map Reduce Architecture to handle these massive amounts of data records.

**Keywords** Big Data · Cancer · Big Data Analytics · Map Reduce

## 1 Introduction

Big data defines the massive collection of data or information which is growing exponentially with time. It is a combination of structured, unstructured, and semi-structured data, and this type of data is very composite and large that none of the traditional software can handle it efficiently. It defines the huge volume of raw data gathered, analyzed, and stored by organizations. The sources and examples of big data are social media, banks, instruments, websites, stock markets, etc. Big data is growing faster day by day. The annual growth of data creation may come to near about 44 trillion zettabytes in the year 2020. Big data is used in so many areas like bioinformatics, healthcare sector, atmospheric science, biomedical, genomics, agriculture, astronomical studies, research, and so on. A massive volume of data is created every day from business, science, and research. So big data is in every place and gives an incredible opportunity to those who can utilize it effectually. Accuracy

S. Priyadarshinee (✉) · M. Panda
G.M.University, Sambalpur, Odisha, India
e-mail: sudiptapatel88@gmail.com

 
S. Patnaik et al. (eds.), *Smart and Sustainable Technologies: Rural and Tribal Development Using IoT and Cloud Computing*, Advances in Sustainability Science and Technology,
https://doi.org/10.1007/978-981-19-2277-0_1

in big data may provide better decisions making capacity and that can provide greater effectual productivity cost-cutting and minimize the risk [1].

Big data characterizes huge data on a colossal number of patients. It contains both organized and unstructured clinical data that can comprise patient qualities, treatment history, and atomic information, genomic and charging history. The difficulty and assorted variety of this gigantic volume of information needs appropriate mastery to process. Legitimate investigation plans and nice systematic techniques are basic to transform enormous information into important data for treatment choices and patient consideration. So large information comprises genuine data, their utilization could traverse holes in translational and clinical examinations. It is exceptionally fundamental in a few oncologic surroundings, where pharmaceutically supported preliminaries regularly utilize problematic control arms and afterward leave a hazy choice between another specialist and the finest existing norm of care routine [2].

Healthcare is a data-intensive sector. The healthcare sector is changing rigorously. In the healthcare industry, there are many processes going on. These processes have an impact on each person's care, as well as assisting medical practitioners and providing care and services. Big data analytics is a strong tool that is rapidly growing in popularity. Big data analytics is playing a main part in the evolving healthcare sector. Using big data-based solutions, a tremendous amount of healthcare data may be gathered and organized. Analytical models depend on the gathered data could help in disease prognosis or health care industry improvement [3]. Big data analytics can be designed using a variety of broad designs and open source software such as Hadoop, Apache Strom, and others [4, 5].

Out of multiple industries of healthcare cancer detection and treatment is the main problem where cancer is the main cause of death of millions of humans every single year. There are a variety of cancers and every cancer has different features. So, depending on the category of cancer and patient diagnosis and treatment is changed [6].

## 2 Literature Review

Among different helpful regions, cancer research area has gathered a huge volume of huge information. This contains information from many patients including gene expression, mutations, amplifications, and proteogenomic data. The essential investigations in cancer are combined into translational medical science to proceed the findings nearer to the clinic [7].

The study of human genomes can aid in the discovery of DNA problems, which can lead to the discovery of the fundamental cause of cancer and a better understanding of the treatment options available to everyone. The human genome is made up of billions of pairs and studying them is a huge undertaking. Such a vast amount of data could be efficiently tackled by big data. As a result, big data can aid with cancer care, diagnosis, prognosis, and treatment [7]. Perfectly maintaining the medical history and for cancer treatment alone, which is a huge undertaking in and of itself, big data is playing a major part [8]. Apart from that, analytical models for cancer prognosis or diagnosis have been developed. There has been work on generating large data sets for lung cancer [6] and developing predictive algorithms for lung cancer detection [9].

Big data is now widely employed in healthcare for disease prognosis. Cancer in the breast is one of the most usual cancers in women. In the United States and Asian countries, it is the second major reason for death for women. There is a better chance of curing this cancer if it is detected in its early stages. The K closest neighbor (KNN) technique is utilized to determine classification accuracy in this study, and it is run on the R tool. It uses the authentic Wisconsin breast cancer data set from the UCI machine learning repository [10].

For the detection and diagnosis of the aforementioned condition, computer vision techniques are crucial. A novel unsupervised skin melanoma extraction and analysis method are given in this paper. The suggested method first removes hairs from dermoscopic pictures before using an easy thresholding-based method to extract skin lesions [11].

Early diagnosis and precise placement of a brain tumor can help with future brain therapy. In this study, a computer vision-based method for pinpointing the correct position of brain tumors using MRI images is established. The procedure begins with the removal of the outer region of the brain and the separation of the white matter. More transmission with local agent (CLA) clustering techniques are used, observed by morphological post-processing approaches for tumor extraction from the white matter area of the brain. A publicly available MRI data set is used to test the approach. When compared to existing methods, quantitative and qualitative metrics demonstrate that the suggested approach reaches an accuracy of 99.64% [12].

An unsupervised computer vision strategy for extracting Meningiomas tumors from Tl-weighted MRI images is provided in this research. In order to effectively extract the tumor mass from the MRI images, the suggested method uses a simple k-means clustering with the thresholding method observed by a variety of morphological methods. The suggested method is validated against many state-of-the-art current computer vision approaches utilizing distinct performance indicators on a broadly obtainable benchmark data set. The results reveal that the proposed approach reaches a data set accuracy of 99.47% [13].

## 3 Proposed Work

In order to manage a vast amount of data records at once, we require an efficient model. Map Reduce Model is best to handle a vast amount of data records of cancer patients. This paper focuses on the Map Reduction Architecture to manage large data.

### 3.1 MapReduce Architecture

Hadoop's MapReduce processing is divided into two stages; first is map stage and second is reduce stage. For managing large data, MapReduce is the best solution. The data is broken into small parts and dispersed across multiple nodes to extract intermediate outcomes; when the data has been processed, the results are pooled to provide final outcomes. A block of data is read and managed at the map stage to generate key-value pairs as intermediate outputs. The output from the map stage is the input to the Reduce stage in the reduce stage. The reducer, on the other hand, gets the key-value pair from many maps. Ultimately, the reducer merges the intermediate key-value pairs into the final output.

Logical View:

The map stage takes one pair of data from one data domain and returns a list of pairings from another domain.

$$\text{Map}(k1, v1) \rightarrow \text{List}(k2, v2)$$

Every pair (keyed by $k1$) in the output data is applied in parallel by the Map function, which returns a list of pairs (keyed by $k2$) for each message. After that, the MapReduce model may accumulate all pairs with the same key ($k2$) from all lists and construct grouping for each key value. The Reduce function is then put into each group that generates a set of values in the same domain at the same time.

$$\text{Reduce}(k2, \text{list}(v2)) \rightarrow \text{List}(v3)$$

Reduce returns either one value $v3$ or an empty result for each call. One call, on the other hand, allows for the return of many values. However, all of these results are compiled into a desirable result list. As a result,

Hadoop MapReduce has parallelism capabilities and can generate output depending on key-value pair analysis. This is excellent for managing a massive volume of growing data in a well-defined fashion [14].

As can be seen from Fig. 1, the system will combine the items having the same key. At last, the system will give the requested output. Patient data is stored in a centralized repository that makes the system cost-effective by reducing the number of storage warehouses as well as eliminating data redundancy, which conducts the system being consistent.

## 4 Conclusion

Big data is an emerging technology which had a great influence on the healthcare sector. No doubt cancer is a very critical disease and there is no specific treatment

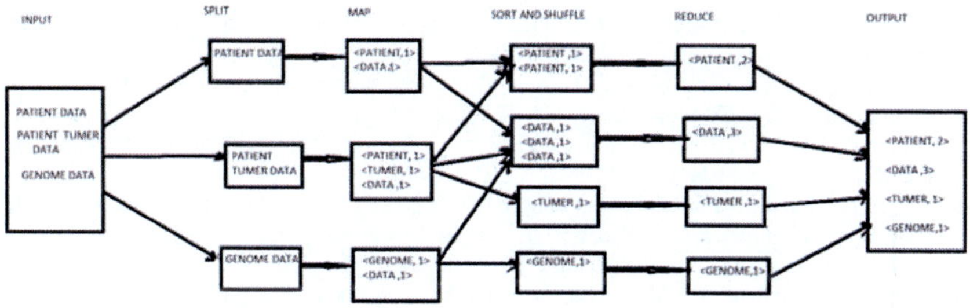


**Fig. 1** Map Reduction of patient data

for cancer. But at the present, with the usage of big data, clinical analysis to forecast cancer recurrence, progression, and response to therapy is no longer a hard task. Big data is a game-changer in cancer research. By the usage of big data, researchers can analyze things like how certain mutations and cancer proteins interrelate with various treatments and discover trends that will cause better patient results. Big data will provide accurate oncology treatment and better treatment for a patient with high accuracy. Ultimately, big data is a boon in to fight against cancer. To handle a massive amount of data records at once we require map reduce architecture which is discussed in this paper. Using this map reduce model, we extract meaningful data from vast amount of healthcare, cancer patient data records. The work can be extended in future using a modified artificial neural network (ANN) classifier technique with a MapReduce framework for the prediction of cancer disease in patients.

## References


1. R. Martiz, M.A. Supaksha, N. Hemalatha, Application of big data in bioinformatics—a survey. Int. J. Latest Trends Eng. Technol. Special Issue SACAIM 2016, 206–212
2. L. Wang, C.A. Alexander, Big data analytics in healthcare systems. Int. J. Math. Eng. Manage. Sci. **4**(1), 17–26 (2019)
3. A. Belle, R. Thiagarajan, S.M. Reza Soroushmehr, F. Navidi, D.A. Beard, K. Najarian, Big data analytics in healthcare. BioMed Res. Int. **2015**, Article ID 370194, 16 pages (2015)
4. V.-D. Ta, C.-M. Liu, G.W. Nkabinde, Big data stream computing in healthcare real-time analytics, in *2016 IEEE International Conference on Cloud Computing and Big Data Analysis*, 978-1-5090-2594-7116, 2016 IEEE
5. R. Vanathi, A. Shaik Abdul Khadir, A robust architectural framework for big data stream computing in personal healthcare real time analytics, World Congress on Computing and Communication Technologies (WCCCT), 978-1-5090-5573-9/16 2016 IEEE. https://doi.org/10.1109/WCCCT.2016.32
6. S.K. Swain, Use of big data analytics in lung cancer data set. Int. J. Comput. Eng. Res. **07**(12) (2017). ISSN 2250-3005
7. A. Makler, R. Narayanan, Big data analytics and cancer. MOJ Proteomics Bioinform. **4**(2), 00115 (2016)
8. Big Data Analysis of Clinical records for Cancer Care, White Paper by Intel
9. R.P. Panda, P.P. Barik, 3P. Alok Kumar Prusty, A review paper on big data in lung cancer big data analytics in lung cancer. Int. J. Trend Res. Dev. **3**(5). ISSN: 2394-9333

10. K. Shailaja, B. Seetharamulu, M.A. Jabba, Prediction of breast cancer using big data analytics. Int. J. Eng. Technol. **7**(4.6), 223–226 (2018)
11. R. Rout, P. Parida, A novel method for melanocytic skin lesion extraction and analysis. J. Discrete Math. Sci. Cryptogr. **23**(2), 461–473 (2020)
12. Sahoo, A. Kumar, P. Parida, Automatic clustering based approach for brain tumor extraction. J. Phys.: Conf. Ser. **1921**(1) (2021)
13. Sahoo, A. Kumar, P. Parida. A clustering based approach for meningioma tumors extraction from brain MRI images, in *2020 IEEE International Symposium on Sustainable Energy, Signal Processing and Cyber Security (iSSSC)* (IEEE, 2020)
14. A.S. Thanuja Nishadi, Healthcare big data analysis using Hadoop MapReduce. Int. J. Sci. Res. Publ. **9**(3) (2019)

# Security Attacks and Its Countermeasures in RPL

Ajay Dilip Kumar Marapatla and E. Ilavarasan

**Abstract** IoT is a rapidly growing network in the present day. With the large collection of applications and ease of deployment in many states of real life, like smart homes, smart cities, smart agriculture, etc. Most of these devices are deployed in distributed way wirelessly. Communication plays a prominent role in IoT, where the communicating devices need to be secured. The standardized protocol that best suits Low power Lossy Networks (LLN) networks is RPL (IPv6 Routing Protocol for Low Power and Lossy Networks). The most important thing that comes in RPL is how to identify the attacks and how to provide the security in LLN. In this paper, we discussed about functioning of RPL and about various RPL attacks that compromised security in LLNs. The review was also conducted on various countermeasures to mitigate the attacks in order to improvise the performance of RPL.

**Keywords** Security attacks in RPL · RPL attacks · Rank attack · Blackhole attack

## 1 Introduction

The IoT concept is to develop an intelligent environment and independent control. Most of the fundamental concept of IoT is inherited from the Wireless Sensor Networks. In today's world, we are in an era called connected smart devices where most of the devices that are connected in this environment are resource-constrained nodes [1]. In IoT, we must provide security to every device and information resource. In some of the networks like LLN Low Power and Lossy Networks where we have very low capabilities for security are prone to internal and external attacks because most of the nodes in LLN are limited with low power, low power processing, and limited memory. The creation of the 6LoWPAN helped to implement the internet

A. D. K. Marapatla (✉) · E. Ilavarasan
Department of Computer Science and Engineering, Pondicherry Engineering College, Puducherry, India
e-mail: dileepmarapatla@pec.edu

E. Ilavarasan
e-mail: eilavarasan@pec.edu



S. Patnaik et al. (eds.), *Smart and Sustainable Technologies: Rural and Tribal Development Using IoT and Cloud Computing*, Advances in Sustainability Science and Technology,
https://doi.org/10.1007/978-981-19-2277-0_2

protocol in small devices. The 6LoWPAN provides the connectivity among point to point enable unlimited applications [2]. 6LoWPAN focus on managing the datagram of IPv6 when transmitting data over the IEEE802.15.4 standard protocol [3]. The growth of internet is very fast and the increase in IoT devices, the standard protocols for routing will accommodate the needs of many added nodes. To handle this kind of routing problem RPL protocol was introduced which gained whose popularity and got standardized for routing in resource-constrained devices [4]. RPL is one of the routing protocols which allow users to take routing topologies logically which are called as Directed Acyclic Graphs (DAGs) in the physical network. These methods support the objective functions (OFs) that are calculated by the user. Several issues are not considered because of many security challenges [5]. In lossy networks, RPL presents with low routing that requires robust nature. This protocol has different group that sends the packet and forward for routing optimization [6].

The paper is organized as follows: Sect. 2 identifies the previous works that are explained the RPL attacks. Section 3 explains the RPL protocol and its control messages. Several attacks descriptions are explained in Sect. 4. In Sect. 5 discussed the analysis in various attack scenarios. In Sect. 6 conclusion and future work are placed.

## 2 Related Work

The author [7] proposed a hierarchical distribution model named Trust-based Neighbor Notification (TN-IDS) strategy for single instance of the RPL. This work tracks malicious nodes in the network using blockchain values for trust report generation that consume less power and computation time. The author [8] proposed (GAIDS) Game models based Anomaly Intrusion Detection System that use stochastic and evolutionary game models by formulating the interaction of players on control packets to measure transmission probability. State transition observations ensure to prevent topology-based attacks by detecting intruders to provide the secure RPL. In [9] proposed a (SRPL-RP) Secure-RPL Routing protocol for version number and attacks on rank to detect and isolate the attacks that support many network topologies. In [10] author proposed a detection technique for both blackhole and version number attacks by implementing them in the COOJA simulator. The proposed work is identified by analyzing the traffic among the nodes. The based traffic analysis mainly focused on throughput along with delay, and PDR value. The author [11] proposed framework to detect sinkhole attacks, black hole attacks, hello flood attacks, and version number attacks using genetic programming. The work was implemented in COOJA simulator in two phases like modulation, and the other phase is the detection of attacks based upon the threshold statements identified in the first phase. This had resulted in higher packet delivery ratio with less delay and data drop. The drawback of this model is the detection technique implemented at each node where only the root node is the resource full node, and the remaining nodes are resource-constrained nodes because of large set of features extracted. In [12] the author proposed the technique called Highest Rank Common Ancestor (HRCA) to identify the node with highest rank among the neighbors to form pair in the network tree. The attack position in the network localizes that makes the process of migration which is lightweight and fast. The simulation result shows a high Packet Delivery Ratio (PDR) and accurate attack detection. This work is limited to True Positive rate (TPR) decreased if the attackers increased in network. To construct the network topology [13], a metric-based Trustworthiness Scheme is introduced. The MRTS calculates the behavior of nodes' trust in their neighborhood by direct and indirect recommendations. The simulation results show that once the detection and isolation of malicious nodes are completed the network topology is more stable with the decrease in the rate of energy consumption. In [14] a time-based trust-aware RPL is proposed which is integrated into RPL protocol for security to prevent attacks. This model detects and isolates the malicious nodes and increases the performance of network and detection ratio. But the drawback of this system is it consumes too much battery power. In [15] Secure-RPL routing protocol is used to migrate the flooding attack by load distribution in the network. Simulation results are done with both static and dynamic networks by comparing the performance of standard RPL and unsecured RPL. The results show a decrease in packet loss ratio and energy consumption. The author [16] studied the various intrusion detection systems that were proposed in recent times. This Work includes identifying various IDS and their design requirements along with the diversity in attacks identified by those IDS. In [17] the author discussed the various IDS placements strategies and analyzed them in the IoT environment and also discussed how to implement ML and DL techniques to detect attacks in IoT networks were discussed. From the related works, it was observed that less research contribution was done on rank attacks and sinkhole attacks.

## 3 About RPL

RPL is the protocol that was developed to work on IEEE 802.15.4 PHY and MAC layers. RPL can illustrate the route solution for the LLN where we have very limited resources of Bandwidth, Power, and Processing which makes the networks lose packets [18]. The networks that are enabled with two features (i) very low data rate. (ii) The communication results in low throughput which is higher data rate. The lossy link is divided by using huge bit error rate with no accessible time which impacts routing design. RPL is intended to be exceptionally versatile and give elective courses at whatever point default courses are difficult to reach [19]. RPL is dependent on the idea called DAG (Directed Acyclic Graph). The DAG represents the tree structure that communicates among the nodes present in the network where RPL has bidirectional correspondence. RPL structure nodes as DODAG (Destination Oriented DAG), where the one giving the course the web is considered as ROOT. Objective Function helps in creating the DODAG, it defines the configurations for routing like Route metrics, computation of rank, and how to elect the parents' selection in a

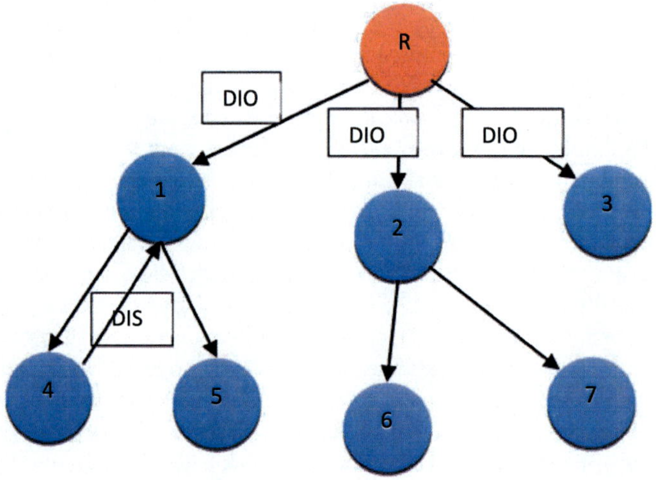


**Fig. 1** A sample DODAG structure [22]

DODAG. A network can have at least one DODAG's, structure together an RPL Instance recognized by a remarkable ID called RPL Instance ID. A node can join numerous RPL occasions yet should just have a place with one DODAG inside each case [20]. Each node will have its rank qualities which expand the descending way and helps in recognizing the situation with different hubs in DODAG. The development of the DODAG occurs at the root node by communicating a control message called DIO which contains arrangement boundaries to assemble the geography. At the point when the neighbor node gets the DIO message from the root, it will add to its rundown of guardians in its table and check the position upsides of every section then it will choose the base position esteem as the parent. Presently the node will communicate the DIO to its neighbors to build up a vertical course to advance traffic to the root node. This steering Protocol upholds three distinct kinds of geographies are point to point, Point to Multipoint, and Multipoint to Point [21]. The sample network architecture of RPL is shown in Fig. 1.

## *3.1 Control Messages in RPL*

RPL introduced the new ICMPv6 control messages. The structures are defined in Fig. 2.

*Solicitation-based DODAG Information*: The representation of DIS message is mapped as $0 \times 00$, which is utilized to refer to the DIO from a node, for example: If any new node joins the network or will take long time to receive the DIO message. In the neighbor DODAG's, the nodes are identified by the network.

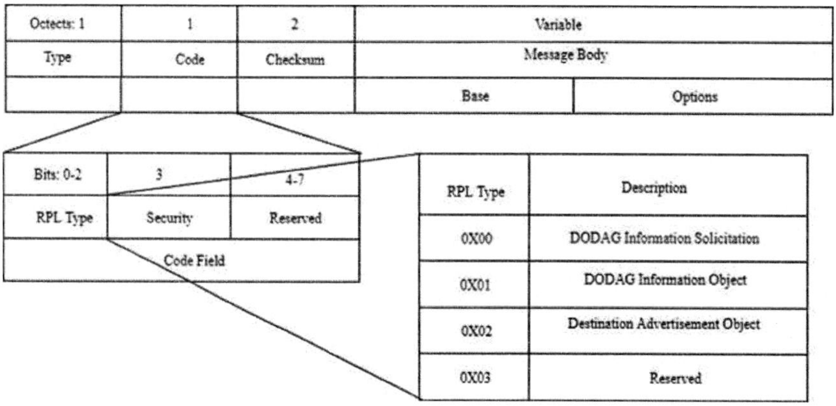


**Fig. 2** RPL control messages

*Object belongs to DODAG Information*: DIO is mapped and it is represented as 0 × 01, for maintaining the message is used in DODAG. DODAG creates the root node with multicasting of DIO messages which consists of values with rank and configures the root node.

*Object belongs to Destination Advertisement*: This is mapped with the 0 × 02 which generates the route information of nodes that are checked with the upgrade path. The message of DAO is sent to each node and this is not sent to root node that creates the routing tables and also the prefixes of the child to their parents. After this step, the overall routing information is generated from the specific node to parent node.

## 3.2 *Traffic and Modes of Operation in RPL*

RPL supports three types of communications:

*Multipoint to Point*: It is called MP2P where the data transfers at the leaf or the child nodes to the sink node.

*Point to Multipoint*: In other words, it is P2MP, from the sink nodes to leaf nodes the communication is established.

*Point to Point*: It is called P2P where the data communication is happening between two nodes.

Modes of Operation

*Non-Storing mode*: This mode mainly focused on not maintaining any routing table. All the network information of nodes is stored at the root node only.

*Storing mode*: In this mode, the Nodes will store the information of its sub-DODAG where it will forward the packets based upon their routing table stored in them.

# 4 Attacks in RPL

The features of security for RPL are defined by ROLL. This model includes confidentiality, integrity, availability, and authentication. Based on the dynamic behavior of the firewall facility is not available to the RPL and these nodes do not have boundaries that are well defined. Based on the centralized nature and cooperation of nodes the cryptography techniques are not applicable.

RPL routing protocols are divided into three categories based on the security attacks.

## *4.1 Attacks Against Resources*

Attacks on the resources aim at the legitimate nodes in the network to perform the unwanted operations to deplete the network resources. This kind of action will consume the memory, computation, and energy of the node. This attack will congest the network and significantly will decrease the network lifetime. The resource attacks are further classified into two types they are direct and indirect attacks. In the direct attack, the attacker (malicious) node will affect the network directly and deplete the network resources by creating the network overhead. In the indirect attack, the malicious node will indirectly affect the network by creating unwanted loops in the RPL network which further produce the network traffic. Figure 3 shows the attacks at the Resource level.

### 4.1.1 Direct Attacks

Resource deplication occurs when the malicious node is attacked by direct attacks. By overloading or by flooding attacks this can occur and also the routing table is active and in storing mode.

**Hello Flood Attacks** To join a network most of the nodes will send HELLO messages to their neighbor nodes. This attack feels like an intelligent, the attacker node merged in the network and acts as a neighbor to most of the nodes in the network and will send requests (HELLO) text to all the nodes. One solution to overcome this attack is to find the distance among nodes in the network and another solution is utilizing the local and global repair of the network. But too much global repair also will lead to the battery drain of the nodes. In [24] studied the various network setups and smart devices setups that affect the battery life hello flood attack scenario. His work identified

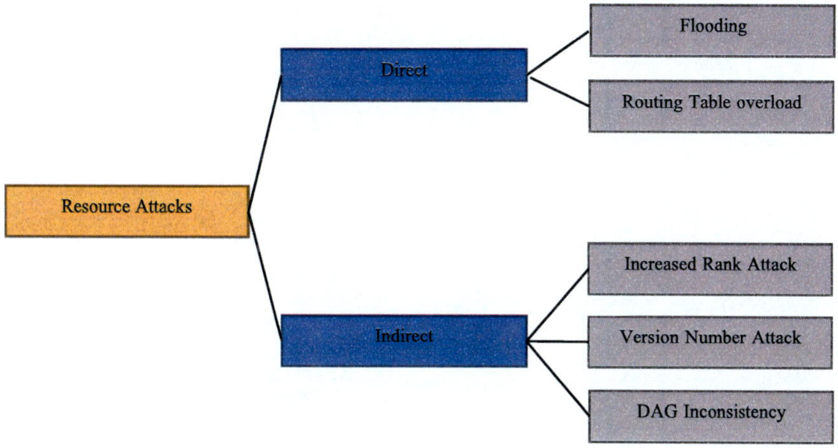


**Fig. 3** Resource Attacks in RPL [23]

the sending of small packets at regular intervals also draining the battery quickly moreover nodes that are close to the malicious node are affected more than the node that is far enough. In [25] the author proposed an ANN Machine Learning approach to accurately detect the Helloflood and rank attacks. Considered the parameters of energy efficiency, network lifetime, and trustworthiness and achieved higher results when compared with other methods.

### 4.1.2 Indirect Attacks

This is one type of attack, that which the malicious node will make the nodes create loops or make the nodes generate the overload on the network.

**Increased Rank Attack** Every node in the RPL network is considered as weighted with ranks that start from root node, if the distance from the root node to the node is far then the rank of that node is very high. The attacker will use this loophole, in this kind of attack the attacker willingly increases its rank value and make the neighboring node think that it is far from the root node and will create a longer path from the root node to its neighboring node which will drain the network resources.

**Version Attack** The root node controls the global repair function by using the version number to ensure all the nodes are updated in DODAG with the routing state. The DIO messages will take the version number. If any nodes receive the outdated version number that can join in new DODAG by improving the value of rank and update the version number [26]. For example, in Fig. 4 we have a malicious node 11 it starts sending DIO messages to its neighbors which are relayed by the other nodes. This attacks version is considered a denial of services.

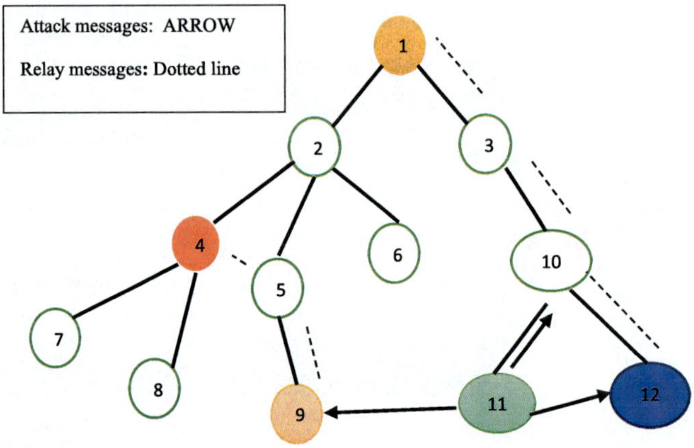


**Fig. 4** Version number attack example [27]

**Version Attack Countermeasures** The authors [28] proposed a hash chain-based authentication security called VeRA. This consists of hash links with version and protected data. In this scheme, nodes will verify other nodes' version numbers are updated by root or not. In version number attacks the attacker will act like a DODAG root node and will start building the routing topology. The DIO messages are validated by the data security which consists of middle nodes that contain new rank and version numbers. In [29] proposed a Trust Anchor interconnection loop called TRAIL. A technique with Digital signature which is root node in the network will go about as the trust anchor when the consistency of positions should be checked by the node in the vertical way a verification message is mentioned to ship off the root node. The difficulty of this technique is neglected to address the sort of attack that manage the cost of the great overhead. Nikravan et al. [30] proposed a lightweight approach providing non-repudiation to mitigate the version number attacks. This signature-based process is separated into online and offline phases. The performance of the node is based on independent algorithm which does not store any information about the nodes. To migrate, the version numbers, the author [31] proposed two new methods. Two phases are present such as (i) the termination method is used to terminate the updated version number that occurs from a strong attack location, it only accepts the updates that come straightly from the parent to the leaf node. (ii) Shield technique utilizes based on the trust strategy where the version is changing frequently with the change in the nodes near the root node. Termination technique requires fewer resources while shield requires high resources.

## 4.2 *Attacks Against Topology*

Attacks against topology are classified into two types, Suboptimization and Isolation. Figure 5 shows the topology attacks.

### 4.2.1 Suboptimization

In suboptimization, the network will not form optimal paths, i.e., in this, the attacker will not allow the nodes to have an optimal path to reach the destination or root node.

**Sinkhole Attack** sinkhole attack affects the routing in the network. When a sinkhole attack is compared with all the routing attacks it is the most destructive. In this attack, a malicious node will broadcast to the nearest nodes that it is the best node to send packets to the destination. Now the nearest nodes start sending the packets through this malicious node thinking that it is the best path to transmit data to the destination. Once all the data passes through this malicious node it will sink the data that it receives from Neighbor nodes [32]. For example, in Fig. 6 we considered an example of nodes communicating with the base station by node 12. Here the node 12 is considered as attacker node and collects packets of data from the neighbors then it drops all the packets.

**Sinkhole attack countermeasures** In [34] author proposed a Neighbor-passive monitoring technique, which is implemented by using the cooja simulator. In this scheme author considered metrics like power consumption and detection accuracy to calculate the performance. By focusing on the re-ranking method, the author achieved the attack detection of the sinkhole. In this proposed detection method using multiple

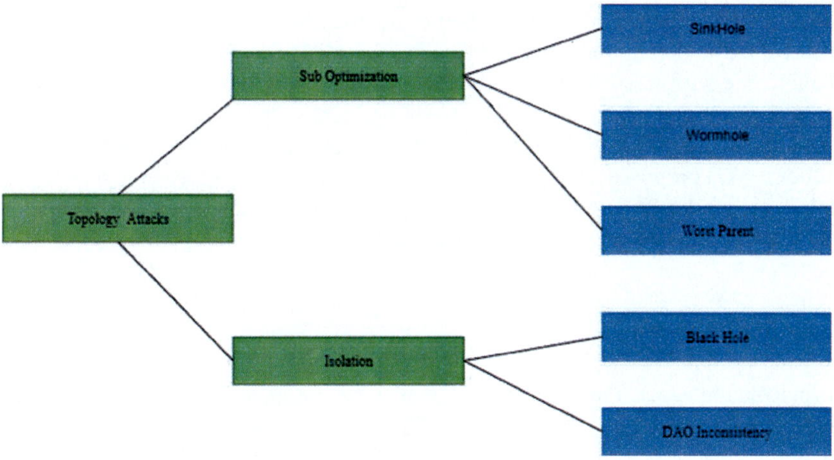


**Fig. 5** Topology attacks in RPL

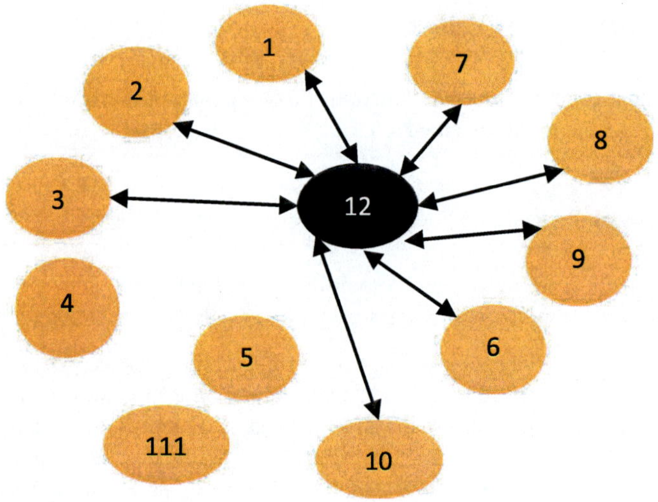


**Fig. 6** Sinkhole attack [33]

sinks means we have more than one sink node called the sinkhole detection method in multi-sinks in IoT (SDMSI). Using the cooja simulator compared the method with traditional attack detection by using single root node topology. Simulation results show that efficiency has been increased in terms of congestion in traffic, transmission, and consumption of resources. The drawback of this method is overhead created when using multi-paths. In [35] a distributed IDS architecture is proposed to detect sinkhole attacks. This approach will detect the attack by introducing network profiles and using data mining techniques. In this approach, the number of false alarms can be significantly reduced by checking the suspicious alarms. The author in [36] proposed the SoS-RPL scheme consists of sinkhole and elimination of sinkhole node. In this method, the malicious nodes are included in a blacklist and whenever a node receives the DIO message it is verified with the list and the detected node details are passed on to the root node and will update the routing table. This scheme is analyzed by NS-3 and achieved a high-level performance in detection rate and throughput. The author in [37] proposed a hybrid technique that monitors the unpredictable behavior in RPL networks. The performance of these networks by using the cooja simulator and identified the decrease in power consumption by 55%.

### 4.2.2 Isolation Attacks

In an isolation attack, the nodes are not able to communicate with the parent or root node.

**Black Hole Attack** The Blackhole attack is the WSN inherited attack in RPL. This is one of the Denial-of-service (DOS) attacks where it denies the services to the leaf

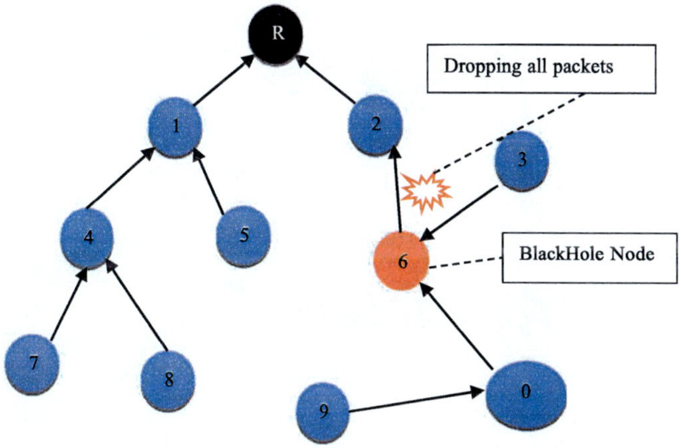


**Fig. 7** Blackhole attack in RPL [38]

nodes from the sink node. The response time is increased because it affects the packet transmission which leads to packet dropping. In this attack, the attacker node extracts the traffic that it receives for forwarding to the sink node. The black hole attack can be combined with the sinkhole attack that stops all the traffic around the black hole. The attack scenario is presented in Fig. 7 here the node 6 acts as a malicious node it attracts the data packets from node 3 and node 0 which is to be forwarded to the sink node but after it collects the data from these nodes it will drop all the packets that pass through it.

**Blackhole attack countermeasures** The author in [39] proposed an IDS that is divided into three modules. The first module is responsible to map the area of the network. The second module is for the detection of possible attacks and the third module is to prevent the attack and act as a firewall to the whole network. In this proposed model it must go through a verification process like the list of neighbors in each node, DIO and DIS message verification along with rank values. In [40] the author implemented the framework with trust values at two levels, i.e., consists of intra level and inter-level DODAG. At intra DODAG level, the trust value is calculated on each node based on the PDR ratio. In the next level, i.e., inter-DODAG level the trust value is same as Intra-DODAG level, but it deals at the central server level in identifying the genuine and malicious servers. The simulation results identified a better throughout, PDR and packet loss percentage when compared with the standard version. Sahay et al. [41] authors proposed an exponential smoothing algorithm to detect malicious node that causes blackhole attack. The proposed algorithm exponential smoothing is a measurable method utilized in time arrangement investigation to decipher dynamic cycles and comprehend the pattern and cyclic segments in time arrangement information. This method eliminates the differences in series data and forecasts the data accurately. In [42] the proposed technique is a trust-based RPL Protocol that is evaluated in the testbed, in this testbed the gathered data were

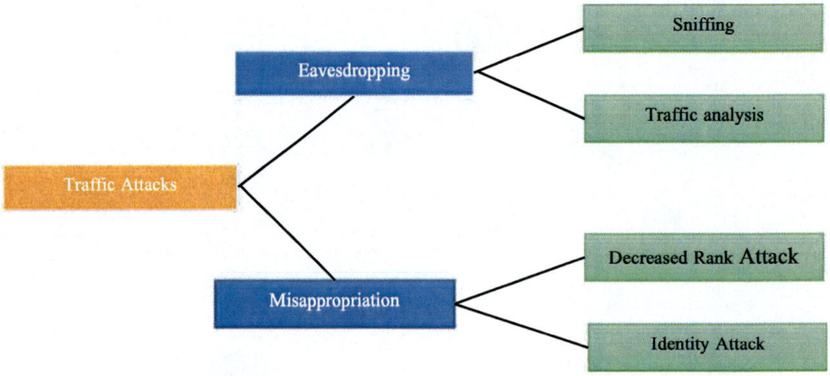


**Fig. 8** Traffic attacks in RPL

analyzed to find the efficiency of the proposed model in mitigating the blackhole attacks against the standard RPL. In [43] authors proposed the AODV based Secure Protocol to detect the Blackhole Attack. Discovery and ID of Blackhole Attack in VANET is a fundamental assignment to keep the organization from imploding. The parcels are dropped by the malevolent hub, and the vindictive hub likewise breaks the course and makes it invalid. The proposed AODV steering calculation effectively recognizes the vindictive hub and helps in its expulsion from the network. Results show that the rate of bundles lost with the proposed AODV is a lot lower when contrasted with the current AODV steering convention. Execution is measured and examined as far as the packet delivery ratio (PDR).

## *4.3 Attacks Against Traffic*

These attacks are mostly concentrated on the traffic of the network. In this type, the attacker creates the traffic in the network and congests the network. In Fig. 8, we have listed the various attacks that are observed against the traffic.

### 4.3.1 Eavesdropping

Eavesdropping attacks are classified into two types: sniffing and traffic analysis. In this attack, the attacker will analyze the traffic or the data transferring between the nodes.

**Sniffing** Using compromised node the attacker attacks the network to damage the packets (data) in the shared medium. With this attack, the attacker not only has packet information but also knows about the topology and addresses in the monitor area just by looking at the starting node and ending addresses. As this is a passive attack it is difficult to identify in the network. To prevent this attack, we must encrypt the messages if the attacker is external.

### 4.3.2 Misappropriation

**Rank Attack** A rank attack is an indirect attack that comes under the network resource category. In this attack, the network performance will be degraded by the malicious node by lower its rank value on purpose, in Fig. 9 we have a malicious node at a deeper level which has reduced its rank value. The child nodes at this level assume this malicious node as the nearest node to the root and they all pass the data packets through this malicious node and this node will drop the data packets. With the rank attack, the path is disturbed to deliver the packets in the network. Another type of rank attack is introduced if the attacker shows the best routing parameters nearest nodes as it is a fake routing that deceives the flow of the network.

**Rank Attack Countermeasures** The specification-based IDS is proposed in [45] that consists of Finite State Machine to detect the attacks early at the monitoring phase only. This architecture monitors the nodes that create the table for the nearest nodes to store the monitoring data. The drawbacks of this scheme are identifying the illegal and allowable node's behavior that depends on system resources. The author in [46] proposed a Sink Based IDS. This is based on the present rank node is compared with the parent rank node. If the present rank node is less than the parent rank node then it is considered as attacker node or mobility of node. Also, among its siblings, the minimum rank is checked and compared with the parent switching threshold. Then, the rank is compared with Node Current rank, and still if it is less we mark that node

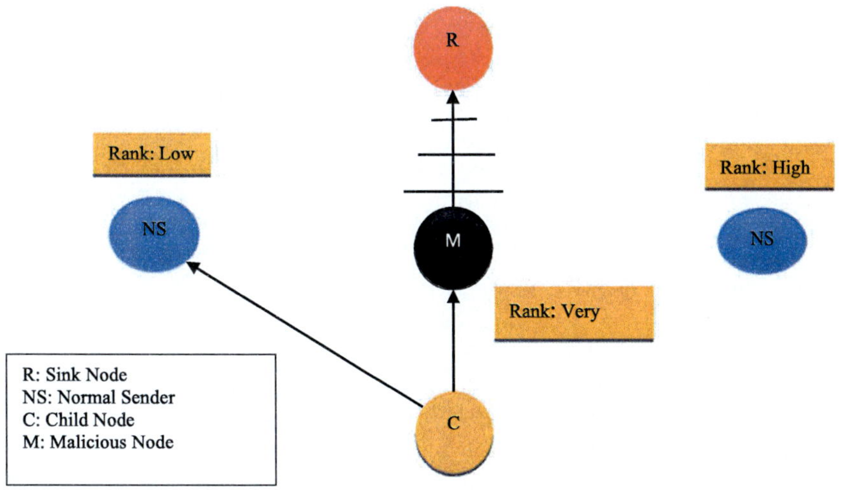


**Fig. 9** Rank attack example [44]

as malicious. In [14] author proposed a Time-Based Trust-Aware RPL protocol. This protocol computes the node trustworthiness based on the recommended and direct values of trust from neighbor nodes. Selection of the secure routing is based on the highest value of trust from neighboring nodes and those who have a lower value of the trust will be considered as malicious or selfish nodes.

## 5 Analysis

In this section, several attacks are discussed with properties and methods along with the techniques that address about the issues and the countermeasures. Table 1 explains about the various attacks against RPL in IoT. The first column explains about the different attacks in the RPL protocol. The attacks are classified into internal attacks and external attacks. Internal attacks are done the internal nodes that are malicious or by some compromised nodes. External attacks are done from the outside nodes that cause the network to degrade its functionality. The second one explains about the methods and techniques that are used by the authors to better explain about the attacks' impact on the networks. The parameters column explains about the various parameters that are considered by the researchers to accomplish their work. The limitations of their work were seen in the fourth column explains about what can be done in the future. The last section explained about the basic tools they used to carry out the experiments. In some attacks like topology, we observed that the attacker must be both the internal and external to carry out the attack.

**Table 1** Summary of attacks against RPL in IoT

| References | Attack | Proposed method | Parameters | Limitations | Tools |
|---|---|---|---|---|---|
| [47] | Sink Hole | DSH-RPL, implemented in four phases<br>This method detects, quarantines the malicious node and data transfer done by encryption | Detection rate, FNR, FPR, PDR | Creating a network overhead by excessive packet transmissions<br>Excessive power consumption.<br>With the increase in the number of nodes the detection rate is decreased | NS-2 |
| [33] | Rank | The low overhead rank attack detection scheme | Accuracy, latency, PDR | Mobility of the nodes is not considered | COOJA |

(continued)

**Table 1** (continued)

| References | Attack | Proposed method | Parameters | Limitations | Tools |
|---|---|---|---|---|---|
| [48] | DIS flooding | Secure-RPL, detection and mitigation of the attacks done in both static and dynamic networks | Packet overhead, Power consumption | Change needs to do in the implementation of RPL. Allows slow DIS flooding, attacker depending on the threshold values | COOJA |
| [49] | Local repair | An IDS implemented using the fuzzy logic, implemented with two phases in which fuzzy based method run on three parameters like distance, RE and ETX. The second fuzzy system identifies whether the attack happens or not | True positive rate, false positive rate | Communication overhead | COOJA |
| [50] | Black Hole | Link-HopValue-based IDS, validates the presence of the attacker node and restrict the attack in the network | PDR, throughput, total packets drop, load analysis | PDR can be improvised, must improve the performance of the network | NS-2 |
| [51] | Rank, worm hole | Intrusion detection system | Energy consumed, power consumption FPR | The accuracy of the network decreases with the increases in the number of nodes | COOJA |
| [52] | Black hole, gray hole | Swarm based artificial bee colony optimization technique | PDR, Throughput | Power consumption is high, the packet drop ratio is high | MATLAB |
| [53] | Hello flood | Gated recurrent unit network model-based deep learning, in this the model was compared with the SVM and logistic regression methods | PDR | Increase in the transmission time with increase in the network, message transmission improves the increase in nodes | COOJA |

(continued)

**Table 1** (continued)

| References | Attack | Proposed method | Parameters | Limitations | Tools |
|---|---|---|---|---|---|
| [54] | Worm hole, black hole, Sinkhole, local repair | DETONAR an IDS capable to deal with multiple RPL Attacks. it is a combination of both signature-based and anomaly-based rules to identify the malicious behavior in the network | Detection rate, false positives, computation time | Identifying the attacks like a sinkhole and local repair is low, Unable to perform in the dynamic networks, Performance in large-scale networks is yet to be done | Machine learning |
| [55] | Version number | ML-LGBM a feature extraction method and LGBM algorithm for optimization | Accuracy, TN rate, FP rate, detection rate | Data is about one type of Attack only one must identify with multiple attack scenarios | COOJA, machine learning |
| [36] | Sinkhole | Distributed IDS architecture | Detection rate, Accuracy | Energy consumption is high because the data collection processing is done at the node end. If the interval between the data collection is low the computation overhead is high | COOJA, machine learning |
| [56] | Rank attack, version number | Blockchain-based framework to analyze the vulnerabilities at each phase of the routing process | Accuracy, precision, F1-score, recall | As IoT devices generate a huge amount of data it is difficult to deal with the growing size of the blockchain, computational and power consumption is high | COOJA, machine learning |
| [57] | Decreased rank | A3-stage process where the data preprocessing, feature extraction and ANN model for detection is done | Accuracy, detection rate, precision, F1-score | Accuracy can be achieved higher with multiclassification The testing phase has lower accuracy when compared with the training phase | Machine learning |

(continued)

**Table 1** (continued)

| References | Attack | Proposed method | Parameters | Limitations | Tools |
|---|---|---|---|---|---|
| [58] | Black hole, gray hole | Lightweight heartbeat protocol | CPU, memory usage, transmission rate, reception rate | Communication overhead was introduced by the protocol because of the additional packets to be sent<br>If a node goes down the detection of that case is not handled in the protocol | COOJA |
| [59] | DAO | Proposed two SecRPL mechanisms one to restrict the number of DAO messages forwarded per child and the other one is to restrict the total number of DAO messages by a specific node | PDR, average power consumption, point to point delay | Less number of nodes are considered a malicious node in the network | COOJA |

## 6 Conclusion and Future Scope

RPL is a well-known network layer protocol for LLN data transmission and communication among nodes in the network. Implementation of security is lacking in LLN networks with resource-constrained nodes. The focus of this article is to explain in detail the working of RPL and various kinds of attacks that can be accomplished on LLN networks. We also investigated the countermeasures on how to identify and restrict the security attacks in RPL. As a future scope, the work can be further implemented on real testbed on restricting the combination of rank, version number, and sinkhole attacks thereby enhancing the performance of RPL protocol.

## References

1. S. Choudhary, N. Kesswani, Cluster-based intrusion detection method for the internet of things, in *2019 IEEE/ACS 16th International Conference on Computer Systems and Applications (AICCSA)* (IEEE, 2019), pp. 1–8
2. B. Farzaneh, M.A. Montazeri, S. Jamali, An anomaly-based IDS for detecting attacks in the RPL-based internet of things, in *2019 5th International Conference on Web Research (ICWR)* (IEEE, 2019), pp. 61–66
3. K. Naito, A survey on the internet-of-things: standards, challenges, and future prospects. J. Inf. Process. **25**, 23–31 (2017)

4. H. Kharrufa, H.A. Al-Kashoash, A.H. Kemp, RPL-based routing protocols in IoT applications: a review. IEEE Sens. J. **19**(15), 5952–5967 (2019)
5. S.Y. Hashemi, F. Shams Aliee, Fuzzy, dynamic and trust based routing protocol for IoT. J. Netw. Syst. Manage. **28**, 1248–1278 (2020)
6. B. Ghaleb, A.Y. Al-Dubai, E. Ekonomou, A. Alsarhan, Y. Nasser, L.M. Mackenzie, A. Boukerche, A survey of limitations and enhancements of the ipv6 routing protocol for low-power and lossy networks: A focus on core operations. IEEE Commun. Surv. Tutor. **21**(2), 1607–1635 (2018)
7. K.N. Ambili, J. Jose, TN-IDS for Network Layer Attacks in RPL based IoT Systems (2020)
8. D.B. Gothawal, S.V. Nagaraj, Anomaly-based intrusion detection system in RPL by applying stochastic and evolutionary game models over IoT environment. Wireless Pers. Commun. **110**(3), 1323–1344 (2020)
9. Z.A. Almusaylim, N.Z. Jhanjhi, A. Alhumam, Detection and mitigation of RPL rank and version number attacks in the Internet of Things: SRPL-RP. Sensors**20**(21), 5997 (2020)
10. M.V.R. Jyothisree, S. Sreekanth, Design and implementation of RPL in Internet of Things, in *Advances in Decision Sciences, Image Processing, Security and Computer Vision* (Springer, Cham, 2020), pp. 707–718
11. K.N. Qureshi, S.S. Rana, A. Ahmed, G. Jeon, A novel and secure attacks detection framework for smart cities industrial internet of things. Sustain. Cities Soc. **61**, 102343 (2020)
12. P. Kaliyar, W.B. Jaballah, M. Conti, C. Lal, LiDL: Localization with early detection of sybil and wormhole attacks in IoT networks. Comput. Secur. **94**, 101849 (2020)
13. N. Djedjig, D. Tandjaoui, F. Medjek, I. Romdhani, Trust-aware and cooperative routing protocol for IoT security. J. Inf. Secur. Appl. **52**, 102467 (2020)
14. D. Airehrour, J.A. Gutierrez, S.K. Ray, SecTrust-RPL: a secure trust-aware RPL routing protocol for Internet of Things. Futur. Gener. Comput. Syst. **93**, 860–876 (2019)
15. E.V. Abhinaya, B. Sudhakar, A secure routing protocol for low power and lossy networks based 6LoWPAN networks to mitigate DIS flooding attacks. J. Ambient Intell. Hum. Comput. 1–12 (2021)
16. G. Simoglou, G. Violettas, S. Petridou, L. Mamatas, Intrusion detection systems for RPL security: a comparative analysis. Comput. Secur. 102219 (2021)
17. A. Thakkar, R. Lohiya, A review on machine learning and deep learning perspectives of IDS for IoT: recent updates, security issues, and challenges. Arch. Comput. Methods Eng. **28**(4), 3211–3243 (2021)
18. A.J. Witwit, A.K. Idrees, A comprehensive review for RPL routing protocol in low power and lossy networks, in *International Conference on New Trends in Information and Communications Technology Applications* (Springer, Cham, 2018), pp. 50–66
19. O. Gaddour, A. Koubâa, RPL in a nutshell: a survey. Comput. Netw. **56**(14), 3163–3178 (2012)
20. A. Raoof, A. Matrawy, C.H. Lung, Enhancing routing security in IoT: performance evaluation of RPL's secure mode under attacks. IEEE Internet Things J. **7**(12), 11536–11546 (2020)
21. H. Bouzebiba, M. Lehsaini, FreeBW-RPL: a new RPL protocol objective function for internet of multimedia things. Wireless Pers. Commun. 1–21 (2020)
22. A. Verma, V. Ranga, Security of RPL based 6LoWPAN networks in the Internet of Things: a review. IEEE Sens. J. **20**(11), 5666–5690 (2020)
23. Z.A. Almusaylim, A. Alhumam, N.Z. Jhanjhi, Proposing a secure RPL based internet of things routing protocol: a review. Ad Hoc Netw. **101**, 102096 (2020)
24. R. Smith, D. Palin, P.P. Ioulianou, V.G. Vassilakis, S.F. Shahandashti, Battery draining attacks against edge computing nodes in IoT networks. Cyber-Phys. Syst. **6**(2), 96–116 (2020)
25. S. Sharma, V.K. Verma, AIEMLA: artificial intelligence enabled machine learning approach for routing attacks on internet of things. J. Supercomput. 1–31 (2021)
26. A. Mayzaud, A. Sehgal, R. Badonnel, I. Chrisment, J. Schönwälder, A study of RPL DODAG version attacks, in *IFIP International Conference on Autonomous Infrastructure, Management and Security* (Springer, Berlin, 2014), pp. 92–104
27. A. Mayzaud, R. Badonnel, I. Chrisment, Detecting version number attacks in RPL-based networks using a distributed monitoring architecture, in *2016 12th International Conference on Network and Service Management (CNSM)* (IEEE, 2016), pp. 127–135

28. A. Dvir, L. Buttyan, VeRA-version number and rank authentication in RPL, in *2011 IEEE Eighth International Conference on Mobile Ad-Hoc and Sensor Systems* (IEEE, 2011), pp. 709–714
29. H. Perrey, M. Landsmann, O. Ugus, T.C. Schmidt, M. Wählisch, TRAIL: Topology authentication in RPL (2013). *arXiv preprint* arXiv:1312.0984
30. M. Nikravan, A. Movaghar, M. Hosseinzadeh, A lightweight defense approach to mitigate version number and rank attacks in low-power and lossy networks. Wireless Pers. Commun. **99**(2), 1035–1059 (2018)
31. A. Arış, S.B.Ö. Yalçın, S.F. Oktuğ, New lightweight mitigation techniques for RPL version number attacks. Ad Hoc Netw. **85**, 81–91 (2019)
32. C. Cervantes, D. Poplade, M. Nogueira, A. Santos, Detection of sinkhole attacks for supporting secure routing on 6LoWPAN for Internet of Things, in *2015 IFIP/IEEE International Symposium on Integrated Network Management (IM)* (IEEE, 2015), pp. 606–611
33. M. Zaminkar, F. Sarkohaki, R. Fotohi, A method based on encryption and node rating for securing the RPL protocol communications in the IoT ecosystem. Int. J. Commun Syst **34**(3), e4693 (2021)
34. M. Alzubaidi, M. Anbar, S.M. Hanshi, Neighbor-passive monitoring technique for detecting sinkhole attacks in RPL networks, in *Proceedings of the 2017 International Conference on Computer Science and Artificial Intelligence* (2017), pp. 173–182
35. M.M. Iqbal, A. Ahmed, U. Khadam, Sinkhole attack in multi-sink paradigm: detection and performance evaluation in RPL based IoT, in *2020 International Conference on Computing and Information Technology (ICCIT-1441)* (IEEE, 2020), pp. 1–5
36. M. Yadollahzadeh Tabari, Z. Mataji, Detecting sinkhole attack in rpl-based internet of things routing protocol. J. AI Data Min. **9**(1), 73–85 (2021)
37. M. Zaminkar, R. Fotohi, SoS-RPL: securing internet of things against sinkhole attack using RPL protocol-based node rating and ranking mechanism. Wireless Pers. Commun. **114**, 1287–1312 (2020)
38. M. Alzubaidi, M. Anbar, Y.W. Chong, S. Al-Sarawi, Hybrid monitoring technique for detecting abnormal behaviour in RPL-based network. J. Commun. **13**(5) (2018)
39. D.B. Gothawal, S.V. Nagaraj, Intrusion detection for enhancing RPL security. Procedia Comput. Sci. **165**, 565–572 (2019)
40. N. Bhalaji, K.S. Hariharasudan, K. Aashika, A trust based mechanism to combat blackhole attack in RPL protocol, in *International Conference on Intelligent Computing and Communication Technologies* (Springer, Singapore, 2019), pp. 457–464
41. R. Sahay, G. Geethakumari, B. Mitra, V. Thejas, Exponential smoothing based approach for detection of blackhole attacks in IoT, in *2018 IEEE International Conference on Advanced Networks and Telecommunications Systems (ANTS)* (IEEE, 2018), pp. 1–6
42. D. Airehrour, J. Gutierrez, S.K. Ray, A testbed implementation of a trust-aware RPL routing protocol, in *2017 27th International Telecommunication Networks and Applications Conference (ITNAC)* (IEEE, 2017), pp. 1–6
43. S. Choudhary, N. Kesswani, Cluster-based intrusion detection method for internet of things, in *2019 IEEE/ACS 16th International Conference on Computer Systems and Applications (AICCSA)* (IEEE, 2019), pp. 1–8
44. M.A. Boudouaia, A. Ali-Pacha, A. Abouaissa, P. Lorenz, Security against rank attack in RPL protocol. IEEE Network **34**(4), 133–139 (2020)
45. A. Le, J. Loo, Y. Luo, A. Lasebae, Specification-based IDS for securing RPL from topology attacks, in *2011 IFIP Wireless Days (WD)* (IEEE, 2011), pp. 1–3
46. U. Shafique, A. Khan, A. Rehman, F. Bashir, M. Alam, Detection of rank attack in routing protocol for Low power and lossy networks. Ann. Telecommun. **73**(7), 429–438 (2018)
47. S. Karmakar, J. Sengupta, S.D. Bit, LEADER: low overhead rank attack detection for securing RPL based IoT, in *2021 International Conference on COMmunication Systems &NETworkS (COMSNETS)* (IEEE, 2021), pp. 429–437
48. A. Verma, V. Ranga, Mitigation of DIS flooding attacks in RPL-based 6LoWPAN networks. Trans. Emerg. Telecommun. Technol. **31**(2), e3802 (2020)

49. B. Farzaneh, M. Koosha, E. Boochanpour, E. Alizadeh, A new method for intrusion detection on RPL routing protocol using fuzzy logic, in *2020 6th International Conference on Web Research (ICWR)* (IEEE, 2020), pp. 245–250
50. G. Soni, R. Sudhakar, A L-IDS against dropping attack to secure and improve RPL performance in WSN aided IoT, in *2020 7th International Conference on Signal Processing and Integrated Networks (SPIN)* (IEEE, 2020), pp. 377–383
51. M.C. Belavagi, B. Muniyal, Multiple intrusion detection in RPL based networks. Int. J. Electr. Comput. Eng. (2088–8708) **10**(1) (2020)
52. P. Rani, S. Verma, G.N. Nguyen, Mitigation of black hole and gray hole attack using swarm inspired algorithm with artificial neural network. IEEE Access **8**, 121755–121764 (2020)
53. S. Cakir, S. Toklu, N. Yalcin, RPL attack detection and prevention in the Internet of Things networks using a GRU based deep learning. IEEE Access **8**, 183678–183689 (2020)
54. A. Agiollo, M. Conti, P. Kaliyar, T. Lin, L. Pajola, 2021. DETONAR: detection of routing attacks in RPL-based IoT. IEEE Trans. Netw. Serv. Manage.
55. M. Osman, J. He, F.M.M. Mokbal, N. Zhu, S. Qureshi, ML-LGBM: a machine learning model based on light gradient boosting machine for the detection of version number attacks in RPL-based networks. IEEE Access (2021)
56. R. Sahay, G. Geethakumari, B. Mitra, A novel blockchain based framework to secure IoT-LLNs against routing attacks. Computing **102**, 2445–2470 (2020)
57. M. Osman, J. He, F.M.M. Mokbal, N. Zhu, Artificial neural network model for decreased rank attack detection in RPL based on IoT networks. Int. J. Netw. Secur. **23**(3), 496–503 (2021)
58. E.G. Ribera, B.M. Alvarez, C. Samuel, P.P. Ioulianou, V.G. Vassilakis, Heartbeat-based detection of blackhole and greyhole attacks in RPL networks, in *2020 12th International Symposium on Communication Systems, Networks and Digital Signal Processing (CSNDSP)* (IEEE, 2020), pp. 1–6
59. I. Wadhaj, B. Ghaleb, C. Thomson, A. Al-Dubai, W.J. Buchanan, Mitigation mechanisms against the DAO attack on the routing protocol for low power and lossy networks (RPL). IEEE Access **8**, 43665–43675 (2020)



**Ajay Dilip Kumar Marapatla** is a research scholar in the department of Computer Science and Engineering at Pondicherry engineering college, Puducherry.

**Dr. E. Ilavarasan** is a Professor of the Department of Computer Science & Engineering at Pondicherry Engineering College, Puducherry. He has more than 25 years of experience in the teaching field. He is an expert in Web service computing.

# An Optimal Policy with Parabolic Demand Carry Forwarded with Three-Parameter Weibull Distribution Deterioration Rate, Scarcity and Salvage Value

**Kailash Chandra Paul, Chandan Kumar Sahoo, and Manas Ranjan Sarangi**

**Abstract** In this work, an EOQ model has been established on behalf of diminishing the goods with parabolic rate of demand and three-parameter weibull deteriorating rate, backlogged partially and allowing scarcity. In deteriorated units, salvage value is included. Our main objective is to reduce the gross relevant value using the optimizing tool during the scarcity, event span and order magnitude. The practicality of the aforementioned model as well as the sensitivity appraisals of the magnificent intention is to communicate a blend of aspect which has been distinguished in view of a mathematical illustration. Ultimately the model significance has been arrived through a wholesome entire inventory cost.

**Keywords** Parabolic rate of demand · Three-factor weibull rate of deterioration · Shortages · Salvage value

## 1 Introduction

Deteriorating import suggested enervate of the product usage in another implementation. Matters concerned to the photoelectric portfolio, radioactive aspect, electronic arena, etc. enfeeble review the inherent value as well as service provision from the time perspective. Hence degradation of warehouse importance considered absolute paramount and consequently warehouse reviewer or inventory analysts recognize this quintessence fact. Most volatile solutions like alcohol, turpentine, gasoline, sprit etc. undergo substantial degradation via evaporation procedure with time elapse. Similarly, radioactive and electronic substances, photographic films, also degrade its value

K. C. Paul (✉) · M. R. Sarangi
GIET University, Gunupur, Raygada, Odisha, India
e-mail: kailash.chandrapaul@giet.edu

M. R. Sarangi
e-mail: manasranjan@giet.edu

C. K. Sahoo
GIET, Ghangapatna, Odisha, India



S. Patnaik et al. (eds.), *Smart and Sustainable Technologies: Rural and Tribal Development Using IoT and Cloud Computing*, Advances in Sustainability Science and Technology,
https://doi.org/10.1007/978-981-19-2277-0_3

with the time elapse. Consequently, substantial degradation in inventory is absolute realistic aspect and henceforth the warehouse analyst experiences the significant usage of this aspect into their deliberation. Ghosh and Chaudhuri [1] presented as well as justified the significance of the worsening object immediate supply, quadratic time-varying order and scarcity. Whitin [2] has been monitored the very degradation of the aforesaid units on completing their contract span. Later on Ghare and Schrader [3] the very primitive scholars who purposefully confirmed the platform upon which the design of warehouse structure concerned to degrading substances has been played out with the aid of differential equation. The duo keenly monitored the classical warehouse design bypassing scarcity with fixing the degradation pace. The researchers namely Shah and Jaiswal [4] as well as Aggarwal [5] fabricated a suitable order intensity warehouse design taking into account the invariable decline pace. Chang [6] developed an inventory frame with deteriorating items corresponding to price rises while trader acknowledgment associated to demand amount on behalf of the aforesaid objects through time-varying order and worsening rates. Alamri and Balkhi [7] have published the consequences of knowledge as well as wavering out the most favorable building up lot amount in favor of diminishing objects considering time changeable order and diminishing rates. Most of the scholars concerned to this portfolio have commenced their research considering fashion substances. In the beginning, Hung [8] has also implemented the most profound sort order, worsening as well as backorder rates. Mishra and Singh [9] have observed a very suitable EOQ warehouse design in favor of slope kind order, and time reliant diminishing objects considering retrieve worth and scarcities and apart from that the same work has also been propounded that an EOQ model for diminishing objects perpetuated with steady refill rate holding power form order as well as cubic polynomial deteriorating rate. Patel [10] has used different demand rate and instantaneous deteriorating items with partial backlogging. Karthikeyan and Santhi [11] have also put forwarded an inventory formulation considering cubic order rate along with time reliant asset value proposing diminishing objects retrieve worth. Singh et al. [12] propounded their observation for diminishing objects considering seasonal and stock reliant order through scarcity. From the routine life scenario, all through the span of scarcity, it has been monitored that the eagerness exhibits via buyers looking toward goods drops with the time elapse. In order to eliminate the raw goods scarcity, work in progress issue or to meet the uncertain demand the very inventory is preferable. Chang et al. [13] being the primitive researchers have successfully monitored the inventory rate which relies upon the time elapse. From the practical aspect, most of the substances concerned to fashionable goods, better technological substances, etc., it has been observed that the more the waiting span, the lower be the inventory rate. Considering storage pace as inconsistent, the same relies on the span of time elapse on the verge of further refilling. Sahoo et al. [14] observed the EOQ model for declining substances with cubic order, variable degradation and unbalanced storage. Sahoo and Paul [15] recognized an EOQ model for cubic degrading substances promoted keeping weibull requisition and lacking scarcity. Sahoo et al. [16] studied the two warehouses EOQ inventory model of declining substances possessing exponential declining order, restricted suspension in cost including salvage values.

In this paper, for deteriorating items have been expressed the rate of demand measuring parabolic utility of time and deterioration rate as three factors weibull distribution function along with retrieve assessment moderately stored and endorsed scarcity. At the unlike sample of time relying requisition, the majority reality procedure is taken as the parabolic form since the same exhibits mutually increasing and decreasing in requisition. Parabolic form of demand is expressed as $R(t) = p+qt-rt^2, p > 0, q \neq 0, r \neq 0$. When $r=0$ and $q=r=0$, for these goods, the three-factor weibull distribution declining rate with time dependent deterioration is used and it is represented by $Z(t) = \theta\beta(t-\mu)^{\beta-1}, \theta(0 < \theta \ll 1), \beta(> 0), \mu(0 < \mu < 1)$. Here $\theta, \beta, \mu$ regarded as the scale, shape and position factor respectively.

## 2 Model Formulation

In accordance with the valuation of the expressed prototype, the following suppositions have been put forward.

### 2.1 *Assumptions*

- Replenishment is infinite.
- Neglect the lead time.
- On time frame the rate of demand is a parabolic function.
- Declining rate is considered as a three-factor weibull distribution utility.
- Scarcity is considered as well as backlogged.
- The salvage value per item in per unit time is calculated.

### 2.2 *Notations*

- $T$: Every ordering cycle particular span.
- $t_1$: The period at which the stage of inventory considered null.
- $I_1(t)$: The inventory on hand at the period $t$, when $t \geq 0$.
- $R(t)$: Considered as parabolic requisition rate i.e. $R(t) = p+qt-rt^2, p > 0, q \neq 0, r \neq 0$.
- $Z(t)$: Being the three factor weibull distribution for the declining pace i.e. $Z(t) = \theta\beta(t-\mu)^{\beta+1}$ $0\langle\theta \ll 1, \beta\rangle 0$ and $0 < \mu < 1$. Here $\theta, \beta, \mu$ regarded as the scale, shape and position factor correspondingly.
- $B(t)$: Backlogging rate i.e. $B(T-t) = \frac{1}{1+\lambda(T-t)}, \lambda >$, where $\lambda$ is considered as a factor of backlogging.
- $\lambda$: The parameter of constant backlogging, where $0 \leq \lambda \leq 1$.

- $\xi$: Salvage cost affiliated during series prosecution considering declining units, where, $0 \le \xi \le 1$.
- $Q_0$: Ordering quantity.
- $A_0$: The order per ordering cost.
- $C_h$: Per unit warehouse holding cost.
- $C_p$: Per unit buying cost.
- $C_b$: Per unit scarcity cost.
- $C_I$: The lost cost sales per unit.
- $\mathrm{TC}(t_1, T)$: The total relevant cost per unit time.

## 2.3 Mathematical Formulation

The system of inventory at time $t = 0$, a lot size of particular unit item launch into the system. In interval, due to the demand factor the inventory level decreases gradually and partially because decay and will disappear with time elapse. Scarcity accepts during gap $[t_1, T]$ along with each requisition within scarcity duration in the interval $[t_1, T]$ is partially backlogged. Suppose be the level of inventory at time, the inventory range in the interval (0, $t_1$) is represented by differential equation:

$$\frac{\mathrm{d}I_1}{\mathrm{d}t} + z(t)I_1(t) = -R(t).$$

where $R(t) = p + qt - rt^2, p > 0, q > 0, r > 0$ and $z(t) = \theta\beta(t-\mu)^{\beta-1}$

$$\frac{\mathrm{d}I_1(t)}{\mathrm{d}t} + \theta\beta(t-\mu)^{\beta-1} I_1(t) = -\left(p + qt - rt^2\right), 0 < t < t_1. \tag{1}$$

$$= -\left[p\int \mathrm{e}^{\theta(t-\mu)^\beta}\mathrm{d}t + q\int t\mathrm{e}^{\theta(t-\mu)^\beta}\mathrm{d}t - r\int t^2\mathrm{e}^{\theta(t-\mu)^\beta}\mathrm{d}t\right] + K. \tag{2}$$

At $I(t_1) = 0, 0 < t < t_1$ then we get

$$I_1(t) = \left[\begin{array}{c} \left(pt_1 + \frac{qt_1^2}{2} - \frac{rt_1^3}{3}\right) \\ +\theta\left\{\begin{array}{c} \frac{p(t_1-\mu)^{\beta+1}}{\beta+1} + q\left(\mu\frac{(t_1-\mu)^{\beta+1}}{\beta+1} + \frac{(t_1-\mu)^{\beta+2}}{\beta+2}\right) \\ -r\left(\mu^2\frac{(t_1-\mu)^{\beta+1}}{\beta+1} + 2\mu\frac{(t_1-\mu)^{\beta+2}}{\beta+2} + \frac{(t_1-\mu)^{\beta+3}}{\beta+3}\right) \end{array}\right\} \\ -\left(pt + \frac{qt^2}{2} - \frac{rt^3}{3}\right) \\ -\theta\left\{\begin{array}{c} \frac{p(t-\mu)^{\beta+1}}{\beta+1} + q\left(\mu\frac{(t-\mu)^{\beta+1}}{(t-\mu)} + \frac{(t-\mu)^{\beta+2}}{\beta+2}\right) \\ -r\left(\mu^2\frac{(t-\mu)^{\beta+1}}{\beta+1} + 2\mu\frac{(t-\mu)^{\beta+2}}{\beta+2} + \frac{(t-\mu)^{\beta+3}}{\beta+3}\right) \end{array}\right\} \end{array}\right]\mathrm{e}^{-\theta(t-\mu)^\beta}. \tag{3}$$

By neglecting the higher power $\theta$ as $0 < \theta \ll 1$. The maximum positive inventory level of the cycle at $t = 0$ is $I_1(0) = I_M$ in Eq. (3), then we get.

$$I_1(o) = I_M = \left[\left[\begin{array}{c}\left(pt_1 + \frac{qt_1^2}{2} - \frac{rt_1^3}{3}\right) \\ +\theta\left\{\begin{array}{c}\frac{p(t_1-\mu)^{\beta+1}}{\beta+1} + q\left(\mu\frac{(t_1-\mu)^{\beta+1}}{\beta+1} + \frac{(t_1-\mu)^{\beta+2}}{\beta+2}\right) \\ -r\left(\mu^2\frac{(t_1-\mu)^{\beta+1}}{\beta+1} + 2\mu\frac{(t_1-\mu)^{\beta+2}}{\beta+2} + \frac{(t_1-\mu)^{\beta+3}}{\beta+3}\right)\end{array}\right\}\end{array}\right]e^{-\theta(-\mu)^\beta}\right.$$
$$\left.-\left[\theta\left\{\begin{array}{c}\frac{p(-\mu)^{\beta+1}}{\beta+1} + q\left(\mu\frac{(-\mu)^{\beta+1}}{(\beta+1)} + \frac{(-\mu)^{\beta+2}}{\beta+2}\right) \\ -r\left(\mu^2\frac{(-\mu)^{\beta+1}}{\beta+1} + 2\mu\frac{(-\mu)^{\beta+2}}{\beta+2} + \frac{(-\mu)^{\beta+3}}{\beta+3}\right)\end{array}\right\}\right]\right]e^{-\theta(t-\mu)^\beta}. \tag{4}$$

When time is $t_1$, the stage of storage attains null, consequently within the range $[t_1, T]$ accepted inventory and the same relies on the part of requisition pace becomes backlogged at the rate $B(T-t) = \frac{1}{1+\lambda(T-t)}$. Hence the behavior of the warehouse unit during the period '$t$' expressed via differential equation:

$$\frac{dI_1(t)}{dt} = -R(t)B(t), t_1 \le t \le T.$$

$B(T-t) = \frac{1}{1+\lambda(T-t)}$, where $\lambda > 0$ then the above equation represented by

$$\frac{dI_1(t)}{dt} = \frac{-(p + qt - rt^2)}{1 + \lambda(T-t)}, t_1 \le t \le T. \tag{5}$$

So the solution of Eq. (5) with the condition $I_2(t_1) = 0$ is written as

$$I_2(t) = \frac{1}{\lambda^3}\left[\begin{array}{c}\{p\lambda^2 + q\lambda(\lambda T + 1) - r(\lambda T + 1)^2\}.\{\{\ln(1 + \lambda(T-t)) \\ - \ln(1 + \lambda(T - t_1))\}\} - \lambda\{q\lambda - 2r(\lambda T + 1)\}(t_1 - t) \\ -\frac{r}{2}\{2\lambda(t_1 - t) + \lambda^2(-2Tt + t^2 + 2Tt_1 - t_1^2)\}\end{array}\right], t_1 \le t \le T. \tag{6}$$

The maximum back order units at $t = T$ is given by

$$I_B = -I_2(T) = \frac{1}{\lambda^3}\left[\begin{array}{c}\{p\lambda^2 + q\lambda(\lambda T + 1) - r(\lambda T + 1)^2\}ln\{1 + \lambda(T - t_1)\} \\ -\lambda\{q\lambda - 2r(\lambda T + 1\}(T - t_1) \\ -\frac{r}{2}\{\lambda^2(T - t_1)^2 + 2\lambda(T - t_1)\}\end{array}\right]. \tag{7}$$

In the time period [0, $T$] order size is written as

$$Q_0 = I_M + I_B = \left[\left[\begin{array}{c} \left(pt_1 + \frac{qt_1^2}{2} - \frac{rt_1^3}{3}\right) \\ +\theta\left\{\begin{array}{c} \frac{p(t_1-\mu)^{\beta+1}}{\beta+1} + q\left(\mu\frac{(t_1-\mu)^{\beta+1}}{\beta+1} \right. \\ \left. +\frac{(t_1-\mu)^{\beta+2}}{\beta+2}\right) - r\left(\mu^2\frac{(t_1-\mu)^{\beta+1}}{\beta+1} \right. \\ \left. +2\mu\frac{(t_1-\mu)^{\beta+2}}{\beta+2} + \frac{(t_1-\mu)^{\beta+3}}{\beta+3}\right) \end{array}\right\} \end{array}\right] e^{-\theta(-\mu)^{\beta}}\right]$$
$$- \left[\theta\left\{\begin{array}{c} \frac{p(-\mu)^{\beta+1}}{\beta+1} + q\left(\mu\frac{(-\mu)^{\beta+1}}{(\beta+1)} + \frac{(-\mu)^{\beta+2}}{\beta+2}\right) \\ -r\left(\mu^2\frac{(-\mu)^{\beta+1}}{\beta+1} + 2\mu\frac{(-\mu)^{\beta+2}}{\beta+2} + \frac{(-\mu)^{\beta+3}}{\beta+3}\right) \end{array}\right\}\right] e^{-\theta(t-\mu)^{\beta}}$$
$$+ \frac{1}{\lambda^3}\left[\begin{array}{c} \{p\lambda^2 + q\lambda(\lambda T + 1) - r(\lambda T + 1)^2\}\ln\{1 + \lambda(T - t_1)\} \\ -\lambda\{q\lambda - 2r(\lambda T + 1\}(T - t_1) \\ -\frac{r}{2}\{\lambda^2(T - t_1)^2 + 2\lambda(T - t_1)\} \end{array}\right]. \quad (8)$$

In a nut shell the model significant worth can be deliberately articulated as the distinct of the summation of the requisition, inventory holding, deterioration, as well as shortage cost on behalf of backlogging and prospect expenditure owing to misplaced sales and salvage assessment of the degraded goods.

The ordering cost per order is

$$CO = A_0. \quad (9)$$

The inventory carrying cost is

$$\mathrm{CCI} = C_h \int_0^{t_1} I_1(t)\mathrm{d}t$$
$$= C_h\left(t_1 - \theta\frac{(t_1-\mu)^{\beta+1} - (-\mu)^{\beta+1}}{\beta+1}\right)$$
$$\left[\begin{array}{c} pt_1 + \frac{qt_1^2}{2} - \frac{rt_1^3}{3} \\ +\theta\left\{\begin{array}{c} p\frac{(t_1-\mu)^{\beta+1}}{\beta+1} + q\left(\frac{(t_1-\mu)^{\beta+2}}{\beta+2} + \mu\frac{(t_1-\mu)^{\beta+1}}{\beta+1}\right) \\ -r\left(\frac{(t_1-\mu)^{\beta+3}}{\beta+3} + 2\mu\frac{(t_1-\mu)^{\beta+2}}{\beta+2} + \mu^2\frac{(t_1-\mu)^{\beta+1}}{\beta+1}\right) \end{array}\right\} \end{array}\right]$$
$$- C_h\left[\begin{array}{c} \left(p\mu + \frac{q\mu^2}{2} - \frac{r\mu^3}{3}\right)t_1 + \frac{(t_1-\mu)^2-(-\mu)^2}{2}(p + q\mu - r\mu^2) \\ +\frac{(t_1-\mu)^3-(-\mu)^3}{3}\left(\frac{q}{2} - r\mu\right) - \frac{r}{3}\frac{(t_1-\mu)^4-(-\mu)^4}{4} \end{array}\right]$$
$$+ \theta C_h\left[\begin{array}{c} \frac{(t_1-\mu)^{\beta+1}-(-\mu)^{\beta+1}}{\beta+1}\left(p\mu + \frac{q\mu^2}{2} - \frac{r\mu^3}{3}\right) \\ +\frac{(t_1-\mu)^{\beta+2}-(-\mu)^{\beta+2}}{\beta+2}(p + q\mu - r\mu^2) \\ +\frac{(t_1-\mu)^{\beta+3}-(-\mu)^{\beta+3}}{\beta+3}\left(\frac{q}{2} - r\mu\right) - \frac{r}{3}\frac{(t_1-\mu)^{\beta+4}-(-\mu)^{\beta+4}}{\beta+4} \end{array}\right]$$

$$-\theta C_h\left[\begin{array}{c} p\frac{(t_1-\mu)^{\beta+2}-(-\mu)^{\beta+2}}{(\beta+1)(\beta+2)}+q\left(\frac{(t_1-\mu)^{\beta+3}-(-\mu)^{\beta+3}}{(\beta+2)(\beta+3)}+\mu\frac{(t_1-\mu)^{\beta+2}-(-\mu)^{\beta+2}}{(\beta+1)(\beta+2)}\right)\\ -r\left\{\begin{array}{c}\frac{(t_1-\mu)^{\beta+4}-(-\mu)^{\beta+4}}{(\beta+3)(\beta+4)}+2\mu\frac{(t_1-\mu)^{\beta+3}-(-\mu)^{\beta+3}}{(\beta+2)(\beta+3)}\\ +\mu^2\frac{(t_1-\mu)^{\beta+2}-(-\mu)^{\beta+2}}{(\beta+1)(\beta+2)}\end{array}\right\}\end{array}\right]. \tag{10}$$

(By neglecting the higher index of $\theta$ as $0 < \theta \ll 1$ and higher power of $(t_1 - \mu)$ i.e. more than $(\beta + 4)$).

The cost of deterioration is

$$\mathrm{DC} = C_p\int_0^{t_1} z(t)I_1(t)\mathrm{d}t = C_p\int_0^{t_1}\theta\beta(t-\mu)^{\beta-1}I_1(t)\mathrm{d}t$$

$$=\theta C_p\left((t_1-\mu)^\beta-(-\mu)^\beta\right)\left[\begin{array}{c} pt_1+q\frac{t_1^2}{2}-r\frac{t_1^3}{3}\\ +\theta\left\{\begin{array}{c} p\frac{(t_1-\mu)^{\beta+1}}{\beta+1}+q\left(\frac{(t_1-\mu)^{\beta+2}}{\beta+2}+\mu\frac{(t_1-\mu)^{\beta+1}}{\beta+1}\right)\\ -r\left(\frac{(t_1-\mu)^{\beta+3}}{\beta+3}+2\mu\frac{(t_1-\mu)^{\beta+2}}{\beta+2}+\mu^2\frac{(t_1-\mu)^{\beta+1}}{\beta+1}\right)\end{array}\right\}\end{array}\right]$$

$$-\theta\beta C_p\left[\begin{array}{c} p\left\{\frac{(t_1-\mu)^{\beta+1}-(-\mu)^{\beta+1}}{\beta+1}+\mu\frac{(t_1-\mu)^{\beta}-(-\mu)^{\beta}}{\beta}\right\}\\ +\frac{q}{2}\left\{\frac{(t_1-\mu)^{\beta+2}-(-\mu)^{\beta+2}}{\beta+2}+2\mu\frac{(t_1-\mu)^{\beta+1}-(-\mu)^{\beta+1}}{\beta+1}+\mu^2\frac{(t_1-\mu)^{\beta}-(-\mu)^{\beta}}{\beta}\right\}\\ -\frac{r}{3}\left\{\begin{array}{c}\frac{(t_1-\mu)^{\beta+3}-(-\mu)^{\beta+3}}{\beta+3}+3\mu\frac{(t_1-\mu)^{\beta+2}-(-\mu)^{\beta+2}}{\beta+2}\\ +3\mu^2\frac{(t_1-\mu)^{\beta+1}-(-\mu)^{\beta+1}}{\beta+1}+\mu^3\frac{(t_1-\mu)^{\beta}-(-\mu)^{\beta}}{\beta}\end{array}\right\}\end{array}\right] \tag{11}$$

(By neglecting the higher index of $\theta$ as $0 < \theta \ll 1$and higher power of$(t_1 - \mu)$ i.e. more than $(\beta + 3)$).

The shortage cost due to backlogging is

$$\begin{aligned}\mathrm{SC} &= C_q\int_{t_1}^{T}\{-I_2(t)\}\mathrm{d}t\\ &= -\frac{C_q}{\lambda^3}\{p\lambda^2+q\lambda(\lambda T+1)-r(\lambda T+1)^2\}.(T\\ &\quad -t_1).\{1-2\ln(1+\lambda(T-t_1))\}-\frac{C_q}{2\lambda^2}(T-t_1)^2\{q\lambda\\ &\quad -2r(\lambda T+1)-\lambda r\}-C_q r(T-t_1)\left\{\frac{2Tt_1}{3}-\frac{1}{3}\left(T^2-t_1^2\right)\right\}.\end{aligned} \tag{12}$$

Again the occasional worth on behalf of missing sales;

$$
\begin{aligned}
\mathrm{COL} &= C_1 \int_{t_1}^{T} [R(t)\{1 - B(T - t)\}]\mathrm{d}t \\
&= C_1 \int_{t_1}^{T} \left[\left(p + qt - rt^2\right)\left\{1 - \frac{1}{1 + \lambda(T - t)}\right\}\right]\mathrm{d}t \\
&= C_1(T - t_1)\left[p + \frac{q}{2}(T + t_1) - \frac{r}{3}\left\{(T + t_1)^2 - Tt_1\right\}\right] \\
&\quad + \frac{C_1}{\lambda^3}\left\{p\lambda^2 + q\lambda(\lambda T + 1) - r(\lambda T + 1)^2\right\}\ln(1 + \lambda(T - t_1)) \\
&\quad - \frac{C_1}{2\lambda}\{2q\lambda - r(3\lambda T + \lambda t_1 + 2)\}(T - t_1).
\end{aligned}
\tag{13}
$$

Considering per unit time, the salvage worth of declining goods becomes:

$$
\mathrm{SV} = \xi C_1\left[Q_0 - \int_0^T R(t)\mathrm{d}t\right]
$$

$$
= \xi C_1 \left[\begin{array}{l}
\left[\begin{array}{c}
p(t_1 - T) + \frac{q}{2}\left(t_1^2 - T^2\right) - \frac{r}{3}\left(t_1^3 - T^3\right) \\
+\theta\left\{\begin{array}{c}
p\frac{(t_1-\mu)^{\beta+1}}{\beta+1} + q\left(\frac{(t_1-\mu)^{\beta+2}}{\beta+2} + \mu\frac{(t_1-\mu)^{\beta+1}}{\beta+1}\right) \\
-r\left(\frac{(t_1-\mu)^{\beta+3}}{\beta+3} + 2\mu\frac{(t_1-\mu)^{\beta+2}}{\beta+2} + \mu^2\frac{(t_1-\mu)^{\beta+1}}{\beta+1}\right)
\end{array}\right\}
\end{array}\right]\mathrm{e}^{-\theta(-\mu)^\beta} \\
-\theta\left\{\begin{array}{c}
p\frac{(-\mu)^{\beta+1}}{\beta+1} + q\left(\frac{(-\mu)^{\beta+2}}{\beta+2} + \mu\frac{(-\mu)^{\beta+1}}{\beta+1}\right) \\
-r\left(\frac{(-\mu)^{\beta+3}}{\beta+3} + 2\mu\frac{(-\mu)^{\beta+2}}{\beta+2} + \mu^2\frac{(-\mu)^{\beta+1}}{\beta+1}\right)
\end{array}\right\}\mathrm{e}^{-\theta(-\mu)^\beta} \\
+\frac{1}{\lambda^3}\left[\begin{array}{c}
\left\{a\lambda^2 + b\lambda(\lambda T + 1) - c(\lambda T + 1)^2\right\}\ln(1 \\
+\lambda(T - t_1)) + \lambda\{b\lambda - 2c(\lambda T + 1)\}(t_1 - T) \\
-\frac{c}{2}\left\{\lambda^2(t_1 - T)^2 - 2\lambda(t_1 - T)\right\}
\end{array}\right]
\end{array}\right].
\tag{14}
$$

Looking toward the retailer, the gross appropriate value per unit time becomes:

$$
TC(t_1,T)=\frac{1}{T}[CO+CCI+DC+SC+COL-SV]=
$$

$$
\frac{1}{T}\left[\begin{array}{l}
A_0+C_h\left(t_1-\theta\frac{(t_1-\mu)^{\beta+1}-(-\mu)^{\beta+1}}{\beta+1}\right)\left[\begin{array}{c} pt_1+\frac{qt_1^2}{2}-\frac{rt_1^3}{3}\\ +\theta\left\{\begin{array}{c} p\frac{(t_1-\mu)^{\beta+1}}{\beta+1}+q\left(\frac{(t_1-\mu)^{\beta+2}}{\beta+2}+\mu\frac{(t_1-\mu)^{\beta+1}}{\beta+1}\right)\\ -r\left(\frac{(t_1-\mu)^{\beta+3}}{\beta+3}+2\mu\frac{(t_1-\mu)^{\beta+2}}{\beta+2}+\mu^2\frac{(t_1-\mu)^{\beta+1}}{\beta+1}\right)\end{array}\right\}\end{array}\right]\\
-C_h\left[\begin{array}{c}\left(p\mu+\frac{q\mu^2}{2}-\frac{r\mu^3}{3}\right)t_1+\frac{(t_1-\mu)^2-(-\mu)^2}{2}(p+q\mu-r\mu^2)\\ +\frac{(t_1-\mu)^3-(-\mu)^3}{3}\left(\frac{q}{2}-r\mu\right)-\frac{r}{3}\frac{(t_1-\mu)^4-(-\mu)^4}{4}\end{array}\right]\\
+\theta C_h\left[\begin{array}{c}\frac{(t_1-\mu)^{\beta+1}-(-\mu)^{\beta+1}}{\beta+1}\left(p\mu+\frac{q\mu^2}{2}-\frac{r\mu^3}{3}\right)\\ +\frac{(t_1-\mu)^{\beta+2}-(-\mu)^{\beta+2}}{\beta+2}(p+q\mu-r\mu^2)\\ +\frac{(t_1-\mu)^{\beta+3}-(-\mu)^{\beta+3}}{\beta+3}\left(\frac{q}{2}-r\mu\right)-\frac{r}{3}\frac{(t_1-\mu)^{\beta+4}-(-\mu)^{\beta+4}}{\beta+4}\end{array}\right]\\
-\theta C_h\left[\begin{array}{c}p\frac{(t_1-\mu)^{\beta+2}-(-\mu)^{\beta+2}}{(\beta+1)(\beta+2)}+q\left(\frac{(t_1-\mu)^{\beta+3}-(-\mu)^{\beta+3}}{(\beta+2)(\beta+3)}+\mu\frac{(t_1-\mu)^{\beta+2}-(-\mu)^{\beta+2}}{(\beta+1)(\beta+2)}\right)\\ -r\left\{\begin{array}{c}\frac{(t_1-\mu)^{\beta+4}-(-\mu)^{\beta+4}}{(\beta+3)(\beta+4)}+2\mu\frac{(t_1-\mu)^{\beta+3}-(-\mu)^{\beta+3}}{(\beta+2)(\beta+3)}\\ +\mu^2\frac{(t_1-\mu)^{\beta+2}-(-\mu)^{\beta+2}}{(\beta+1)(\beta+2)}\end{array}\right\}\end{array}\right]\\
+\theta C_p((t_1-\mu)^{\beta}-(-\mu)^{\beta})\left[\begin{array}{c} pt_1+q\frac{t_1^2}{2}-r\frac{t_1^3}{3}\\ +\theta\left\{\begin{array}{c} p\frac{(t_1-\mu)^{\beta+1}}{\beta+1}+q\left(\frac{(t_1-\mu)^{\beta+2}}{\beta+2}+\mu\frac{(t_1-\mu)^{\beta+1}}{\beta+1}\right)\\ -r\left(\frac{(t_1-\mu)^{\beta+3}}{\beta+3}+2\mu\frac{(t_1-\mu)^{\beta+2}}{\beta+2}+\mu^2\frac{(t_1-\mu)^{\beta+1}}{\beta+1}\right)\end{array}\right\}\end{array}\right]\\
-\theta\beta C_p\left[\begin{array}{c} p\left\{\frac{(t_1-\mu)^{\beta+1}-(-\mu)^{\beta+1}}{\beta+1}+\mu\frac{(t_1-\mu)^{\beta}-(-\mu)^{\beta}}{\beta}\right\}\\ +\frac{q}{2}\left\{\frac{(t_1-\mu)^{\beta+2}-(-\mu)^{\beta+2}}{\beta+2}+2\mu\frac{(t_1-\mu)^{\beta+1}-(-\mu)^{\beta+1}}{\beta+1}+\mu^2\frac{(t_1-\mu)^{\beta}-(-\mu)^{\beta}}{\beta}\right\}\\ -\frac{r}{3}\left\{\begin{array}{c}\frac{(t_1-\mu)^{\beta+3}-(-\mu)^{\beta+3}}{\beta+3}+3\mu\frac{(t_1-\mu)^{\beta+2}-(-\mu)^{\beta+2}}{\beta+2}\\ +3\mu^2\frac{(t_1-\mu)^{\beta+1}-(-\mu)^{\beta+1}}{\beta+1}+\mu^3\frac{(t_1-\mu)^{\beta}-(-\mu)^{\beta}}{\beta}\end{array}\right\}\end{array}\right]\\
-\frac{C_q}{\lambda^3}\{p\lambda^2+q\lambda(\lambda T+1)-r(\lambda T+1)^2\}.(T-t_1).\{1-2ln(1+\lambda(T-t_1))\}\\
-\frac{C_q}{2\lambda^2}(T-t_1)^2\{q\lambda-2r(\lambda T+1)-\lambda r\}-C_q r(T-t_1)\left\{\frac{2Tt_1}{3}-\frac{1}{3}(T^2-t_1^2)\right\}\\
+C_1(T-t_1)\left[p+\frac{q}{2}(T+t_1)-\frac{r}{3}\{(T+t_1)^2-Tt_1\}\right]\\
+\frac{C_1}{\lambda^3}\{p\lambda^2+q\lambda(\lambda T+1)-r(\lambda T+1)^2\}ln(1+\lambda(T-t_1))\\
-\frac{C_1}{2\lambda}\{2q\lambda-r(3\lambda T+\lambda t_1+2)\}(T-t_1)\\
-\xi C_1\left[\begin{array}{l}\left[\begin{array}{c} p(t_1-T)+\frac{q}{2}(t_1^2-T^2)-\frac{r}{3}(t_1^3-T^3)\\ +\theta\left\{\begin{array}{c} p\frac{(t_1-\mu)^{\beta+1}}{\beta+1}+q\left(\frac{(t_1-\mu)^{\beta+2}}{\beta+2}+\mu\frac{(t_1-\mu)^{\beta+1}}{\beta+1}\right)\\ -r\left(\frac{(t_1-\mu)^{\beta+3}}{\beta+3}+2\mu\frac{(t_1-\mu)^{\beta+2}}{\beta+2}+\mu^2\frac{(t_1-\mu)^{\beta+1}}{\beta+1}\right)\end{array}\right\}\end{array}\right]e^{-\theta(-\mu)^{\beta}}\\
-\theta\left\{\begin{array}{c} p\frac{(-\mu)^{\beta+1}}{\beta+1}+q\left(\frac{(-\mu)^{\beta+2}}{\beta+2}+\mu\frac{(-\mu)^{\beta+1}}{\beta+1}\right)\\ -r\left(\frac{(-\mu)^{\beta+3}}{\beta+3}+2\mu\frac{(-\mu)^{\beta+2}}{\beta+2}+\mu^2\frac{(-\mu)^{\beta+1}}{\beta+1}\right)\end{array}\right\}e^{-\theta(-\mu)^{\beta}}\\
+\frac{1}{\lambda^3}\left[\begin{array}{c}\{a\lambda^2+b\lambda(\lambda T+1)-c(\lambda T+1)^2\}\ln(1+\lambda(T-t_1))\\ +\lambda\{b\lambda-2c(\lambda T+1)\}(t_1-T)-\frac{c}{2}\{\lambda^2(t_1-T)^2-2\lambda(t_1-T)\}\end{array}\right]\end{array}\right]
\end{array}\right]. \quad (15)
$$

Ultimately, the very target of the present model is to reduce the appropriate value taking time as a factor:

$$\frac{\partial(TRC(t_1, T))}{\partial t_1} = \frac{1}{T}[P_1 + P_2 + P_3 + P_4 + P_5 - P_6]. \quad (16)$$

where $P_1 = 0$

$$P_2 = C_h\left\{1 - \theta(t_1 - \mu)^\beta\right\}$$

$$\left[\begin{array}{c} pt_1 + \frac{qt_1^2}{2} - \frac{rt_1^3}{3} \\ +\theta\left\{\begin{array}{c} p\frac{(t_1-\mu)^{\beta+1}}{\beta+1} + q\left(\frac{(t_1-\mu)^{\beta+2}}{\beta+2} + \mu\frac{(t_1-\mu)^{\beta+1}}{\beta+1}\right) \\ -r\left(\frac{(t_1-\mu)^{\beta+3}}{\beta+3} + 2\mu\frac{(t_1-\mu)^{\beta+2}}{\beta+2} + \mu^2\frac{(t_1-\mu)^{\beta+1}}{\beta+1}\right) \end{array}\right\} \end{array}\right]$$

$$+ C_h\left(t_1 - \theta\frac{(t_1 - \mu)^{\beta+1} - (-\mu)^{\beta+1}}{\beta + 1}\right)$$

$$\left[\begin{array}{c} p + qt_1 - rt_1^2 \\ +\theta\left\{\begin{array}{c} p(t_1 - \mu)^\beta + q\left((t_1 - \mu)^{\beta+1}\right) \\ +\mu(t_1 - \mu)^\beta - r\left((t_1 - \mu)^{\beta+2}\right. \\ \left.+2\mu(t_1 - \mu)^{\beta+1} + \mu^2(t_1 - \mu)^\beta\right) \end{array}\right\} \end{array}\right]$$

$$+C_h\left[\begin{array}{c} \left(p\mu + \frac{q\mu^2}{2} - \frac{r\mu^3}{3}\right)\left(\theta(t_1 - \mu)^\beta - 1\right) \\ +\left(p + q\mu - r\mu^2\right)\left((t_1 - \mu)^{\beta+1} - (t_1 - \mu)\right) \\ +\left(\frac{q}{2} - r\mu\right)\left((t_1 - \mu)^{\beta+2} - (t_1 - \mu)^2\right) \\ -\frac{r}{3}\left((t_1 - \mu)^{\beta+3} - (t_1 - \mu)^3\right) \end{array}\right]$$

$$-\theta C_h\left[\begin{array}{c} \frac{p}{\beta+1}(t_1 - \mu)^{\beta+1} + q\left(\frac{(t_1-\mu)^{\beta+2}}{\beta+2} + \mu\frac{(t_1-\mu)^{\beta+1}}{\beta+1}\right) \\ -r\left(\frac{(t_1-\mu)^{\beta+3}}{\beta+3} + 2\mu\frac{(t_1-\mu)^{\beta+2}}{\beta+2} + \mu^2\frac{(t_1-\mu)^{\beta+1}}{\beta+1}\right) \end{array}\right]$$

$$P_3 = \theta\beta C_p(t_1 - \mu)^{\beta-1}\left[\begin{array}{c} pt_1 + \frac{qt^2}{2} - r\frac{t_1^3}{3} \\ +\theta\left\{\begin{array}{c} p\frac{(t_1-\mu)^{\beta+1}}{\beta+1} + q\left(\frac{(t_1-\mu)^{\beta+2}}{\beta+2} + \mu\frac{(t_1-\mu)^{\beta+1}}{\beta+1}\right) \\ -r\left(\frac{(t_1-\mu)^{\beta+3}}{\beta+3} + 2\mu\frac{(t_1-\mu)^{\beta+2}}{\beta+2} + \mu^2\frac{(t_1-\mu)^{\beta+1}}{\beta+1}\right) \end{array}\right\} \end{array}\right]$$

$$+\theta C_p\left((t_1 - \mu)^\beta - (-\mu)^\beta\right)\left[\begin{array}{c} p + qt_1 - rt_1^2 \\ +\theta\left\{\begin{array}{c} p(t_1 - \mu)^\beta + q((t_1 - \mu)^{\beta+1} \\ +\mu(t_1 - \mu)^\beta) - r\left((t_1 - \mu)^{\beta+2}\right. \\ \left.+2\mu(t_1 - \mu)^{\beta+1} + \mu^2(t_1 - \mu)^\beta\right) \end{array}\right\} \end{array}\right]$$

$$-\theta\beta C_p\left[\begin{array}{c} p\left\{(t_1 - \mu)^\beta + \mu(t_1 - \mu)^{\beta-1}\right\} \\ +\frac{q}{2}\left\{(t_1 - \mu)^{\beta+1} + 2\mu(t_1 - \mu)^\beta\right. \\ \left.+\mu^2(t_1 - \mu)^{\beta-1}\right\} - \frac{r}{3}\left\{(t_1 - \mu)^{\beta+2} + 3\mu(t_1 - \mu)^{\beta+1}\right. \\ \left.+3\mu^2(t_1 - \mu)^\beta + \mu^3(t_1 - \mu)^{\beta-1}\right\} \end{array}\right]$$

$$P_4 = \frac{C_q}{\lambda^3}\left\{p\lambda^2 + q\lambda(\lambda T + 1) - r(\lambda T + 1)^2\right\}$$
$$\left\{\frac{1 + 3\lambda(T - t_1) - 2\ln(1 + \lambda(T - t_1))}{1 + \lambda(T - t_1)}\right\}$$

$$+\frac{C_q}{\lambda^2}(T - t_1)\{q\lambda - 2r(\lambda T + 1) - \lambda r\} - C_q r\left\{\left(T^2 - t^2\right) - \frac{2Tt_1}{3}\right\}$$

$$P_5 = -C_1\left[p + \frac{q}{2}(T + t_1) - \frac{r}{3}\left\{(T + t_1)^2 - Tt_1\right\}\right]$$
$$+ C_1(T - t_1)\left\{\frac{q}{2} - \frac{r}{3}(T + 2t_1)\right\}$$

$$- \frac{C_1}{\lambda^3(1 + \lambda(T - t_1))}\left\{a\lambda^2 + b\lambda(\lambda T + 1) - c(\lambda T + 1)^2\right\}$$
$$- \frac{C_1}{\lambda}\{r(\lambda T + \lambda t_1 + 1) + q\lambda\}$$

$$P_6 = \xi C_1\left[\begin{array}{c} p + qt_1 - rt_1^2 \\ +\theta\left\{\begin{array}{c} p(t_1 - \mu)^\beta + q\left((t_1 - \mu)^{\beta+1} + \mu(t_1 - \mu)^\beta\right) \\ -r\left((t_1 - \mu)^{\beta+2} + 2\mu(t_1 - \mu)^{\beta+1} + \mu^2(t_1 - \mu)^\beta\right) \end{array}\right\} \end{array}\right] e^{-\theta(-\mu)^\beta}$$

$$+\frac{\xi C_1}{\lambda^3}\left[\begin{array}{c} -\frac{a\lambda^2 + b\lambda(\lambda T + 1) - c(\lambda T + 1)^2}{1 + \lambda(T - t_1)} \\ +\lambda\{q\lambda - 2r(\lambda T + 1)\} - r\lambda^2(t_1 - T) + r\lambda \end{array}\right],$$

and

$$\frac{\partial(TRC(t_1, T))}{\partial T} = [q_1 + q_2 + q_3 + q_4 + q_5 - q_6], \tag{17}$$

where $q_1 = -\frac{A_0}{T^2}$

$$
q_2 = -\frac{1}{T^2}\left[\begin{aligned}
& C_h\left(t_1-\theta\frac{(t_1-\mu)^{\beta+1}-(-\mu)^{\beta+1}}{\beta+1}\right)\\
& \left[\begin{aligned} & pt_1+\frac{qt_1^2}{2}-\frac{rt_1^3}{3}\\
& +\theta\left\{\begin{aligned} & p\frac{(t_1-\mu)^{\beta+1}}{\beta+1}+q\left(\frac{(t_1-\mu)^{\beta+2}}{\beta+2}+\mu\frac{(t_1-\mu)^{\beta+1}}{\beta+1}\right)\\
& -r\left(\frac{(t_1-\mu)^{\beta+3}}{\beta+3}+2\mu\frac{(t_1-\mu)^{\beta+2}}{\beta+2}+\mu^2\frac{(t_1-\mu)^{\beta+1}}{\beta+1}\right)\end{aligned}\right\}\end{aligned}\right]\\
& -C_h\left[\begin{aligned} & \left(p\mu+\frac{q\mu^2}{2}-\frac{r\mu^3}{3}\right)t_1+\frac{(t_1-\mu)^2-(-\mu)^2}{2}\left(p+q\mu-r\mu^2\right)\\
& +\frac{(t_1-\mu)^3-(-\mu)^3}{3}\left(\frac{q}{2}-r\mu\right)-\frac{r}{3}\frac{(t_1-\mu)^4-(-\mu)^4}{4}\end{aligned}\right]\\
& +\theta C_h\left[\begin{aligned} & \frac{(t_1-\mu)^{\beta+1}-(-\mu)^{\beta+1}}{\beta+1}\left(p\mu+\frac{q\mu^2}{2}-\frac{r\mu^3}{3}\right)\\
& +\frac{(t_1-\mu)^{\beta+2}-(-\mu)^{\beta+2}}{\beta+2}\left(p+q\mu-r\mu^2\right)\\
& +\frac{(t_1-\mu)^{\beta+3}-(-\mu)^{\beta+3}}{\beta+3}\left(\frac{q}{2}-r\mu\right)\\
& -\frac{r}{3}\frac{(t_1-\mu)^{\beta+4}-(-\mu)^{\beta+4}}{\beta+4}\end{aligned}\right]\\
& -\theta C_h\left[\begin{aligned} & p\frac{(t_1-\mu)^{\beta+2}-(-\mu)^{\beta+2}}{(\beta+1)(\beta+2)}\\
& +q\left(\frac{(t_1-\mu)^{\beta+3}-(-\mu)^{\beta+3}}{(\beta+2)(\beta+3)}\right.\\
& \left.+\mu\frac{(t_1-\mu)^{\beta+2}-(-\mu)^{\beta+2}}{(\beta+1)(\beta+2)}\right)\\
& -r\left\{\frac{(t_1-\mu)^{\beta+4}-(-\mu)^{\beta+4}}{(\beta+3)(\beta+4)}\right.\\
& +2\mu\frac{(t_1-\mu)^{\beta+3}-(-\mu)^{\beta+3}}{(\beta+2)(\beta+3)}\\
& \left.+\mu^2\frac{(t_1-\mu)^{\beta+2}-(-\mu)^{\beta+2}}{(\beta+1)(\beta+2)}\right\}\end{aligned}\right]
\end{aligned}\right]
$$

$$
q_3 = -\frac{1}{T^2}\left[\begin{array}{l}
\theta C_p\left((t_1 - \mu)^{\beta} - (-\mu)^{\beta}\right)\left[\begin{array}{c} pt_1 + q\frac{t_1^2}{2} - r\frac{t_1^3}{3} \\
+\theta\left\{\begin{array}{l} p\frac{(t_1-\mu)^{\beta+1}}{\beta+1} + q\left(\frac{(t_1-\mu)^{\beta+2}}{\beta+2} + \mu\frac{(t_1-\mu)^{\beta+1}}{\beta+1}\right) \\ -r\left(\frac{(t_1-\mu)^{\beta+3}}{\beta+3} + 2\mu\frac{(t_1-\mu)^{\beta+2}}{\beta+2} + \mu^2\frac{(t_1-\mu)^{\beta+1}}{\beta+1}\right)\end{array}\right\}\end{array}\right] \\
-\theta\beta C_p\left[\begin{array}{l}
p\left\{\frac{(t_1 - \mu)^{\beta+1} - (-\mu)^{\beta+1}}{\beta+1} + \mu\frac{(t_1 - \mu)^{\beta} - (-\mu)^{\beta}}{\beta}\right\} \\
+\frac{q}{2}\left\{\frac{(t_1 - \mu)^{\beta+2} - (-\mu)^{\beta+2}}{\beta+2} + 2\mu\frac{(t_1 - \mu)^{\beta+1} - (-\mu)^{\beta+1}}{\beta+1} + \mu^2\frac{(t_1 - \mu)^{\beta} - (-\mu)^{\beta}}{\beta}\right\} \\
-\frac{r}{3}\left\{\begin{array}{l}\frac{(t_1-\mu)^{\beta+3} - (-\mu)^{\beta+3}}{\beta+3} \\ + 3\mu\frac{(t_1 - \mu)^{\beta+2} - (-\mu)^{\beta+2}}{\beta+2} \\ + 3\mu^2\frac{(t_1 - \mu)^{\beta+1} - (-\mu)^{\beta+1}}{\beta+1} \\ + \mu^3\frac{(t_1 - \mu)^{\beta} - (-\mu)^{\beta}}{\beta}\end{array}\right\}
\end{array}\right]
\end{array}\right]
$$

$$
q_4 = -\frac{C_q}{\lambda^3}\left(1 - \frac{t_1}{T}\right)\left\{q\lambda^2 - 2r\lambda(\lambda T + 1)\right\}\{1 - 2\ln(1 + \lambda(T - t_1))\}
$$

$$
-\frac{C_q t_1}{\lambda^3 T^2}\left\{a\lambda^2 + b\lambda(\lambda T + 1) - c(\lambda T + 1)^2\right\}\{1 - 2\ln(1 + \lambda(T - t_1))\}
$$

$$
+ 2\frac{C_q}{\lambda^2}\left\{\frac{a\lambda^2 + b\lambda(\lambda T + 1) - c(\lambda T + 1)^2}{1 + \lambda(T - t_1)}\right\}\left(1 - \frac{t_1}{T}\right)
$$

$$
- \frac{C_q}{2\lambda^2}\left(1 - \frac{t_1^2}{T^2}\right)\{q\lambda - 2r(\lambda T + 1) - \lambda r\}
$$

$$
+\frac{C_q r}{\lambda}\left(T + \frac{t_1^2}{T} - 2t_1\right) - C_q r\left(\frac{2t_1^2}{3T} - \frac{t_1}{3} + \frac{t_1^3}{3T^2}\right) + \frac{C_q r}{T}(T - t_1)^2
$$

$$
q_5 = \frac{C_1 t_1}{T^2}\left[p + \frac{q}{2}(T + t_1) - \frac{r}{3}\left\{(T + t_1)^2 - Tt_1\right\}\right]
$$

$$
+ C_1\left(1 - \frac{t_1}{T}\right)\left\{\frac{q}{2} - \frac{r}{3}(2T + t_1)\right\}
$$

$$
- \frac{C_1}{\lambda^3 T^2}\left\{a\lambda^2 + b\lambda(\lambda T + 1) - c(\lambda T + 1)^2\right\}\ln\{1 + \lambda(T - t_1)\}
$$

$$
\begin{aligned}
&+\frac{C_1}{\lambda^3 T}\left\{q\lambda^2 - 2r\lambda(\lambda T + 1\right\}\ln(1+\lambda(T-t_1)) \\
&+\frac{C_1}{\lambda^2 T}\left[\frac{a\lambda^2 + b\lambda(\lambda T+1) - c(\lambda T+1)^2}{1+\lambda(T-t_1)}\right] \\
&-\frac{C_1 t_1}{2\lambda T^2}\{2q\lambda - r(3\lambda T + \lambda t_1 + 2+)\} + \frac{3C_1 r}{2}\left(1-\frac{t_1}{T}\right)
\end{aligned}
$$

$$
\begin{aligned}
q_6 = -\xi C_1 &\left[\begin{array}{c} p\frac{t_1}{T^2} + \frac{q}{2}\left(\frac{t_1^2}{T^2}+1\right) - \frac{r}{3}\left(\frac{t_1^3}{T^2}+2T\right) \\ +\frac{\theta}{T^2}\left\{\begin{array}{c} p\frac{(t_1-\mu)^{\beta+1}}{\beta+1} + q\left(\frac{(t_1-\mu)^{\beta+2}}{\beta+2} + \mu\frac{(t_1-\mu)^{\beta+1}}{\beta+1}\right) \\ -r\left(\frac{(t_1-\mu)^{\beta+3}}{\beta+3} + 2\mu\frac{(t_1-\mu)^{\beta+2}}{\beta+2} + \mu^2\frac{(t_1-\mu)^{\beta+1}}{\beta+1}\right) \end{array}\right\} \end{array}\right] e^{-\theta(-\mu)^\beta} \\
&+\frac{\xi C_1\theta}{T^2}\left\{\begin{array}{c} p\frac{(-\mu)^{\beta+1}}{\beta+1} + q\left(\frac{(-\mu)^{\beta+2}}{\beta+2} + \mu\frac{(-\mu)^{\beta+1}}{\beta+1}\right) \\ -r\left(\frac{(-\mu)^{\beta+3}}{\beta+3} + 2\mu\frac{(-\mu)^{\beta+2}}{\beta+2} + \mu^2\frac{(-\mu)^{\beta+1}}{\beta+1}\right) \end{array}\right\} e^{-\theta(-\mu)^\beta} \\
&-\frac{\xi C_1}{\lambda^3 T^2}\left\{a\lambda^2 + b\lambda(\lambda T+1) - c(\lambda T+1)^2\right\}\ln(1+\lambda(T-t_1)) \\
&+\frac{\xi C_1}{\lambda^3 T}\left\{q\lambda^2 - 2r\lambda(\lambda T+1\right\}\ln(1+\lambda(T-t_1)) \\
&+\frac{\xi C_1}{\lambda^2 T}\left\{\frac{a\lambda^2 + b\lambda(\lambda T+1) - c(\lambda T+1)^2}{1+\lambda(T-t_1)}\right\} - \frac{\xi C_1 t_1}{\lambda^2 T^2}\{q\lambda - 2r(\lambda T+1)\} \\
&+\frac{2\xi C_1 r}{\lambda}\left(1-\frac{t_1}{T}\right) + \frac{\xi C_1 r}{2\lambda^3 T^2}\{\lambda^2(t_1-T)^2 - 2\lambda(t_1-T)\} \\
&+\frac{\xi C_1 r}{\lambda^2 T}\{\lambda(t_1-T) - 1\}.
\end{aligned}
$$

Thus, the required constrains for minimizing the total appropriate cost $TC^*(t_1, T)$ per unit time are

$$\frac{\partial^2(TRC(t_1,T))}{\partial t_1^2} < 0, \tag{18}$$

$$\frac{\partial^2(TRC(t_1,T))}{\partial T^2} < 0, \tag{19}$$

and provided that it must justify the adequate circumstances,

$$\frac{\partial^2(TRC(t_1,T))}{\partial t_1^2}\frac{\partial^2(TRC(t_1,T))}{\partial T^2} - \left(\frac{\partial^2(TRC(t_1,T))}{\partial t \partial T}\right)^2 < 0. \tag{20}$$

Subsequently, to obtain the best possible assessment of the finest span of ordering succession, the duration of most select scarcity instant, and ultimately the most favorable amount of order. Then we can calculate optimal total relevant cost using Eqs. (7) and (14).

## 3 Numerical Examples

To illustrate the Model, an example is demonstrated considering the underneath, factors. Let $p = 1300$, $q = 200$, $r = 120$ $\xi = 0.1$, $\beta = 4$, $C_h = 15$, $\mu = 0.5$, $Cp = 90$, $\theta = 0.2$, $C_1 = 30$, $\lambda = 0.4$, $A_0 = 220$ in appropriate unit. By applying the subroutine find root in MATHEMATICA version 9. We obtain the optimum solution for $t_1$, $T$ and $Q_0$ of equation as $t_1^* = 5.2481$, $T^* = 6.5112$, $Q_0^* = 15.5623$, and optimum average cost as TC* $(t_1, T)$ = 22,908.00 (Table 1).

### 3.1 Sensitivity Analysis

Under this resolution sensitivity analysis against the system factors; $p, q, r, A_0, \theta, \beta, \mu, \lambda, c_1$, upon the most suitable entire cost, the weight of alteration in each factors being investigated considering sensitivity study via 50%, 25%, −25%, −50% and again holding a single factor at once keeping another factor unaltered has been discussed below with outcomes.

The arguments basically rely upon the mathematical pattern followed by the results reflected in the chart revealed above.

1. Along with the rise in the matter of cost for factor '$p$', the value of $t_1^*$, $T^*$ and $Q_0^*$ increases. In this case, it is observed that while '$p$' rises, the optimal total relevant cost $TC^*$ also increases.
2. From parameter '$q$' and '$r$' shows that with increase in the amount of factor '$q$' and '$r$,' there is no affect on $t_1^*$, $T^*$ and $Q_0^*$ with change in '$q$' and '$r$'. There is affect on $TC^*$ with change in '$q$' and '$r$'. This is clear that if the demand is high the total relevant cost will be higher.
3. With increase in the amount of factor '$\theta$', amount of $t_1^*$, $T^*$ and $Q_0^*$ decreases while $TC^*$ increases. As a result of this raises $TC^*$. The formulation then reduces both optimum quantity and cycle time.
4. With the rise in the amount of factor '$\beta$', the value of $t_1^*$, $T^*$ and $Q_0^*$ decreases. In this case, it is observed that when '$\beta$' increases, the requirement for that particular also rises keeping the total relevant cost increased.
5. With the increase in the amount of factor '$\mu$', the amount of $t_1^*$, $T^*$ and $Q_0^*$ increases. Here, when '$\mu$' increases the total relevant cost also increases.
6. With the increase in the amount of the factor '$A_0$', the values of $t_1^*$, $T^*$ and $Q_0^*$ decrease. Here, when '$A_0$' rises, the total relevant cost of the system decreases.

**Table 1** Sensitivity values

| Changing parameter | % change | $t_1^*$ | $T^*$ | $Q_0^*$ | $TC^*\ (t_1, T)$ |
|---|---|---|---|---|---|
| $p$ | +50 | 6.0513 | 8.0915 | 31.7314 | 89,317.907 |
| | +25 | 5.6542 | 7.1333 | 20.5074 | 42,109.903 |
| | −25 | 4.8321 | 5.9603 | 12.4167 | 12,198.607 |
| | −50 | 4.4064 | 5.4554 | 10.3172 | 6123.713 |
| $q$ | +50 | 5.2481 | 6.5112 | 15.5623 | 34,360.61 |
| | +25 | 5.2481 | 6.5112 | 15.5623 | 28,634.20 |
| | −25 | 5.2481 | 6.5112 | 15.5623 | 17,181.80 |
| | −50 | 5.2481 | 6.5112 | 15.5623 | 11,455.92 |
| $r$ | +50 | 5.2474 | 6.5112 | 15.5623 | 22,909.00 |
| | +25 | 5.2474 | 6.5112 | 15.5623 | 22,908.50 |
| | −25 | 5.2474 | 6.5112 | 15.5623 | 22,907.51 |
| | −50 | 5.2474 | 6.5112 | 15.5623 | 22,907.50 |
| $\Theta$ | +50 | 4.7105 | 5.9725 | 15.2541 | 23,339.52 |
| | +25 | 4.9624 | 6.2067 | 15.3614 | 23,073.10 |
| | −25 | 5.6631 | 6.9283 | 15.8774 | 22,004.80 |
| | −50 | 6.2045 | 7.5576 | 16.3546 | 22,170.80 |
| $\beta$ | +50 | 3.4475 | 6.3317 | 11.7557 | 98,618.50 |
| | +25 | 4.2471 | 6.3966 | 13.1625 | 21,290.31 |
| | −25 | 5.7421 | 6.7846 | 20.9487 | 26,771.40 |
| | −50 | 6.8473 | 7.1574 | 27.6654 | 30,196.10 |
| $\mu$ | +50 | 5.7965 | 7.4733 | 23.8116 | 41,160.40 |
| | +25 | 5.5454 | 7.0004 | 19.3276 | 31,156.00 |
| | −25 | 4.8823 | 6.9637 | 12.2116 | 15,718.40 |
| | −50 | 4.4145 | 5.2958 | 9.1158 | 9594.55 |
| $A_O$ | +50 | 5.2471 | 6.5112 | 15.5604 | 22,925.90 |
| | +25 | 5.2471 | 6.5112 | 15.5604 | 22,906.50 |
| | −25 | 5.2506 | 6.5145 | 15.5837 | 22,920.30 |
| | −50 | 5.2781 | 6.5536 | 15.8313 | 23,020.80 |
| $C_h$ | +50 | 4.2485 | 4.5115 | 15.5626 | 35,360.60 |
| | +25 | 4.9247 | 5.5126 | 15.5626 | 28,634.20 |
| | −25 | 6.2436 | 7.5119 | 15.5624 | 18,181.80 |
| | −50 | 7.2489 | 8.5117 | 15.5623 | 11,445.90 |
| $C_p$ | +50 | 6.7967 | 8.4738 | 23.8116 | 45,170.40 |
| | +25 | 5.5456 | 7.0003 | 19.3277 | 33,146.00 |
| | −25 | 4.8828 | 4.9634 | 12.2118 | 16,718.40 |
| | −50 | 3.4143 | 3.2953 | 10.1154 | 8584.56 |

(continued)

**Table 1** (continued)

| Changing parameter | % change | $t_1^*$ | $T^*$ | $Q_0^*$ | TC* $(t_1, T)$ |
|---|---|---|---|---|---|
| $C_1$ | +50 | 7.7966 | 9.4735 | 25.8116 | 48,170.40 |
| | +25 | 6.5454 | 7.0007 | 21.3275 | 35,166.00 |
| | −25 | 4.8723 | 4.9638 | 14.2117 | 14,718.40 |
| | −50 | 3.5147 | 2.2955 | 10.1159 | 9593.558 |
| λ | +50 | 9.0516 | 3.0916 | 10.7318 | 89,316.90 |
| | +25 | 7.6545 | 4.1338 | 20.5073 | 42,119.90 |
| | −25 | 4.8323 | 6.9607 | 32.4164 | 12,188.60 |
| | −50 | 2.4064 | 8.4555 | 45.3175 | 6143.72 |

7. As of shortages cost corresponding to the increase in the amount of the factor '$C_h$', the amount of $t_1^*$, $T^*$ reduces while $Q_0^*$ rises. In this case when '$C_h$' increases, total relevant cost increases.
8. It is observed that with increase in the amount of the factor '$Cp$ and $C1$', the values of $t_1^*$, $T^*$ and $Q_0^*$ increase. Here, the total relevant cost of the arrangement rises with the rise in purchase cost.
9. With the rises in the amount of the factor '$\lambda$', then the amount of $T^*$ and $Q_0^*$ reduces whereas $t_1^*$ rises. Here, the total relevant cost of the arrangement rises with the rise in the parameter '$\lambda$'.

## 4 Conclusion

In this paper, an inventory model has been developed that determines the deterioration rate as three parameter Weibull function and shortages are permitted which are partially backlogged. The demand rate is assumed as parabolic function of time. The backlogging rate is dependent on the waiting time for the next replenishment. This model can be used for items like cosmetic, high-tech products, fruits and vegetables. The objective of this paper is to minimize the relevant cost per unit time. Sensitivity analysis of the optimal solution with respect to parameters is also carried out to check the stability of the model.

## References


1. S.K. Ghosh, K.S. Chaudhuri, An order level inventory model for a deteriorating item with Weibull deterioration time-quadratic demand and shortages. Adv. Model. Optim. **6**(1), 21–35 (2004)
2. T.M. Within, *The Theory of Inventory Management* (Princeton University Press, Princeton, 1953), pp. 62–72

3. P.N. Ghare, G.F. Schrader, A model for exponentially decaying inventories. J. Ind. Eng. **15**, 238–243 (1963)
4. Y.K. Shah, M.C. Jaiswal, An order-level inventory model for a system with constant rate of deterioration. OPESEARCH **14**, 174–184 (1997)
5. S.P. Aggarwal, A note on an order-level inventory model for a system with constant rate of deterioration. OPESEARCH **15**, 184–187 (1987)
6. C.T. Chang, An EOQ model with diminishing objects under inflation when supplier credits linked to order quantity. Int. J. Prod. Econ. **188**, 307–316 (2004)
7. A.A. Alamri, Z.T. Balkhi, The effects of learning and forgetting on the optimal production lot size for diminishing objects with time-varying demand and deterioration rates. Int. J. Prod. Econ. **10**(7), 125–138 (2007)
8. K. Hung, An inventory model with generalized type demand deterioration and backorder rates. Eur. J. Oper. Res. **208**(3), 239–242 (2011)
9. S.S. Mishra, P.K. Singh, A computational approach to EOQ model with power-form stock-dependent demand and cubic deterioration. Am. J. Oper. Res. **1**(1), 5–13 (2011)
10. H. Patel, Pricing model for instantaneous deteriorating items with partial back logging and different demand rates. Uncertain Supply Chain Manage. **7**, 97–10 (2019)
11. K. Karthikeyan, G. Sarthi, An inventory model for constant deteriorating items with cubic demand and salvage values. Int. J. Appl. Eng. Res. **10**(55), 3723–3728 (2015)
12. V. Mishra, L. Singh, Inventory model for stock type demand, on hand inventory deteriorates with shortages. Int. J. Appl. Math. Stat. **23**(D11), 84–91 (2016)
13. H.J. Chang, C.Y. Dye, An EOQ model for deteriorating items with time varying demand and partial backlogging. J. Oper. Res. Soc. **50**, 1176–1182 (1999)
14. C.K. Sahoo, K.C. Paul, A. Kalam, An EOQ representation for declining matters with cubic order, inconsistent declination and inequitable backlogging. AIP Conf. Proc. **2253**, 020010 (2020). https://doi.org/10.1063/5.0018991
15. C.K. Sahoo, K.C. Paul, EOQ model for cubic deteriorating items carry forward with Weibull demand and without shortages. Int. J. Res. Eng. Innov. **5**(5), 285–290 (2021)
16. C.K. Sahoo, K.C. Paul, S. Kumar, Two warehouses EOQ inventory model of degrading matter having exponential decreasing order, limited suspension in price including salvage value. SN Comput. Sci. **1**(334) (2020). https://doi.org/10.1007/s42979-020-00346-1

# Information Actions Use for System Activity: Action Modeling Schemas

**Alexander Geyda**

**Abstract** Main results of the review of research in digitalization, digital economy, society, organizations digitalization are provided. The problem considered is mathematical research of information in the action of systems. Main white spots and gaps in the study of information in the action of systems described. As a result, the author points to the existing multidisciplinary gap between the need to solve the use of information for the further action of various systems as mathematical, system theoretic problems and available theoretic and mathematical means to solve such problems. Candidates for the mathematical theory of using the information in system action are suggested. Role of information actions for system action is considered. As a result, the author considers families of alternative stochastic actions networks as an example of modeling information actions use for system actions. The novelty of formalism is in the dynamical nature of networks alternations, modeled as a tree of possible networks. Schemas to model information actions use in system actions are provided.

**Keywords** Digitalization · Information · Information actions · Information technology · Modeling · Methods · Schemas

## 1 Introduction

Digital transformation, digital economy, production, digital platforms, education, and other "digital phenomenons" are well-known "buzzwords" nowadays. But it is well known that there is no shared agreement about their exact meaning, not about conceptual and especially—formal models, frameworks, or theories that allow to measure, predict results of such "digital phenomenons" and design them.

Moreover, such phenomenons, their appearance, use, and value are not yet clearly specified conceptually, especially for new phenomenons, that is—phenomenons under construction.

A. Geyda (✉)
St. Petersburg Federal Research Center of the Russian Academy of Sciences, St. Petersburg, Russia 198188
e-mail: geida@iias.spb.su



S. Patnaik et al. (eds.), *Smart and Sustainable Technologies: Rural and Tribal Development Using IoT and Cloud Computing*, Advances in Sustainability Science and Technology,
https://doi.org/10.1007/978-981-19-2277-0_4

I conducted a systematic literature review of current conceptual and formal methods, theories, and trends in "digital phenomenons" research [1]. A study was made to answer research questions:

Q1. What are the main white spots or gaps in research of the "digital phenomenons," and why do they exist?

Q2. What are the possible research directions to overcome existing white spots or gaps, and by which means?

Q3. Which theories suited best to overcome existing white spots or gaps?

Scopus and Web of Science core collection articles at least for five years were reviewed by keywords "digital transformation," "digital economy," and other "digital phenomenon" alike.

We consider the main results of our research as a plan to create a theory of information actions used for system activity and schemas to model information actions use for system actions.

## 2 The Need in Information Actions Use for System Activity

Results [1] show a fast-growing number of articles in European Community (EC) countries, China, and former USSR countries but not in the US, which is strange given the massive difference in financing. Our research shows that most probably it is the result of differences in schemas of financing research. EC, China, Russia have financed a relatively small number of themes related to digital transformation using government public programs of various kinds. However, the US funded many themes of research in the various fields related to digitalization through universities. For this reason, we obtained linked keywords, which describe specific areas of digitalization. Keywords cloud because of additional searches for research shown in Fig. 1.

There were a few hundred keywords that were used to find research articles of interest.

Articles found by these keywords were used to find out answers to research questions. As a result of our research to see white spots, it became clear that there are many white spots and gaps in "digital phenomenons," but the main are:

1. Common enough formal models of mechanisms of "digital phenomenons" use are absent. This is the case because of the exceptionally multidisciplinary nature of such phenomenons research. An appropriate theory of using the information in further activity—based on mathematical models and methods—was not yet created. However, various theoretical means exist to explain mechanisms of "digital phenomenons" use and results.
2. The absence of general enough theory to explain "digital phenomenons" to explain the use of information results leads to the gap. The gap is between the need to formalize and predict outcomes of possible new "digital phenomenons," the need to evaluate effects of using information, the need to design future

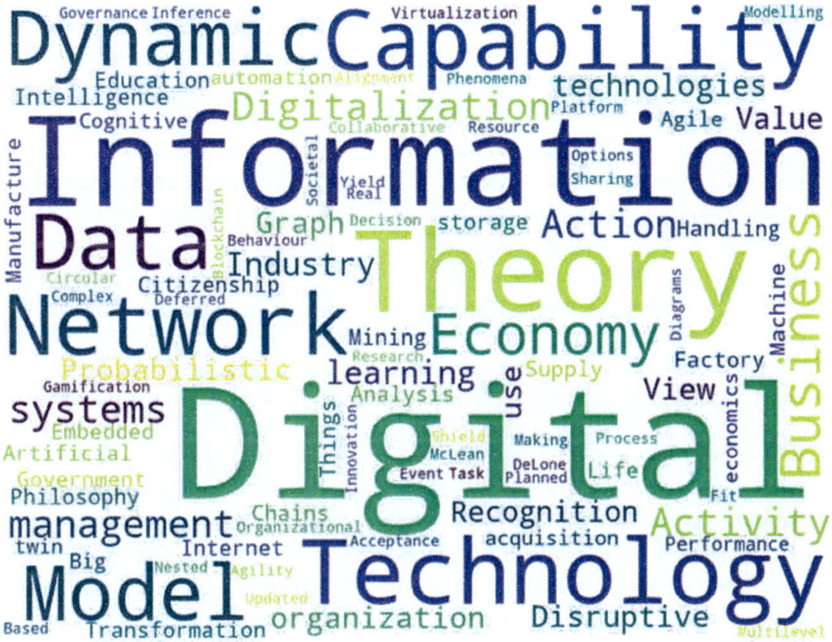


**Fig. 1** Keywords cloud obtained in the research

new "digital phenomenons" for better results—based on formal methods—from one hand—and available conceptual and based on them formal frameworks, mathematical models and methods suitable for such formalization, prediction, and design—from another hand.

To overcome this gap, I researched articles found for theories to use as part of *Theory of Using Information for System Action*. The total number of references found was hundreds of thousands. We used statistical software R and package bibliometrics [2] to find patterns in data obtained, to find keywords co-occurrences and authors collaboration networks. As a result of the literature review for theories to use, candidate theories that can be used to build "Theory of Using Information for System Action" suggestions were formed.

Main candidate theories: action theory [3–5], complexity theory [6], deferred action theory [7, 8], system potential theory [9], complex and dynamic networks theories [10–13], network science, and probabilistic/fuzzy dynamic graphs theories [14–16], dynamic decision making [17–19].

The hypothesis is: by creating a unified conceptual model of "digital phenomenon's" based on action theory, deferred action theory, system potential theory, dynamic decision making, and by utilizing mathematical models and methods of theories mentioned above, it is possible to create mathematical models and methods to overcome the existing gap. The theory of using information for system action is possible because of the theory of system potential [20]. I have already built models, which can estimate system capability regarding characteristics of information technology use. But such models were created for the limited case of a one-of-a-kind information operation to alter project fulfillment.

Such models are based on suggested families of alternative stochastic action networks (FASAN) [21]. Such families are complex of basic graph-theoretic objects and mappings between them. They describe functioning in terms of the alternative action networks, which can be interrupted and alternated as a whole, depending on the system's states and environment. They can be considered as alternative stochastic network expansion. Alternative Stochastic Networks [22] alter networks locally, depending on local conditions. Families of alternative stochastic action networks alternate networks as a whole, depending on the state of the environment and the system and depending on managerial information operation used to alternate system functioning.

Such alternations form families of alternative stochastic action networks. It is worth noticing, to build the FASAN model, the model of environment and reaction of the information operation on environment impact must be created.

A corresponding alternative dynamical data structure is suggested to represent possible system changes under the condition of given environment changes, representing sequences of possible network fragments of different types (cuts, networks parts between cuts) in time (Table 1). This data structure includes rules implemented by information operation in the form of state–network rules.

It can be considered a timed multidimensional array or multidimensional cyclogram of the graph-theoretic model of system functioning under a fixed environment functioning scenario model.

The researcher can use such data structure to represent Log files corresponding to complex system changes over time under the given environment scenario. Such representation could help the researcher automatically build FASAN models from

**Table 1** Dynamic alternative data structure

| # | $\langle c_n^s \rangle / T_n$ | $N_{in}^v, N_{in}^b$ | $T_0$ | $T_1$ | $T_2$ | … | $T_m$ | … |
|---|---|---|---|---|---|---|---|---|
| 1 | $Null, \langle c_1^s \rangle$ | | | | | | | |
| 2 | $Null, \langle c_2^s \rangle$ | | | | | | | |
| … | … | | | | | | | … |
| n | $Null, \langle c_n^s \rangle$ | | | | | | | |
| … | … | | | | | | | … |
| m | $(C5)\langle c_1^s \rangle$ | | | | | | | |
| … | $(C5), \langle c_2^s \rangle$ | | | | | | | |
| | … | | | | | | | … |
| … | $(C6), \langle c_1^s \rangle$ | | | | | | | |
| … | … | | | | | | | … |

Big Data collected in Log files and automate FASAN model creation, learning, and usage.

## 3 Action Modeling Schemas for the Theory of Information Actions Use for System Activity

We propose to describe kinds of information operations by sets named by identifiers, for example—between «Alpha» and «Omega.» «Alpha» group contains actions that start with the material world and ends with information. «Omega» group has steps that begin with information and ends with the result of information use. An example of such classification was provided in [23].

Among obtaining information, kinds of actions («Alpha») are Sensor operations, Situation awareness operations. We consider «Beta» kinds of operations as getting information from the source of information. These kinds of actions are learning, receiving.

Dynamic decision making [17–19] provides a good description of the types of decisions that require information action.

Action theory [3–5] explains various aspects of human activity types and their relations and a way of understanding the change and development of human practice. Therefore, it is well suited to change and development contexts and complements sociological theories of practice by supplying a well-developed model of the dynamics of human activity [24]. According to this theory, activities form an activity system. Its nodes are individual, mediating artifact, object, community, rules, and division of labor. Systemic-structural activity theory (SSAT) introduced qualitative and quantitative stages of activity analyses during task performance [25]. The most important one is the functional analysis that considers activity as a complex self-regulative system. The SSAT model of self-regulation of orienting activity and its application to task analysis are presented in [25].

Based on activity theory, information actions may be considered a kind of artifact, object, community, rules, division of labor information actions, and all possible combinations of actions.

As stated in [26]: «activity theory is more in sync with science and technology studies, which reason that actors shape technologies while being continuously shaped by technologies. Tools are extensions of human agency, but at the same time, affect other tools. For instance, an IS (Information System) solution can be used to produce another IS, and tools used in one activity can also be used in different ways in associated and unrelated activities».

«Technologies are increasingly taking hold of all the aspects of the activity system: technology can compose elements of the subject (e.g., humans working with robots; or possibly technology as the subject), the object, community and form hybrids of these. This is an opportunity for the greater theorization of technologies and for the central notion of "tools" to be unpacked in IS studies (largely concerned with the interaction between technology and human activity).»

The goal of obtaining, processing the type of information actions [23] is information itself. Under information, we mean the form of the existence of ideas or thoughts in the material world (that is, outside of mind).

It is urgent to note that we state that human ideas and thus—humans are required for information to exist with such a definition. We do not consider information kinds of reflections—for example, in quantum mechanics explanations of the material world and reflections/information roles in it [27]. However, human ideas may be created and stored as reflections automatically (for example, because of the sensor's activities).

Based on the articles revised, I propose three *types of information action* use schema and directions to combine suggested schema. The purpose of schemas is to start to formalize various information actions use for system action. Information use schemas are shown in Figs. 2, 3 and 4.

*The first type of information action* $ia_1$ use schema for system action $ma$ consists in information action used to correct $ma$ *after it finished and a new instance of* $ma$ planned, effects of the action obtained and effects compliance to known demands $ec$ measured (as the probability of compliance, for example).

Information ($is$) about the system and information ($ie$) about environment system action $ma$ may be altered to suit better obtained. After alternation of and before one, the difference may serve as an indicator $\varphi_{aec} = w'_{ec} - w_{ec}$ of information action $ia$ quality for system action.

*The second type of information action* use schema $U_3(ia_3)$—for the set of (changing in the course of functioning) possible system actions $ma'_2$ depending on information about the system and environment states from the sets and, and depending information operation $ia_2$ consists in information action use to select elements of the set $ma'$ in the course *of action fulfillment depending changing conditions.* Set of

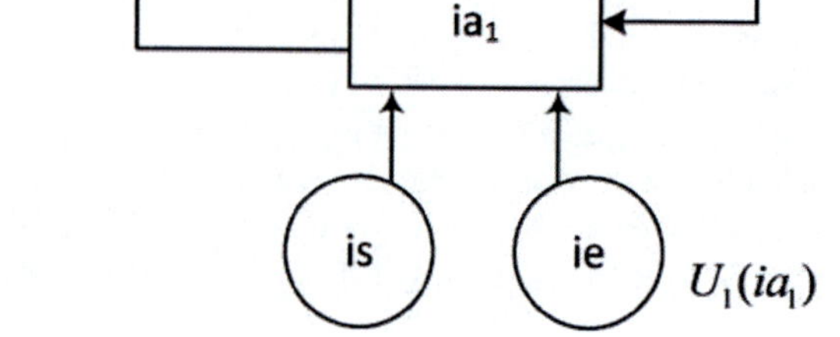


**Fig. 2** The first type of information action use schema for system action

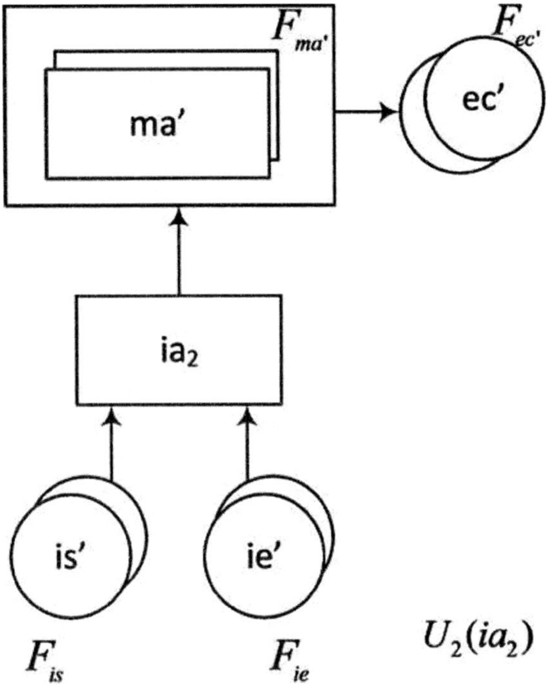


**Fig. 3** The second type of information action use schema for system action

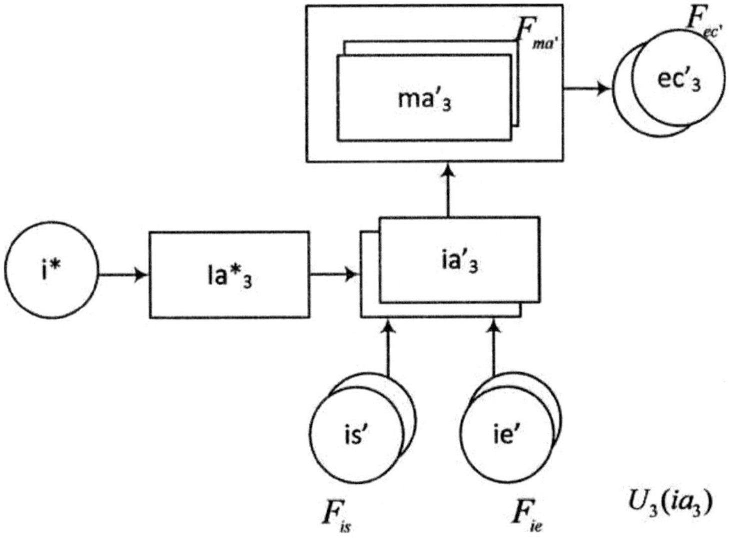


**Fig. 4** The third type of information action use schema for system action

possible effects of the action obtained as a result of such selection and effects compliance to the evolving demands from the set measured as distribution of probabilities of compliances $w'_{ec}$ on possible states of, $ie'$ and taking into account possibility information operation will select particular. Based on distribution $F_{w'_{ec}}$, which is obtained based on possible information ($is'$) about the system and possible information ($ie'$) about the environment and information action characteristics, the quality of information action $ia_2$ results can be measured. Such measures, for example, can be in the form of $F_{w'_{ec}}$ moments $\mu_{w'_{ec}}$. Then, the difference between possible alternations of and in the case $ma$ was not altered (which can be one point value) may serve as an indicator $\varphi_{aec} = F_{w'_{ec}} - w_{ec}$ of information action $ia_2$ quality for system action.

*The third type of information action* $ia_3$ use schema $U_3(ia_3)$—for the set of changing in the course of functioning possible system actions from the set of possible actions $ma'_3$, depending from possible information action depending information about the system and environment states from the sets and $ie'$, and depending information operation $ia_{3i} \in ia_3$ selected by meta-operation consists in (cnnnnn by $ia_3^*$) in the course *of information action fulfillment depending changing conditions.* Information action is used to determine elements of the set in the system *of material action fulfillment depending on changing conditions.* Set of possible effects of the action obtained as a result of such selections (information and material actions) and effects compliance to the evolving demands from the set measured as distribution of probabilities of compliances $w'_{ec3}$ on possible states of $is'$, $ie'$, $i^*$ (where $i^*$—information, available for meta-operation) and taking in account possibility meta-operation will select information action $ia_{3i}$, which, in turn, will choose particular $ma'_{i3} \in ma'_3$. Such possibilities form nested probability distributions of effects compliance given probabilities of selection $ia_{3i}$, $ma'_{i3}$: $F_{w'_{ec3}}$. Still, the measure needed can be in the form of $F_{w'_{ec3}}$ moments $\mu_{w'_{ec3}}$. Other actions may be constructed from the main kinds of actions listed above, such as building complexes of actions.

Then again, the difference between possible alternations of $ma'_3$ and $F_{w_{ce3}}$ in the case $ma$ was not altered (or altered by $ia_2$ schema) may serve as indicator $\varphi_{aec}$ of information action by $ia_3$ schema quality for system action compared to $ia_2$ and $ia_1$ information operation schemas, respectively.

The action was not considered a possible composition of actions (of various kinds) at the schemas above. To overcome this issue, I suggest some procedures to create composite actions. Examples of such compositions are shown in Figs. 5 and 6.

When sequence composition is considered, possible successor's actions variants usually depend on predecessors' information and material actions. As a result, the effects of sequential operations depend too. Examples of sequential operations are preparation and fulfillment of action or actions in the same workplace with the same item. An example of possible alternations is shown at the bottom of Fig. 5 as a typical business process diagram fragment with XOR branches.

When the parallel composition is considered, possible material actions variants are usually independent of other parallel material actions variants. However, they are dependent on information action—if expected (like in the case with FASAN, when the whole network is altered), and they are dependent on common predecessors. As a result, the effects of parallel actions (operations) depend too. Effects of a few parallel

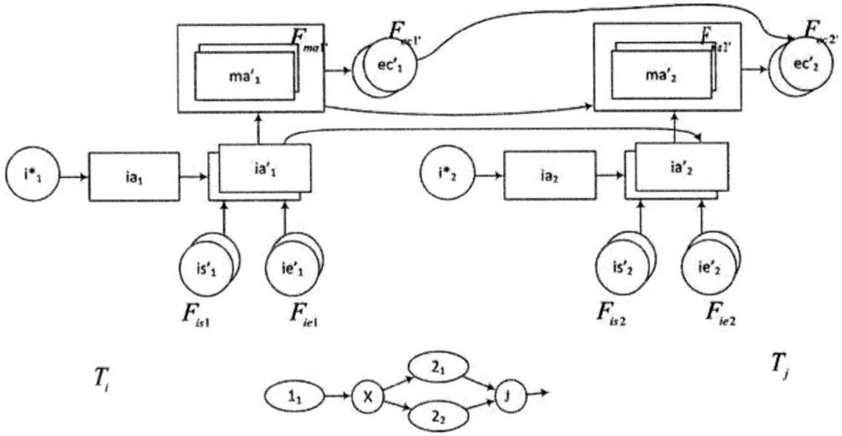


**Fig. 5** Composition of actions. Sequence composition example

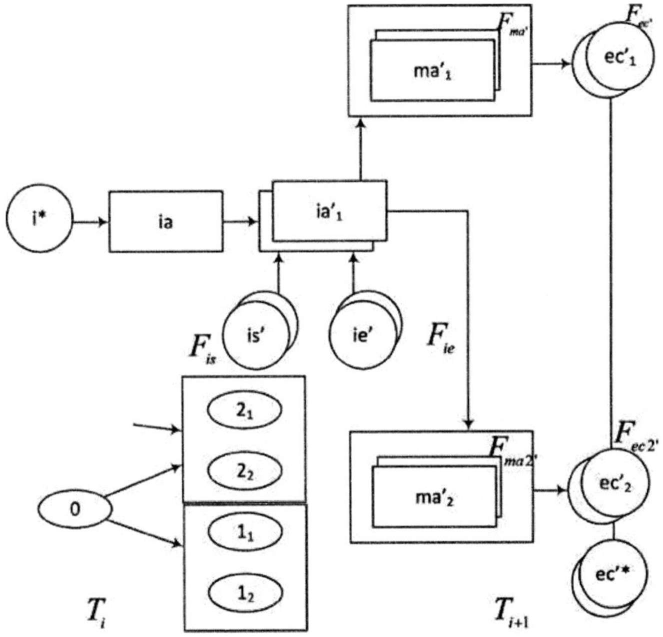


**Fig. 6** Composition of actions. Parallel composition example

operations are formed by the composition of each particular parallel action. Examples of parallel operations are the fulfillment of actions on the different workplaces with different parts or check of the same item at a few workplaces. An example of possible alternations, in this case, is shown at the bottom of Fig. 5.

## 4 Conclusion

As a result of the suggested schemas, they build mathematical models for quantitative estimation of system capability and other properties concerning information technology use in changing environments. Researchers can implement such assessments depending on the parameters and variables of the information operations. Predicting and prescriptive problems of systems behavior quality depending on information operations characteristics will be possible as a result.

The limitations of the suggested models and methods are their discrete nature reflected by graph-theoretical models used. To overcome this limitation marked by functional dependencies, graph-theoretic models can be used.

Further research should be done to utilize theories discovered, describe and model a wide range of various kinds of information operations and system functioning using candidate theories results found because of literature review.

**Acknowledgements** The reported study was funded by RFBR, project numbers 20-08-00649 and 19-08-00989.

## References

1. A.S. Geyda, T.N. Gurieva, V.N. Naumov, Conceptual and mathematical models, methods, and technologies for studying the digital transformation of economic and social systems: a literature review and research agenda. Adm. Consult. **11**, 12 (2021)
2. H. Derviş, Bibliometric analysis using bibliometrix an R package. JSCIRES **8**, 156–160 (2020). https://doi.org/10.5530/jscires.8.3.32
3. G. Goos, J. Hartmanis, J. van Leeuwen, M. Thielscher, *Challenges for Action Theories* (Springer, Berlin Heidelberg, 2000)
4. B. Dick, E. Stringer, C. Huxham, Theory in action research. Action Res. **7**, 5–12 (2009). https://doi.org/10.1177/1476750308099594
5. J. Aranzadi, (ed.), *Human Action, Economics, and Ethics*. SpringerBriefs in Economics (Springer International Publishing, Cham, 2018). https://doi.org/10.1007/978-3-319-73912-0
6. E. Angela, *Complexity Approach to Sustainability, A. Theory, and Application* (Imperial College Press, 2011)
7. N.V. Patel, The Theory of Deferred Action: Purposive Design as Deferred Systems for Emergent Organisations, in *Information Systems Theory. Integrated Series in Information Systems*, ed. by Y.K. Dwivedi, M.R. Wade, S.L. Schneberger, vol. 28 (Springer New York, New York, 2012), pp. 125–149. https://doi.org/10.1007/978-1-4419-6108-2_7
8. N.V. Patel, *Organization and Systems Design. Theory of Deferred Action* (Springer, 2006)

9. A. Geyda, I. Lysenko, System potential estimation with regard to digitalization: main ideas and estimation example. Information **11**, 164 (2020). https://doi.org/10.3390/info11030164
10. F. Ghanbarnejad, R. Saha Roy, F. Karimi, J.-C. Delvenne, B. Mitra, *Dynamics on and of Complex Networks III. Machine Learning and Statistical Physics Approaches*, ed. by F. Ghanbarnejad, R.S. Roy, F. Karimi, J.-C. Delvenne, B. Mitra (Springer, Cham, 2019)
11. A. Barrat, M. Barthelemy, A. Vespignani, *Dynamical Processes on Complex Networks* (Cambridge University Press, Leiden, 2008)
12. X. Fu, M. Small, G. Chen, *Propagation Dynamics on Complex Networks. Models, Methods, and Stability Analysis*, ed. by X. Fu, M. Small, G. Chen (Wiley/Higher Education Press, Chichester, 2014)
13. A. Zinilli, Competitive project funding and dynamic complex networks: evidence from Projects of National Interest (PRIN). Scientometrics **108**, 633–652 (2016). https://doi.org/10.1007/s11192-016-1976-4
14. M. Szelągowski, *Dynamic Business Process Management in the Knowledge Economy. They Are Creating Value from Intellectual Capital*, ed. by M. Szelągowski (Springer, Cham, 2019)
15. S.S. Hashemin, S.M.T. Fatemi Ghomi, Constrained consumable resource allocation in alternative stochastic networks via multi-objective decision making. J. Ind. Eng. Int. **8**, 207 (2012). https://doi.org/10.1186/2251-712X-8-18
16. K. Neumann, *Stochastic Project Networks. Temporal Analysis, Scheduling and Cost Minimization* (Springer Berlin Heidelberg, Berlin, 1990)
17. J. Fox, R.P. Cooper, D.W. Glasspool, A canonical theory of dynamic decision-making. Front. Psychol. **4**, 150 (2013). https://doi.org/10.3389/fpsyg.2013.00150
18. C. Gonzalez, P. Fakhari, J. Busemeyer, Dynamic Decision Making: Learning Processes and New Research Directions. Hum. Factors **59**, 713–721 (2017). https://doi.org/10.1177/0018720817710347
19. J.M. Hotaling, P. Fakhari, J.R. Busemeyer, *Dynamic Decision Making International Encyclopedia of the Social & Behavioral Sciences*, vol. 8 (Elsevier, 2015), pp. 708–713. https://doi.org/10.1016/B978-0-08-097086-8.43040-0
20. Гейда, А.С., Ашимов, А.А., Лысенко, И.В., Юсупов, Р.М.: Эффективность функционирования и другие операционные свойства систем: задачи и метод оцениван ия. Research Problems, Evaluation Method. Труды СПИИРАН [SPIIRAS Proceedings] **5**, 241–270 (2018)
21. A. Geyda, Families of alternative stochastic action networks: use for process science, in *Conference of Open Innovation FRUCT 28*, ed. by Balandin (FRUCT Oy, 2021), p. 9347589
22. D. Golenko-Ginzburg, A. Gonik, Project planning and control by stochastic network models, in *Managing and Modelling Complex Projects*, ed. by T.M. Williams (Springer Netherlands, Dordrecht, 1997), pp. 21–45. https://doi.org/10.1007/978-94-009-0061-5_4
23. A.S. Geyda, System capability estimation for various information operations used. MATEC Web of Conference, vol. 346, 2021, in *International Conference on Modern Trends in Manufacturing Technologies and Equipment (ICMTMTE)* (2021)
24. R. Miettinen, S. Paavola, P. Pohjola, From habituality to change: contribution of activity theory and pragmatism to practice theories. J. Theory Soc. Behav. **42**, 345–360 (2012). https://doi.org/10.1111/j.1468-5914.2012.00495.x
25. G.Z. Bedny, I. Bedny, *Work activity studies within the framework of ergonomics, psychology, and economics* (Taylor & Francis, Boca Raton, 2019)
26. S. Karanasios, Toward a unified view of technology and activity. ITP **31**, 134–155 (2018). https://doi.org/10.1108/ITP-04-2016-0074
27. C. Marletto, Constructor theory of information, in *Information and Interaction. The Frontiers Collection*, ed. by I.T. Durham, D. Rickles, vol. 33 (Springer International Publishing, Cham, 2017), pp. 103–111. https://doi.org/10.1007/978-3-319-43760-6_6

# High Security Object Integrity and Manipulation of Conceal Information by Hiding Partition Technique

**Girish Padhan and Rout Ranjita**

**Abstract** In the current period, because of the far reaching accessibility of the Internet, it is incredibly simple for individuals to convey and share interactive media substance with one another. Nonetheless, simultaneously, secure exchange of individual and protected material has turned into a basic issue. Thusly, secure method for information move are the most dire need of the time. Steganography is the science and specialty of shielding the restricted information from an unapproved access. The steganographic approaches hide restricted information into a cover record of type sound, video, text or potentially image. The genuine test in steganography is to accomplish high strength and limit without dealing on the subtlety of the cover document. In this article, a proficient steganography strategy is proposed for the exchange of privileged information in computerized images utilizing number hypothesis. For this reason, the proposed technique addresses the cover image utilizing the Fibonacci grouping. The portrayal of an image in the Fibonacci succession permits expanding the bit planes from 8-bit to 12-bit planes. The test aftereffects of the proposed strategy in examination with other existing steganographic strategies display that our technique accomplishes high installing of privileged information as well as gives top caliber of stego images as far as pinnacle signal-to-clamor proportion (PSNR). Besides, the power of the technique is likewise assessed within the sight of salt and pepper clamor assault on the cover images.

**Keywords** Security · Object · Integrity · Manipulation · Information · Hiding · Partition · Technique

G. Padhan (✉)
Vikash Institute of Technology, Bargarh, Odisha, India
e-mail: drgirishpadhan@gmail.com

R. Ranjita
GIETU, Gunupur, Odisha, India

 
S. Patnaik et al. (eds.), *Smart and Sustainable Technologies: Rural and Tribal Development Using IoT and Cloud Computing*, Advances in Sustainability Science and Technology,
https://doi.org/10.1007/978-981-19-2277-0_5

# 1 Introduction

Presently a day's normal correspondence of information transmission is done effectively by online media like images, recordings, and sounds for digital correspondence. Sharing, stockpiling, business, eservices like aircraft reservations, Internet business, media communications charging, electronic banking, exchange preparing, Visas, capital stock exchanging, life sciences research, medical care claims handling, and a lot more exercises are acted in our everyday existence schedule. Expanding conditions of digital correspondence force a few difficulties in ensuring and overseeing information as gotten correspondence. Altering becomes simpler because of effectively accessible programming. A significant proportion of confirmation assortment, stockpiling, and verification in criminological sciences, which chooses the wellbeing and security of any framework records, can be either in versatile archive organizes or filtered images. To accumulate proof, or plan a measurable examination, digital images are gotten with various present day strategies. Information hiding, achieved by taking advantage of a PC's record framework and different other working framework qualities, can take on many structures. Much of the time, information hiding is a purposeful movement that an individual utilizes to store away touchy information trying to make it imperceptible to every other person. Notwithstanding, there are a few special cases, for example, digital watermarking, that are utilized for fitting purposes. Some normal techniques for information hiding include: stowed away documents, erased records, stowed away partitions, substitute information streams, steganography, and slack space hiding. There are numerous PC criminology toolboxes accessible that permit a client to recognize different sorts of information hiding.

Taking a more top to bottom look to how these tool stash identify the kinds of information hiding referenced gives a more profound comprehension of how the information hiding was cultivated. The most generally utilized document frameworks for Windows working frameworks are the record allotment table (FAT) record frameworks and the new advances document framework (NTFS). These record frameworks both play out similar fundamental errands, yet for the motivations behind information hiding, their unpretentious contrasts change the manners in which that a few techniques for information hiding are refined. In this manner, a short outline of each document framework, alongside a little conversation of their disparities is justified and will be introduced preceding the conversation of the different techniques for information hiding. Digital image investigation incorporates image recuperation and reconnaissance for image information improvement. The objective of fraud location is to amplify the extraction of information from controlled images, especially uproarious and post-handled images. Since digital image preparing is becoming famous with many benefits in logical and designing applications, the imitation techniques are likewise developing at a quick rate. In any case, the validness of images and recordings becomes shifty information as addressed by Garfinkel in Digital image scientific methodologies beat image altering and expect to further develop image quality for the present digital world by featuring the requirement for new strategies for creativity and its quality. Information on the image input source is significant since it permits gadget information for agents' necessities. The most well-known sort of duplicate move altering was generally used in a digital exchange where a piece or some segment of an image messed with comparable provisions of a similar image, and that made it hard to gauge the exact area of the image imitations.

The different categories of information security systems are depicted in Fig. 1. The cryptography and information hiding are security systems that are used to protect data from deceivers, crackers, hackers, and spies. Commonly, most of the malicious users want to leave traces from cuts, manipulations, and infections [1]. The cryptography scrambles a plain text into ciphertext which is reversible without data loss. The goal of cryptography is to prevent unauthorized access to the secret information by scrambling the content of information. On the other hand, information hiding is a powerful security technique which hides a secret data in a cover media (e.g., text, image, audio, or video) so that the trace of embedding hidden data is completely unnoticeable. The cryptography and information hiding are similar in a way that both are utilized to protect sensitive information. However, the imperceptibility is the difference between both techniques; that is, information hiding concerns how to hide information unnoticeably. Generally, the information hiding can be further categorized into steganography and watermarking. The aim of steganography is to hide a secret message in a cover media in order to transmit the secret information; therefore, the main concern is how to conceal the secret information without raising suspicion; that is, steganography needs to conceal the fact that the message is hidden. Watermarking is concerned with hiding a small data in digital files such that the hidden data is robust to alterations and adjustments. In other words, watermarking aims to protect intellectual property of digital media against unauthorized copy or access by embedding a watermark (visible or invisible) in the cover media which can remain beside the data, and it can be used whenever there is any query about the originality of media (e.g., the hidden watermark refers to the original owner) [1–9].

**Fig. 1** Block diagram of Information security system

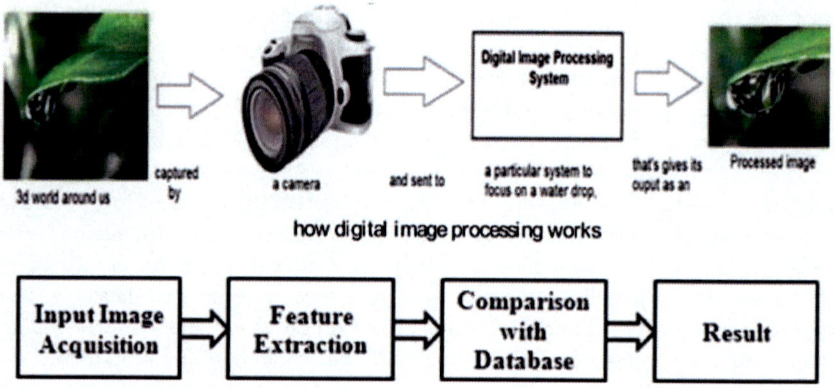


**Fig. 2** Digital image processing system

The image size of altered areas is likewise a significant boundary for the recognition of falsification. At the point when post-handling is applied then the outcome will be an inferior quality factor that will make calculations troublesome. Digital legal is an assortment of logical strategies for distinguishing proof, investigation, understanding, content validation, characterization, documentation from digital hotspots for the reproduction of unique information, which helps in fraud location in this manner recognizing who, what and why in such conditions. Measurable science is vital for image examination in which factual twofold examples were investigated utilizing distinctive change techniques (Fig. 2).

## 2 Requirements of Hiding Information Digitally

There are various conventions and inserting techniques that empower us to shroud information in a given object. In any case, the entirety of the conventions and techniques should fulfill various necessities so steganography can be applied accurately.

Coming up next is a rundown of primary necessities that steganography techniques should fulfill: The integrity of the secret information after it has been implanted inside the stego object should be right.

1. The secret message should not change at all, for example, extra information being added, loss of information or changes to the privileged data after it has been covered up. In the event that restricted Intel is changed during steganography, it would nullify the entire purpose of the cycle.
2. The stego object should stay unaltered or almost unaltered to the unaided eye. In the event that the stego object changes fundamentally and can be seen, an outsider might see that information is being covered up and subsequently could endeavor to extricate or to obliterate it.

3. In watermarking, changes in the stego object must have no impact on the watermark. Suppose you had an unlawful duplicate of an image that you might want to control differently.
4. These manipulations can be straightforward cycles, for example, resizing, managing or pivoting the image. The watermark inside the image should endure these manipulations, in any case the assailants can without much of a stretch eliminate the watermark and the place of steganography will be broken.
5. Finally, we generally accept that the aggressor knows that there is covered up information inside the stego object.

## 3 Digital Image Forensic Techniques for Feature Extraction

Change is vital in a few image preparing applications like Image examination, Image separating, Image upgrade, and Image pressure. Change having sinusoidal as the essential capacity is called Fourier Transform. Change having nonsinusoidal as an essential capacity is called Haar-Transform (most straightforward), Walsh Transform, Hadamard-Transform, and Inclination Transform. Change whose essential cycle relies upon measurements of the info signal is KL Transform (best direct Transform as far energy compaction) and Singular Value Decomposition. Change which addresses directional information of an image signal incorporates Hough Transform, Radon change.

## 4 Spatial Transform Techniques

This methodology extricates highlights dependent on Scale Invariant Feature Transform and Speed up Reduced Features, second, power, obscure, Zernike approach, and so forth.

Training and Testing Procedure of Passive Forgery Detection Techniques: This incorporates the course of Image Processing and Feature Extraction wherein highlights from a specific set are taken out for each class which helps in distinctive that from additional classes while staying invariant to recognizing modifications inside the class from input counterfeit information. Extricated edifying elements and picked highlights should be unpretentious to image manipulation and low measurement diminishes the computational multifaceted nature of grouping and preparing without decreasing AI execution. Classifier Selection helps the classifier in separating highlights from preparing image sets and Feature Pre Processing Classification and post-handling is associated with manufactured district examination.

## 5 Conclusion

Duplicate moving falsification recognition has been tried by numerous systems for the extraction of components, division, procurement, histogram, change, and so forth. Subsequently, the confirmation of digital images is a vital region in the field of scientific examination on image preparing. This examination assists with distinguishing new approaches and thoughts for future specialists working in the field of electronic fraud recognizable proof. Duplicate moving phony discovery calculation is executed with all image limitations and works with no speculated image subtleties like a digital steganography or digital mark.

## References

1. O. Mayer, M.C. Stamm, Forensic similarity for digital images. IEEE Trans. Inf. For. Secur. **15**(1), 1331–1346, 3456 (2017)
2. J. Fridrich, M. Goljan, R. Du, Invertible authentication, in *Proceedings of SPIE Secur Watermarking Multimedia Contents III*, vol. 4314, 2001, pp. 197–208
3. J. Fridrich, M. Goljan, R. Du, Lossless data embedding New paradigm in digital watermarking. EURASIP J. Adv. Signal Process. **2002** (2002)
4. M.U. Celik, G. Sharma, A.M. Tekalp, E. Saber, Lossless generalized-LSB data embedding. IEEE Trans. Image Process **14**(2), 253–266 (2005)
5. J. Tian, Reversible data embedding using a difference expansion. IEEE Trans. Circuits Syst. Video Technol. **13**(8), 890–896 (2003). Show in Context View Article Full Text PDF (544KB) Google Scholar
6. J. Tian, Wavelet-based reversible watermarking for authentication, in *Proceedings of SPIE Secur Watermarking Multimedia Contents IV*, vol. 4675, 2002, pp. 679–690
7. F. Matern, C. Riess, M. Stamminger, Gradient-based illumination description for image forgery detection. IEEE Trans. Inf. For. Secur. IEEE **2341**, 1303–1317 (2020)
8. S. Ryu, et al., Rotation invariant localization of duplicated image regions based on Zernike moments. IEEE Trans. Inf. For. Secur. **762**, 1355–1370 (2018)
9. S. Joshi, S. Member, N. Khanna, Single classifier-based passive system for source printer classification using local texture features. IEEE Trans. Inf. For. Secur. **4217**, 1603–1614 (2019)

**Dr. Girish Padhan** a young lad of 38 years from a well-educated rural family, working as an associate professor in Vikash Institute of Technology, Bargarh, Odisha having 15 numbers of national and international papers goes ahead on the area of image processing.

# Performance Evaluation of FSO Under Different Atmospheric Conditions

**Bibhu Prasad, Krishna Chandra Patra, Nalinikanta Barpanda, Subham Dey, and Somya Ranjan Pradhan**

**Abstract** To fulfil the ever-increasing demand of the modern generation, various technologies are developed so that the communication becomes faster and efficient. Free Space Optics has proved to be a great boon to the modern world. FSO is the basis behind the idea of safe optical wireless communication. It offers an unsurpassed reliability and high-speed connectivity. FSO indicates to the transmission of beams of modulated signal in free space (atmosphere) to obtain high-speed communications with a high data rate. FSO communication has advanced a lot in few years, giving the freedom of portability and reliance to wireless communication. The most challenging factor in FSO communication is the attenuation which increases with respect to extreme weather conditions. There are a wide range of climate and weather conditions that FSO has to deal with. In this work a FSO communication system is analyzed and simulated under various weather conditions like clear air, fog, haze, etc. with variable power, wavelength and attenuation using Optisystem software and the performances are compared in accordance with Q Factor and Bit error rate.

**Keywords** FSO · Weather conditions · Q Factor · BER · OptiSystem18

## 1 Introduction

Communication is very necessary in the process of making human life easier. The process of transmitting information where optical fiber acts as a medium is known as fiber-optic communication and here the original information is converted to optical pulses before transmission. Optical fiber is made up of plastic or glass like material,

B. Prasad (✉) · K. C. Patra · N. Barpanda
Sambalpur University, Sambalpur, Odisha 768019, India
e-mail: bibhu.lect@gmail.com

N. Barpanda
e-mail: nkbarpanda@suniv.ac.in

S. Dey · S. R. Pradhan
Department of ECE, GIET University, Gunupur, Odisha 765022, India



S. Patnaik et al. (eds.), *Smart and Sustainable Technologies: Rural and Tribal Development Using IoT and Cloud Computing*, Advances in Sustainability Science and Technology,
https://doi.org/10.1007/978-981-19-2277-0_6

because light can easily pass through it. TIR (Total Internal Reflection) is the principle behind optical communication.

FSO (Free space optical communication) is a method where the modulated input signal is transmitted in free space, that may be air, space or vacuum [1–4]. In comparison to various technologies of fiber-optic communication FSO is much more flexible in designing complex optical communication system [5–7]. Also, there is no interference between other radio signals and other FSO systems [8]. It's desired to achieve great speed at a high data rate in case of FSO [9–11]. However environmental factors like haze, fog, clouds bring the phenomenon of attenuation into picture [12]. As attenuation is considered as the most important bottle neck of optical communication, to eradicate it we varied the wavelength, input power of our proposed system. To analyze the performance of the whole system Bit Error Rate (BER), Quality-Factor (Q-factor) and Eye Diagram is taken into consideration.

## 2 Simulation Setup

A designed system of FSO in OptiSystem18 is shown in Fig. 1. The simulation set up consists of PRBS (Pseudo-Random Bit Sequence Generator) which creates random data like real time scenario, NRZ Pulse Generator creates a train of pulses which follows non-return to zero principle, Continuous Wave Laser which generates an optical wave which is continuous in nature, Mach-Zehnder Modulator is used for the modulation of the given signal, FSO channel is used for wireless connectivity between two points, EDFA is used to reduce the loss of an optical fiber due to long range fiber-optic communication, Photodetector APD is present as a receiver which improves the signal to noise ratio, Low Pass Bessel filter allows specific wavelength or set of wavelength of light, BER Analyzer is a visualizer which gives the bit error rate (BER) along with the Q-factor of the transmitted signal at the receiver's side.

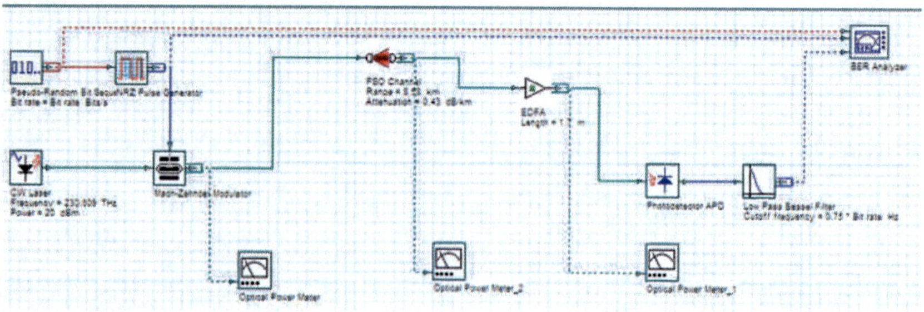


**Fig. 1** Simulation setup using OptiSystem18

## 2.1 Simulation Result and Analysis

While keeping the Q-Factor and BER value close to 5.5 and $10^{-9}$ respectively, so that we could maximize the distance of communication range, here we have varied the power from 0 to 20 dB and the results are as follows.

Table 1, shows the value of attenuation at 0, 5, 10, 20 db input for clear air, rain (heavy and moderate), haze, light fog, fog (heavy and moderate). For clear air the attenuation value is 20 at 20 db input power, for heavy rain 2.8 at 20 db input power and for heavy fog the attenuation is 0.347 at 20 db input power.

Figure 2 gives information about the eye diagram of radiated pulse in FSO system for clear air at 20 db input power and the BER value is 5.88.

**Table 1** Results of attenuation and communication range for weather conditions with varied power range

| Weather conditions | Attenuation | Distance (km) 0 db | Distance (km) 5 db | Distance (km) 10 db | Distance (km) 15 db | Distance (km) 20 db |
|---|---|---|---|---|---|---|
| Clear air | 0.43 | 4.4 | 6.95 | 10.4 | 14.9 | 20 |
| Haze | 4.2 | 2.05 | 2.65 | 3.356 | 4.2 | 4.98 |
| Moderate rain (12.5 mm/h) | 5.8 | 1.7 | 2.2 | 2.75 | 3.33 | 3.92 |
| Heavy rain (25 mm/h) | 9.2 | 1.32 | 1.666 | 2.02 | 2.41 | 2.8 |
| Light fog | 20 | 0.91 | 0.98 | 1.165 | 1.355 | 1.535 |
| Moderate fog | 42.2 | 0.49 | 0.57 | 0.665 | 0.755 | 0.85 |
| Heavy fog | 125 | 0.217 | 0.247 | 0.28 | 0.314 | 0.347 |

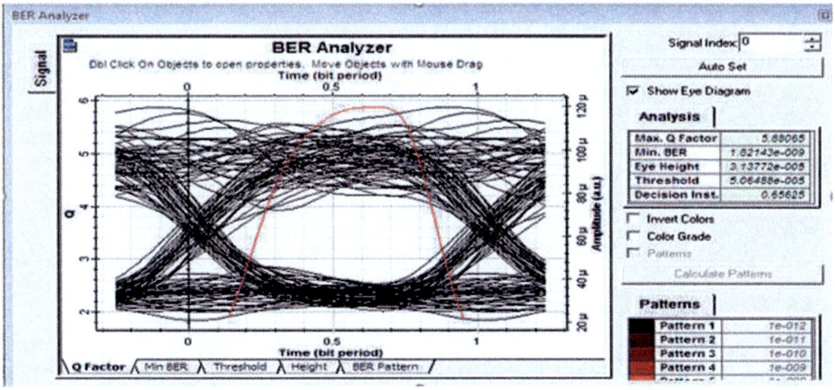


**Fig. 2** Eye diagram for clear air at 20 db power

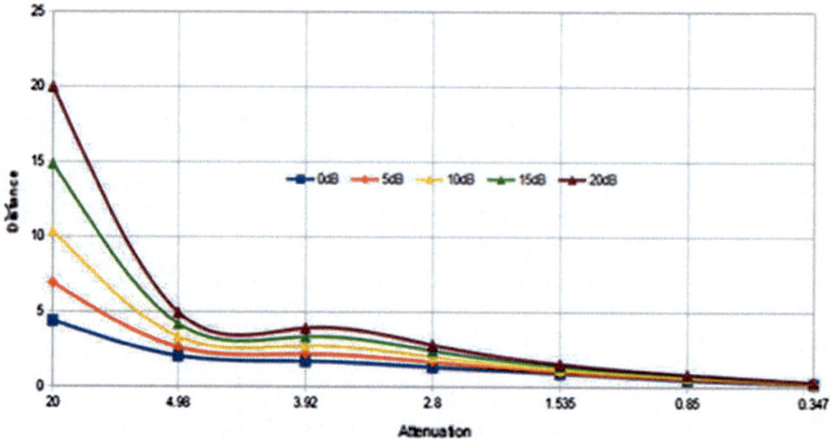


**Fig. 3** FSO with variable power and attenuation

Figure 3 shows the attenuation value for different atmospheric factors which are responsible for attenuation loss in FSO, where the distance of communication varies from 0 to 20 km (Table 2).

Figure 4 shows the attenuation value for atmospheric factors at 850, 1300, 1550, 1760 nm wavelength at 10db input power. For clear air the attenuation value is 10.4, for heavy rain 2.02 and for heavy fog the attenuation is 0.28 at 1550 nm wavelength. The best attenuation value is obtained for clear air and the worse is in case of heavy fog.

The previous Fig. 5 shows the attenuation value for atmospheric factor at 850, 1300, 1550, 1760 nm wavelength at 20 db input power. For clear air the attenuation value is 20, for heavy rain 2.8 and for heavy fog the attenuation is 0.34 at 1550 nm wavelength. The worse attenuation value is obtained in heavy fog and the best value is obtained in case of clear air as there is less obstacle.

Figure 6 gives the information for the eye diagram of radiated pulse in FSO system in heavy fog at 20db input power and the BER value is 5.72.

**Table 2** Results of attenuation and communication ranges for weather conditions with power range of 10 db and 20 db

| Weather conditions | Attenuation | 10 dB | | | | 20 dB | | | |
|---|---|---|---|---|---|---|---|---|---|
| | | 850 nm | 1300 nm | 1550 nm | 1760 nm | 850 nm | 1300 nm | 155 0 nm | 1760 nm |
| Clear air | 0.43 | 3.77 | 3.48 | 10.4 | 5.6 | 9.165 | 8.58 | 20 | 12.61 |
| Haze | 4.2 | 1.848 | 1.76 | 3.356 | 2.359 | 3.143 | 3.02 | 4.98 | 3.775 |
| Moderate rain (12.5 mm/h) | 5.8 | 1.575 | 1.5 | 2.75 | 1.972 | 2.573 | 2.48 | 3.92 | 3.051 |
| Heavy rain (25 mm/h) | 9.2 | 1.225 | 1.17 | 2.02 | 1.499 | 1.903 | 1.84 | 2.8 | 2.22 |
| Light fog | 20 | 0.763 | 0.74 | 1.165 | 0.904 | 1.107 | 1.08 | 1.53 | 1.262 |
| Moderate fog | 42.2 | 0.461 | 0.45 | 0.66 | 0.533 | 0.63 | 0.45 | 0.85 | 0.71 |
| Heavy fog | 125 | 0.207 | 0.2 | 0.28 | 0.233 | 0.27 | 0.27 | 0.34 | 0.29 |

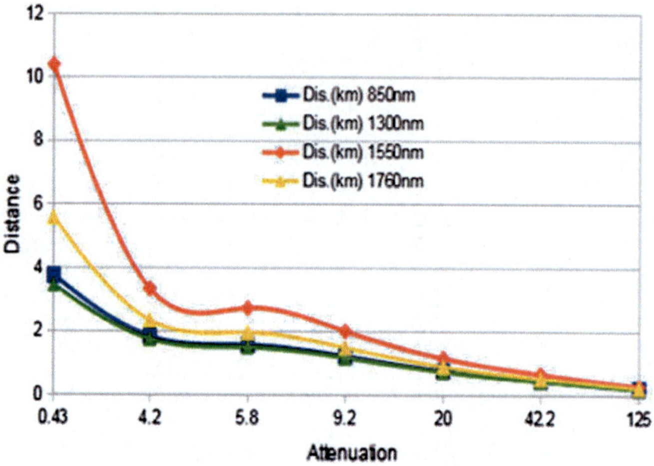


**Fig. 4** FSO with 10 db power at variable attenuation and wavelength

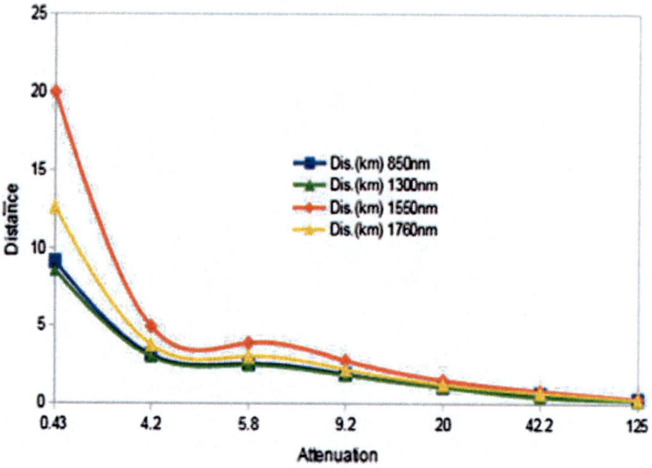


**Fig. 5** FSO with 20 db power at variable attenuation and wavelength

## 3 Conclusion

From the variable power observation tables and graphs we can observe that FSO is best suited for low attenuation area so that it can provide us with the best data rate, suitable for high attenuation areas so that it would be easily accessible to all even if the weather is harsh or not suitable for proper communication, data loss will be reduced with higher security but still being cost effective which we need in our daily life. We can increase the range of the FSO communication even with high attenuation

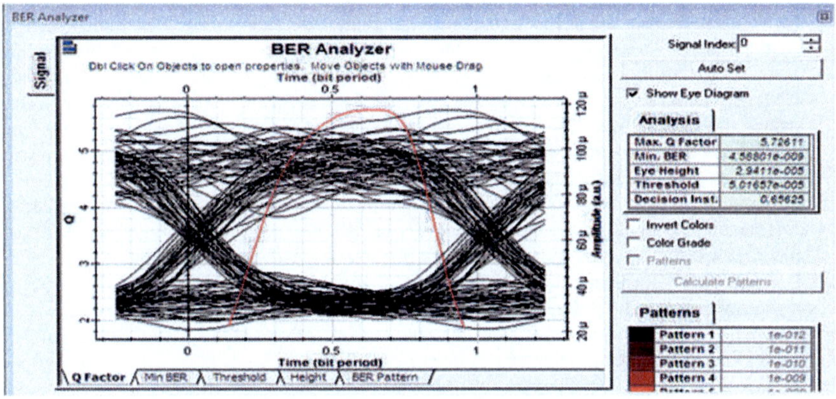


**Fig. 6** Eye diagram at heavy fog with1550 nm wavelength

by installing multiple optical amplifiers (which isn't very cost effective). The other way to do so is by increasing the power of the source signal so that it can travel longer distances. From the variable wavelength observation tables and graphs we can conclude that FSO requires a basic wavelength of 1550 nm to perform best at any given condition.

## References


1. S. Burdah, R. Alamtaha, O.N. Samijayani, S. Rahmatia, A. Syahriar, Performance analysis of Q factor optical communication in free space optics and single mode fiber. Univ. J. Electr. Electron. Eng. **6**(3), 167–175 (2019). http://www.hrpub.org. https://doi.org/10.13189/ujeee.2019.06031
2. D. Jain, R. Mehra, Dept. of ECE, Govt. Engineering College, Ajmer Ajmer (Raj.), Performance analysis of free space optical communication system for S, C and L band, in *2017 International Conference on Computer, Communications and Electronics (Comptelix) Manipal University Jaipur*, Malaviya National Institute o/Technology Jaipur & IRISWORLD, 01–02 July 2017
3. S. Mahajan, D. Prakesh, H. Singh, Scholar, Department of ECE Professor, Department of ECE Assistant Professor, Department of ECE Chandigarh University Chandigarh University Chandigarh University Gharuan, Mohali—Punjab, India Gharuan, Mohali—Punjab, India Gharuan, Mohali—Punjab, India, Performance analysis of free space optical system under different weather conditions, in *2019 6th International Conference on Signal Processing and Integrated Networks (SPIN)*
4. G. Sharma, L. Tharani, Performance evaluation of WDM-FSO based hybrid optical amplifier using Bessel filter, in *International Conference on Communication and Signal Processing*, 3–5 Apr 2018, India

5. M. Prajapat, Department of Electronics and Communication Engineering Government Women Engineering College Ajmer, Rajasthan, India ermanisha22@gmail.com Chetan Selwal Department of Electronics and Communication Engineering Government Women Engineering College Ajmer, Rajasthan, India selwal.chetan@gmail.com, Free space optical link performance simulation under different atmospheric conditions and diversity, in *2017 International Conference on Computer, Communications and Electronics (Comptelix)*. Manipal University Jaipur, Malaviya National Institute of Technology Jaipur & IRISWORLD, 01–02 July 2017
6. M. Prajapat, Department of Electronics and Communication Engineering Government Women Engineering College Ajmer, Rajasthan, India ermanisha22@gmail.com Chetan Selwal Department of Electronics and Communication Engineering Government Women Engineering College Ajmer, Rajasthan, India selwal.chetan@gmail.com, Free space optical link performance evaluation with different atmospheric conditions at different power and diversity
7. D.A. Luong, T.C. Thang, A.T. Pham, Average capacity ofmimo/fso systems with equal gain combining over log-normal channels, in *Fifth International Conference on Ubiquitous and Future Networks (ICUFN 2013)*, 7 (2013)
8. A.C. Motlagh, V. Ahmadi, Z. Ghassemlooy, K. Abedi, The effect of atmospheric turbulence on the performance of the free space optical communications, in *2008 6th International Symposium on Communication Systems, Networks and Digital Signal Processing*, 2008, pp. 540–543. https://doi.org/10.1109/CSNDSP.2008.4610725
9. N. Dayal, P. Singh, P. Kaur, Long range cost-effective WDM-FSO system using hybrid optical amplifiers. Wireless Pers. Commun. **97**, 6055–6067 (2017). https://doi.org/10.1007/s11277-017-4826-7
10. F. Nadeem, V. Kvicera, M.S. Awan, E. Leitgeb, S.S. Muhammad, G. Kandus, Weather effects on hybrid FSO/RF communication link. IEEE J. Sel. Areas Commun. **27**(9), 1687–1697 (2009). https://doi.org/10.1109/JSAC.2009.091218
11. M. Usman, H. Yang, M. Alouini, Practical switching-based hybrid FSO/RF transmission and its performance analysis. IEEE Photonics J. **6**(5), 1–13 (2014), Art no. 7902713. https://doi.org/10.1109/JPHOT.2014.2352629
12. M. Alzenad, M.Z. Shakir, H. Yanikomeroglu, M. Alouini, FSO-based vertical backhaul/fronthaul framework for 5G+ wireless networks. IEEE Commun. Mag. **56**(1), 218–224 (2018). https://doi.org/10.1109/MCOM.2017.1600735

# Smart Communication

# Design and Performance Analysis of High Reliability QOS-Oriented Adaptive Routing Protocol for MANET

**K. Venkatesulu, V. Gajendrakumar, and B. Nancharaiah**

**Abstract** Hybrid wireless networks designing are important part of routing protocols. MANET routing protocols QoS performance is mostly affected by the network mobility conditions. Because of the open nature network of MANET, there is chance of attacking various types of threats like DoS attacks, malicious attack and black hole attack etc. Design and performance analysis of high reliability QOS-oriented adaptive routing protocol for MANET is proposed in this paper based on QoS routing parameters named as routing throughput, routing delay, PDR (packet delivery ratio) in terms of mobility nodes. The Adaptable opportunistic routing algorithm based on data transmission broadcasting nature is used to minimize the security issues and challenges of the MANET networks. QoS-Routing parameters are improved with the introduction of QoS-Oriented Adaptive Expedient Broadcast Routing (OOS-oriented AEBR). Advance opportunistic schemes are used in forwarding the packets in routes and failure nodes are detected with this framework in hybrid wireless network. The QoS Routing performance and secure communication with user's authentication can be provided by employing the optimization scheme called Secure-QMAA (QoS Mobility Aware ACO) (SQMAA). From the simulation results it is clear that QoS performance is improved with SQMAA routing protocol than other routing protocols.

**Keywords** A mobile ad hoc network (MANET) · QoS performance · Secure-QMAA · Routing analysis

---

K. Venkatesulu (✉) · V. Gajendrakumar
Research scholar, Department of ECE, GIET University, Gunupur, Odisha, India
e-mail: k.venkatesulu@giet.edu

K. Venkatesulu
Department of ECE, Swarnandhra Institute of Engineering and Technology, Narsapur, AP, India

B. Nancharaiah
Department of ECE, Usha Rama College of Engineering and Technology (A), Telaprolu, Vijawada, AP, India



S. Patnaik et al. (eds.), *Smart and Sustainable Technologies: Rural and Tribal Development Using IoT and Cloud Computing*, Advances in Sustainability Science and Technology,
https://doi.org/10.1007/978-981-19-2277-0_7

# 1 Introduction

One of the infrastructures less network is Mobile Ad hoc Network (MANET) which is self-healing, self-configured and self-coordinated when the demand of communication arises [1]. Continuous mobility of nodes creates a routing problem in MANETs, this is because of topology changing which results the breaking of routes frequently so re-establishment is required. Finding and selecting a path is very complex that can exist as long as possible in MANET reliable and optimal. A nontrivial mobile scenario used for holding guaranteeing quality in terms of parameters as maximum end -to-end delay, requirement of minimum bandwidth and loss permit of maximum data packet etc. for any data application or flow such as file transfer, data stream of video or voice [2].

Cell hubs gathering incorporated by a portable system which works on multi-jump way does not require any consistent framework. Therefore a suitable and flexible framework less network is required and MANET is utilized in this situation [3]. However, limited quality activity, ability of variable connections and requirement of transmission capacity are in corporate with some stand-out constraints [4]. More than one jump courses are comprised because moderate hubs are utilized as switches so chatting with hubs are out of its broadcast assortment. Because of node reasons, active topology visits the intemperate mistakes costs and interface disappointments. Therefore bearer (QoS) desired tasteful safeguard is hardly maintained in the network. Among neighbor hubs WiFi network is pooled and time-delicate measurements (video parcels) communication uses the modification of organize topology as hubs course as more noteworthy troublesome [5].

MANET fundamentals are developed in addition with mixed media prevalence applications, and Quality of Service (QOS) instruments and a green steering are used in bundles [6]. Multicast administrations advancement uses the extreme significance and ensured the QoS arrangement. MANETs are directed by the QoS on the grounds troublesome because of changing the system topology continuously. These changes are resulted from powerful steering changes of accessible state data and hub portability. In MANET QoS steering is trouble and directing changes uses the accessible state data. Novel routing protocol designing is proposed in this paper which makes the MANET QoS improvement and considered two main issues as reliable data communication and optimal path selection. MANET algorithm of this paper uses the QoS-Oriented adaptive routing which uses the advance opportunistic schemes for packets transferring. So complexity and overhead are maintained as low. Retransmissions of unwanted packets are avoided by introducing the QoS-Oriented Adaptive Expedient Broadcast computing hence MANET throughput is also improved. This paper presents a SQMAA routing protocol for solving the different security attacks and selection of routing path which affects the QoS performance. ACO (Ant Colony Optimization) based path selection approach is presented. Next hop-based energy level and their nodes are selected by the ACO algorithm.

## 2 MANET

From last two decades there is a huge development in Ad hoc networks because of its wireless communication fast progress [7]. Without any infrastructure a group of mobile nodes are called as Mobile Adhoc network (MANET) and there is no existence of central controller. MANET each node does not have any restrictions to move freely. A router with multiple hosts is present at each node. Limited physical security, constrained operation of energy, variable capacity links, constrained bandwidth and dynamic topologies are owned by the MANETs [8]. Different applications present with MANETs as Military Applications, rescue and search operations.

### *2.1 MANETS Characteristics*

- **Dynamic Topology**: Each and every node is mobile in MANETs. So at any time the communication may leave or break then difficult problem of routing is raised. In the range of transmission, node is acted as both host as well as router [9].
- **Multi hop structure**: In the given radio range no availability of destinations then multi hop routing came into action and no central control in the MANETs.
- Wireless links bandwidths, security and reliability are relatively lower in MANETs than wired ones.
- Battery power is limited so every node energy is precious and entire network is collapsed if there is overload on nodes.

It is useful for small and as well as large networks because these are scalable [10].

### *2.2 Application Areas of MANETS*

- **Military battlefield**: Military operations with ARPANET are the first populated Adhoc networks. The information can be gathered among tracking vehicles and soldiers.
- **Search and Rescue operations**: In the rescue operations as fire operations, floods, earthquake situations these Adhoc networks are widely used to help the people. Rescue operations are made easy with ship to ship communication [11].
- **Personal Data Networks**: Present days each person is having several gadgets. The data can be transferred in between these gadgets by using MANETs.
- **Sensor Networks**: Home automation systems uses the sensor networks widely. Adhoc networks developed many sensors.

## 2.3 Link Stability

Topology changes are results phenomenon of breaking links in MANETs. Wired networks gives more strength to systems than wireless links. Therefore there is requirement of study the link stability. Limited bandwidth is maintained in the wireless networks. So a better bandwidth is utilized with the increment of link stability. Therefore networks QOS (Quality of Service) is enhanced. By using the link lifetime, link formation relative speed, nodes remaining battery power, ABR (Associatively Based Routing) based pilot signals and signal strengths, the link stability is estimated.

Link breakage can be detected by three methods in MANETs.

1. **Route Time Out**: The routes in which the packets do not flow then that situation of route is termed as timeout. This method is supported by AODV as Reactive routing protocol.
2. **Neighbor discovery**: There is a constant flow of hello packets from neighbor nodes. Hello packet not receiving links are treated as link breaks and it is supported by OLSR as proactive routing. Demand protocols adapts this proactive routing.
3. **Notification from link layer to network layer**: If there is a break then link break information is received by link layer and it is informed by upper network layer. Therefore network blocking will be avoided.

Two types of Routing Protocols based on link stability are classified:

**Distance based routing protocols**: The distance will be minimized in this protocol in between node and it's before node.

**Mobility based routing protocols**: Mobility of nodes based parameters are considered in this protocol as any node to remain in radio range probability, node movement and speed, link stability.

# 3 Literature Survey

Different advances and challenges of QoS Routing protocols are proposed by Raghavendran et al. [12] in MANET. QoS aware routing protocols survey is also proposed by the author in the MANET.

QoS aware adaptive multipath routing protocol is proposed by the Pillai et al. [13]. Node disjoint routes are finding by using the on demand multipath routing protocol and according to the QoS metrics, route breakage uses the periodic route maintenance prior. A RREQ message is transmitted by the source node to all its neighbors when there is a requirement of communication in between the source and destination. Then the information is transmitted to destination by the intermediate nodes with first hop information. A RREP message is forwarded as the reply message to the neighbors by the destination node. Therefore different routes are reaching the source with multiple next hops from destination. From multiple routes best routes selection and alternative route prior switching for route breakage by using QoS metric by the dynamic route maintenance method is proposed by the author. Remaining battery energy, link stability and signal strength are the QoS metrics. These metrics are calculated by the every node accordingly.

In MANET, the Quality of service is improved by using the AODV modified variant which is proposed by the Sedrati et al. [14] with the discovering of source to destination multiple paths. During the phase of route discovery, multiple node disjoint routes are calculated in this method. Therefore there is existence of multiple routes in between the source to destination. Communication is not disturbed in the system if the any one of the node fails. Multiple paths providing from source to destination in Modified-AODV helps in decreasing the reduce packet because of node failure or path failure. Availability or consistency of link advancement is known by the control message of 'HELLO' transmission which is included in the route maintenance phase. After detection of failure, packets overflow or overload is reduced by the route maintenance phase. In network, overload and packet loss is reduced by the Modified-AODV.

Routing Protocol (ARA) based on Ant colony Optimization was proposed by Gunes et al. [15]. It is on demand routing protocol because of simple ant colony metaheuristic algorithm used in this routing protocol. Failure handling, maintenance and discovery phase of route are the three phases of this algorithm. The creation of routes is done in the route discovery phase. Forward ant and backward ant based Net algorithms are required in the new routes creation. Sender broadcast the forward Ants (FANT). Unique number is acquired with small packet is called as FANT. According to sequence number, duplicate packets are differentiated by the nodes and intermediate nodes delete these duplicate nodes. Backward Ants (BANT) are created by the destination after it receives the FANT and these BANT sends back to the source node. If a node recognizes failure it first sends the pheromone value to zero so that that link may not be active. An alternate link will be searched after it. If this route also fails, then it informs its neighbors. After discovering the alternative route this process is terminated automatically.

## 4 QOS-Oriented Adaptive Routing Protocol for 'MANET'

Light weight security is improved by eliminating the stable routes problems. The security system is proposed in this paper from state-of-art solutions. Data transmission broadcasting nature based Opportunistic routing is presented. During data transmission, node behavior is not considered in this scheme which has some limitations as minimum throughput and high overhead. Advance opportunistic schemes are used in transferring the packets which are allowed by Hybrid Wireless Network (QoS-Oriented AEBR) algorithm and it uses the QoS-Oriented AEBR. So complexity and overhead are maintained as low. Retransmissions of unwanted packets are avoided by introducing the QoS-Oriented Adaptive Expedient Broadcast computing hence

MANET throughput is also improved. Failure node detection is also described in this framework. Distributed Adaptive Opportunistic Routing Algorithm is analyzed in this paper. A framework of QoS-oriented adaptive routing protocol for MANET is represented in Fig. 1.

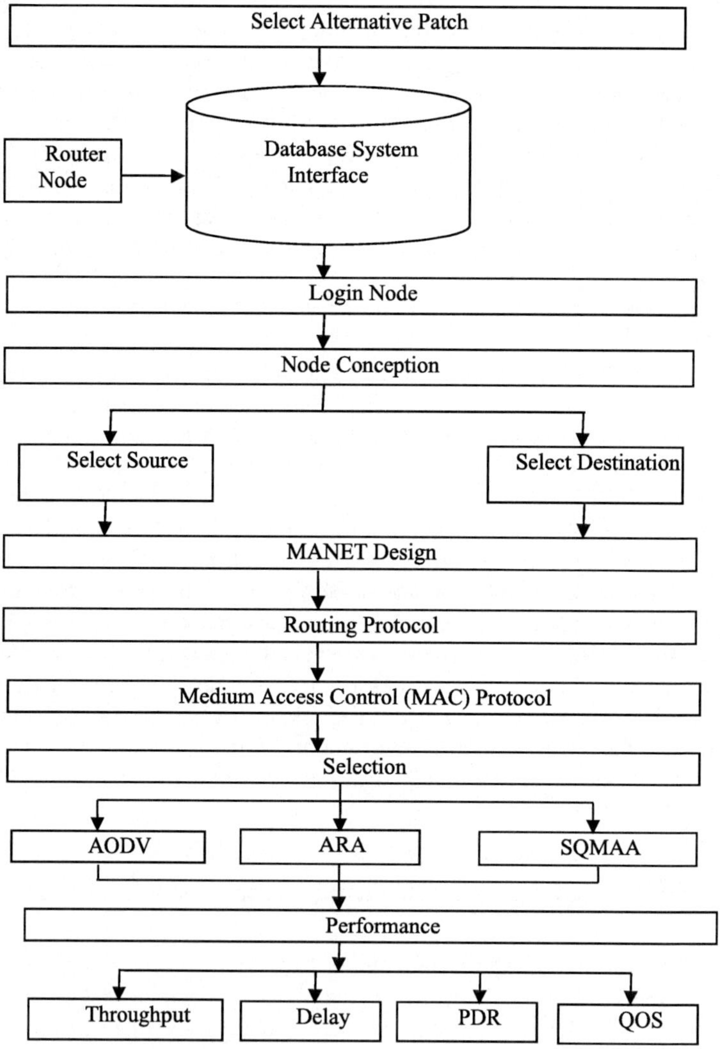


**Fig. 1** Framework of QoS-oriented adaptive routing protocol for MANET

## 4.1 Formulas

Efficient packet transmission is performed by opportunistic scheduling algorithm overall effect with the irregular wireless medium than the EDF (Earliest Deadline First) priority algorithm. Assume that set of neighbors as N and a node as $S$ in this algorithm. Accepting nodes group is denoted as $D$ and it is because of node $S$ transmission. Result acceptance is based on any time of instance routing decision and termination node or intermediate node selection, re-transmission is included. Three handshaking methods adaption make the conclusions in broadcast mode in between node $S$ and its neighbor nodes.

1. A packet was sent by node $S$ at time $t$. Acknowledgment (ACK) packet is sent again to the node $S$ after by successful received packets group to nodes $N$.
2. Command message is combined with the node acknowledgment packet and it is named as Approximated best grade (ABG).
3. Termination is declared by node $S$ after declaration of next sender as node $k$ and this is mentioned in forwarding packets.

In this paper SQMAA routing protocol is presented for achieving QoS efficiency performance and data security. SQMAA routing protocol with their algorithms designing is proposed in this section. The algorithm is consisting with five major essentials as:

1. **Initialization State**: LSTM (Long Short Term Memory) Network used all nodes are initialized in this step and these are as random set of nodes $N$, network $M$, destination $D$, packet, routing table and source $S$ etc.
2. **Packet Transmission State**: If node S has packets for transmission then this stage is processed at time $t$.
3. **Response and Recognition State**: Nodes random set is denoted by $N$ which accepts the packets from node $S$. Acknowledgment is received by the node $S$ from all successful packet acceptance of nodes in $M$. Small duration is took by acknowledgment state delay. In time $t$, $M$ is derived by node $S$ for all the nodes. Node $k$ ACK packet is combined with ABG message. Increased the random variable $N_n$ based on recognition and response measuring.
4. **Intermediate State**: Based on randomized method the process of optimum routing is selected by the node $S$ in this state. Routing decision information is present in control packet which is send by node $S$ in between two intervals at some time. Variable is measured after the routing process selection. In this paper, the Secure QMAA routing protocol method is used for taking the optimum and secure routing decision for data communication between the sender nodes to the destination nodes in the network $M$. ARA and AODV routing protocols are also evaluated along with SQMAA routing protocol. QMAA algorithm method performs the path selection based on ACO and it is the main task of SQMAA. Light weight decryption and encryption algorithms are based on PKI is used in the process of data security at intermediate nodes for avoiding the malicious attackers into data.

5. **Adaptive Computing State**: The score of transmitter can update at time ($t$ + 1) by node $S$, after sending the packets at time $t$. ABG message is also updated by node $S$ for next acknowledgment. Result acceptance is based on any time of instance routing decision and termination node or intermediate node selection, re-transmission is included. Such decision is made by our algorithm in a broadcast manner in different ways as follows. Node $S$ is handshakes with its neighbors $N(S)$.
1. A packet is transmitted by node $S$ at time $t$.
2. Acknowledgment (ACK) packet is sent again to the node $i$ after by successful received group of nodes $D$. Command message is combined with the node acknowledgment packet and it is named as ABG. It is referred by JK max.
3. Termination $T$ is declared by node $i$ after declaration of next sender as node $k$ and this is mentioned in forwarding packets.

## 4.2 Algorithm

| | | |
|---|---|---|
| Input: | Source $S$, packet, Routing table, destination $D$, $P$ threshold $E$ threshold | |
| Output: | Selection of Optimal path and Secure Data Transmission | |
| Step 1: | Initialization | |
| | 1 | Let Network $M$ with Nodes $N$ randomly deployed in size $X$ and $Y$ dimensions |
| Step 2: | Set Packet Transmission form $S$ to $D$ | |
| Step 3: | Response and Recognition of Packet transmission from $S$ | |
| Step 4: | Intermediate State of packet transmission | |
| | 1 | Extract optimal path using QoS Mobility Aware ACO |
| | 2 | Set RT = route discovery ($S$, $D$) |
| | 3 | Start secure data communication |
| Step 5: | At node $S$ | |
| | 1 | Broadcasting source node ID to the set of nodes $N$ |
| | 2 | Broadcast RREQ Packet |
| | 3 | Set of nodes $D_m^i$ accept the packet, All $K \epsilon N_m^i$ communicate their id |
| | 4 | Node $S$ declare routing conclusion $D_m^i$ |
| | 5 | Applying the algorithms of light weight encryption and decryption at intermediate nodes |
| | 6 | Update the routing table with new session key along with Public Key and Private Keys |
| | 7 | Broadcasting finally the authenticated RREQ (route request packet) |
| Step 6: | At Intermediate Node | |

(continued)

(continued)

| | | | |
|---|---|---|---|
| | 1 | received packets are verified with group private key | |
| | 2 | Transmit acknowledgment (ACK) packets to the Node $S$ | |
| | | If | The details can be extracted from the received packet after successful packet verification |
| | | else | |
| | | | From malicious node current received packet is marked and drops it |
| | 3 | Forward it to next hope by performing the operation of encryption with generated keys | |
| | 4 | Source node is signed with its group private key | |
| | 5 | Status '$P$' is setup and entry of routing table is updated with present route | |
| | 6 | The task will be repeated till destination node reaches for each intermediate node | |
| Step 7: | At Destination node | | |
| | 1 | By using keys data is verified | |
| | 2 | Decrypt the original data by applying Decryption operation | |
| | 3 | Send acknowledgment to $S$ | |

## 5 Results

Different performance metrics are considered for evaluating the SQMAA protocol performance and comparing the results with ARA and AODV. From source to destination packet loss is caused by mobile ad hoc network including with black hole nodes existence. Application throughput is also affected and it is in between source to destination. One performance metric is considered for application throughput selection and another metric is considered for data packet loss by the packet delivery ratio (PDR). Application data end-to-end delay considered as another performance metric. Next the QoS of the network were analyzed by the above performance metrics. NS3 simulator performs the QMAA routing protocol evaluation and simulation. Considering the QMAA algorithm, state-of-art routing protocol ARA for evaluating the SQMAA protocol with respect to varying mobility speed under the presence of 10% malicious or black hole nodes over 60 number of sensor nodes (Table 1).

**Throughput ratio**: In one second, number of bytes transmitted or received is called as throughput. $T$ represents throughput, and defined as:

$$T = \frac{\sum_{i=1}^{n} T_i^r}{\sum_{i=1}^{n} T_i^s} \times 100\% \tag{1}$$

**Table 1** Simulation parameters

| Parameter | Value |
|---|---|
| Network size | 1500 × 500 m |
| Number of sensors | 60 |
| Packet size | 512B |
| Speed of mobile nodes | 5, 10, 15, 20, 25, 30 mps |

where, application numbers is denoted as *n*, for the *i*th application, average receiving throughput is denoted as $T_i^r$ and for the *i*th application, average sending throughput is denoted as $T_i^s$.

**Average end-to-end delay**: The application data packets end-to-end delay average is represented as *D*, and formulated as:

$$D = \frac{\sum_{i=1}^{n} d_i}{n} \tag{2}$$

where, *i*th application data packets end-to-end delay average is denoted by $d_i$, constant Bit Rate applications number is denoted by $n$.

**Packet Delivery Ratio**: Data packet delivery ratio, (PDR) is calculated as follows:

$$\text{PDR} = \frac{\sum_{i=1}^{n} \left(N_i^s - N_i^r\right)}{\sum_{i=1}^{n} N_i^s} \times 100\% \tag{3}$$

where, for the *i*th application, the sender sending of application data packets is denoted by $N_i^s$, the receiver receiving of application data packets number is denoted by $N_i^r$.

Delay and throughput are two important parameters to be evaluated for MANET QoS. So these two parameters are required for evaluation of proposed approach. Proposed approach obtained delay is minimum.

Throughput performance comparative analysis in kbps is represented in Fig. 2. The proposed SQMAA better performs as big margin among AODV, ASR and QMAA. PDR performance of proposed approach is represented in Fig. 3. From results it is clear that, there is a requirement of throughput with increment in mobility speed because of introduction of attackers in MANET. Therefore the linear relationship is present in between the mobility speed and throughput. The performance of QMAA routing protocol is degraded with malicious attackers because of inexistence of any security approach. But its performance is better than the state-of-art ARA routing protocol because of QoS efficient paths selection.

Comparative delay performance is represented in Fig. 4. Best delay characteristics are obtained by using proposed SQMAA than other methods of protocols. Because shortest path based path is selected with highest throughput neighbor node.

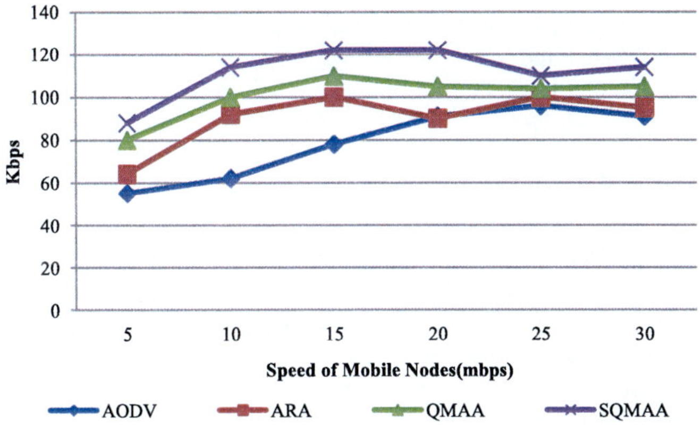


**Fig. 2** Comparative throughput results

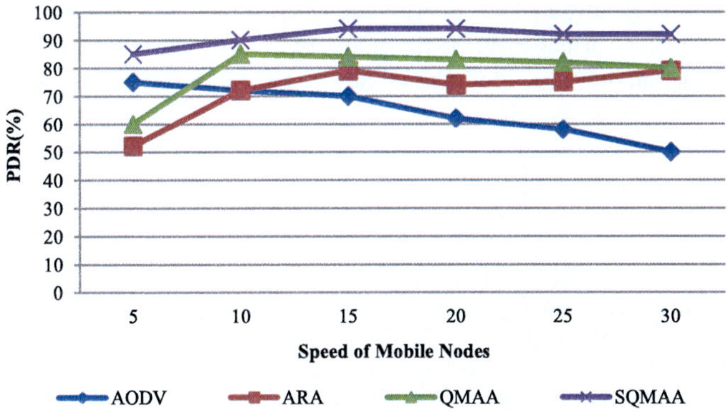


**Fig. 3** Comparative 'PDR' results

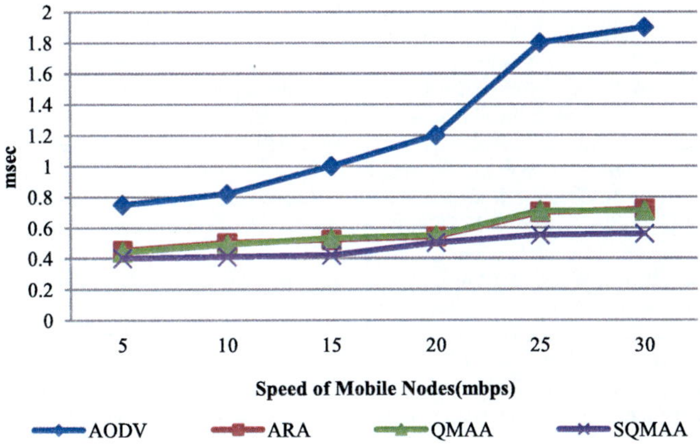


**Fig. 4** Comparative delay results

Highest delay is acquired by AODV and almost similar delay is obtained from routing techniques of QMAA and ARA.

## 6 Conclusion

Secure data communication and QoS efficiency are the two main problems of MANET. Nodes location information is used in efficient routing algorithm based on ACO with secure QoS is presented in this paper for handling the MANETs mobility issues. Security issues of MANETs are also discussed. The packets are routing with minimum overhead in Opportunistic Routing Algorithm. The packets routing framework is provided by algorithm with improved packet delivery ratio and network performance. In presence of malicious attackers, SQMAA routing protocol is simulated against the QMAA method and ARA routing protocol. From the simulation result it is clear that, data security and QoS efficiency is improved against the malicious attackers in network. In future a new routing protocol based on greedy forwarding can be developed for throughput energy aware multipath routing.

## References


1. Y. Kirsal, V.V. Paranthaman, P. Shah, H.X. Nguyen, G. Mapp, Exploiting resource contention in highly mobile environments and its application to vehicular ad-hoc networks. IEEE Trans. Veh. Technol. **68** (2019)
2. Y.-H. Lo, Y. Zhang, J. Li, F. Shu, Achieving maximum reliability in deadline-constrained random access with multiple-packet reception. IEEE Trans. Veh. Technol. **68** (2019)
3. P. Chatterjee, U. Ghosh, W.S. Alnumay, A trust-based predictive model for mobile ad hoc network in Internet of Things. Sensors **19** (2019)
4. H.V. Poor, M. Debbah, M. Bennis, Ultrareliable and low-latency wireless communication: tail risk and scale. Proc. IEEE Proc. IRE **106** (2018)
5. D. Dash, U.N. Kar, D. Guha, S. Chattopadhyay, D.K. Sanyal, A survey of topology-transparent scheduling schemes in multi-hop packet radio networks. IEEE Commun. Surv. Tutor. **19** (2017)
6. J. Fan, V.C.M. Leung, H. Zhou, S. Xu, C. Zhu, Predicting temporal social contact patterns for data forwarding in opportunistic mobile networks. IEEE Trans. Veh. Technol. **66** (2017)
7. M. Ismail, W.A. Jabbar, S. Arif, R. Nordin, Power-efficient routing schemes for MANETs: a survey and open issues. Wireless Netw. **23** (2017)
8. N.B. Bhople, P.N. Chatur, S.R. Deshmukh, AODV-based secure routing against blackhole attack in MANET, in *Proceedings of IEEE International Conference on Recent Trends in Electronics, Information and Communication Technology*, 2016
9. C. Li, Y. Chen Z. Wang, PSR: a lightweight proactive source routing protocol for mobile ad hoc networks. IEEE Trans. Veh. Technol. **63** (2014)
10. G. Cardone, P. Bellavista, L. Foschini, A. Corradi, Convergence of MANET and WSN in IoT urban scenarios. IEEE Sensors **13** (2013)
11. B.E. Wang, M.X. Zhang, K.D. Sung, J.J. Xia, A neighbor coverage-based probabilistic rebroadcast for reducing routing overhead in mobile ad hoc networks. IEEE Trans. Mob. Comput. **12** (2013)
12. C.V. Raghvendran, G. Naga Satish, Challenges and advances in QoS routing protocols for mobile ad hoc networks. IJARCSSE **3**(8) (2013)

13. M. Pillai, M.P. Sebastian, S.D. Madhukumar, *Dynamic Multipath Routing for MANET—A QoS Adaptive Approach* (IEEE, 2013)
14. M. Balachandra, K.V. Prema, M. Krishnamoorthy, Enhancing the quality of service in MANET by improving the routing technique. Int. J. Comput. Appl. **51**(7) (2012)
15. M. Gunes, U. Sorges, I. Bouazizi, ARA-the ant-colony based routing algorithm for MANETs, in *Proceedings of International Conference on Parallel Processing, Workshops* (2002), pp. 79–85

# Circular Patch Antenna with Perturbed Slots for Various Wireless Applications

**Ribhu Abhusan Panda, Nishit Mohapatra, and Subudhi Sai Susmitha**

**Abstract** A conformal circular patch antenna is designed with perturbed slots. These slots are the combination of L-shape and circular shape. The perturbed slots makes the design a novel one. The S-parameter ($S_{11} < -10$ dB) has been considered to verify the frequency at which the proposed patch resonates. The bandwidth and other essential parameters are determined from the simulation by the HFSS (High Frequency Structure Simulator) software. The proposed antenna has been intended for WBAN and WLAN application that includes 2.4 and 3.6 GHz. This proposed antenna has multiple resonant frequency with appropriate bandwidth so it can be used for different C-Band applications.

**Keywords** Antenna gain · Circular patch · C-Band · Directivity · Perturbed slots · WLAN

## 1 Introduction

Recent developments in planar antennas have provided different structures. To enhance the bandwidth different type of slots can be implemented. In the year 2012, a unique antenna was designed for WBAN (wireless body area network) [1]. Few unique antennas have been designed for WBAN in recent years [2–5]. Perturbation of the circular shape of the patch has been done in recent years for definite applications [6–16]. Here, a conventional shaped circular patch is designed with perturbed strips. High resistive surface has been considered for reducing the surface waves in the patch

R. A. Panda · N. Mohapatra (✉) · S. S. Susmitha
Department of Electronics and Communication Engineering, GIET University, Gunupur, Odisha, India
e-mail: 18ece098.nishitmohapatra@giet.edu

R. A. Panda
e-mail: ribhuabhusanpanda@giet.edu

S. S. Susmitha
e-mail: 18ece106.subudhisaisusmitha@giet.edu



S. Patnaik et al. (eds.), *Smart and Sustainable Technologies: Rural and Tribal Development Using IoT and Cloud Computing*, Advances in Sustainability Science and Technology,
https://doi.org/10.1007/978-981-19-2277-0_8

antenna in past few years [17]. Proposed conformist circular patch is designed with line feed. Perpendicular perturbed shaped slots have been implemented with a high resistive substrate including silicon material which is having a dielectric constant 11.9. A circular slot has also been implemented to make the antenna operable for multiband application like WBAN at 2.4 GHz, WLAN (wireless local area network) at 3.6 GHz and other C-Band applications (4–8 GHz).

## 2 Parameters for the Proposed Design

The proposed structure is an assemble of three parts, ground, patch and dielectric substrate. For the ground plane the dimension is 80 mm × 80 mm × 0.01 mm and for the substrate it is 80 mm × 80 mm × 1.6 mm. The dimensions have been provided in Table 1.

Two mutually perpendicular slots resembling a shape of L have been implemented with perturbation on the conventional circular patch. The patch structure has been illustrated in Fig. 1 and the design using HFSS has been shown in Fig. 2. The novelty of the proposed design is included in the perpendicular slots which adds another capacitive element to the circuit for the proposed patch that enhances the bandwidth. The material used for the substrate is the high resistive silicon which has a dielectric constant of 11.9 that support the frequency up to 60 GHz. The high resistive silicon

**Table 1** Design parameters of proposed antenna

| Symbol | Value (mm) |
|---|---|
| sw 1 | 1 |
| sw 2 | 1 |
| sw 3 | 1 |
| sw 4 | 1 |
| sw 5 | 1 |
| sw 6 | 1 |
| sw 7 | 1 |
| sw 8 | 1 |
| sw 9 | 1 |
| sw 10 | 1 |
| D | 4 |
| W | 80 |
| L1 | 42 |
| L2 | 42 |
| L3 | 40 |
| L4 | 40 |
| FW | 3 |

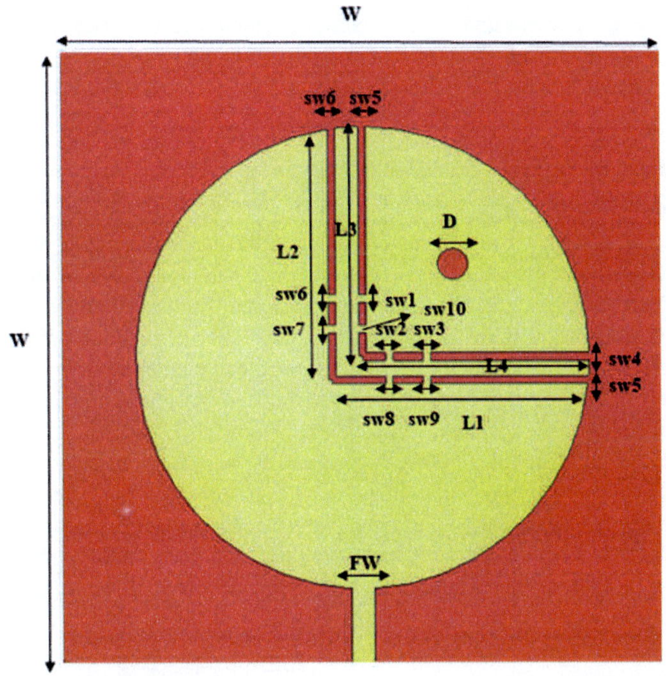


**Fig. 1** Proposed patch with perturbed slots

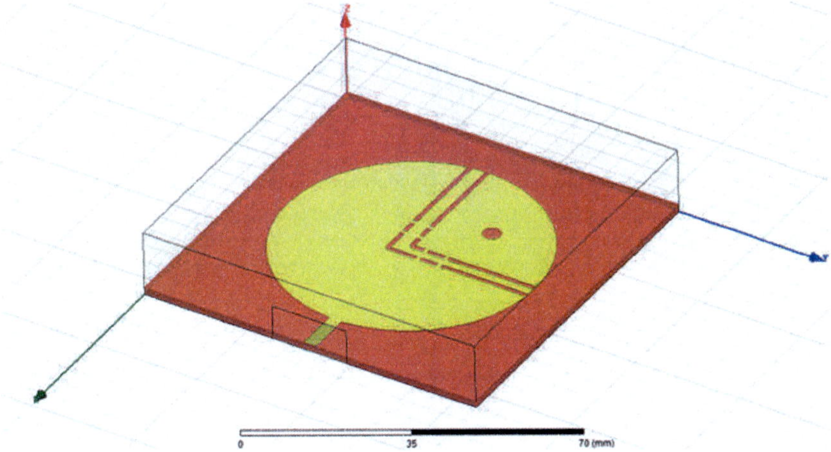


**Fig. 2** Design of antenna

prevents the leaky waves and increases the radiation efficiency. The implementation of circular slot leads to further enrichment of bandwidth.

## 3 Results of the Proposed Antenna

To verify the resonant frequencies and bandwidth the S-Parameter plot has been taken into consideration. Figure 3 includes the S-Parameter plot. From the plot it is clear that the antenna can be operated at different frequency bands has multiple resonant frequencies. Figure 4 indicates the VSWR (Voltage Standing Wave Ratio) at different resonant frequencies for the proposed patch antenna. From Fig. 4 it has been observed that the proposed antenna has the VSWR closely equal to 1, which is the desired value at the resonant frequencies.

Figures 5 and 6 includes both 2D gain pattern and 3D gain pattern. From that pattern it is clear that the radiation pattern is almost uniform. The E-Plane and H-Plane pattern has been shown in Figs. 7 and 8 respectively.

The resonant frequencies with corresponding return loss and bandwidths have been listed in Table 2. Other simulated parameters have been listed in Table 3.

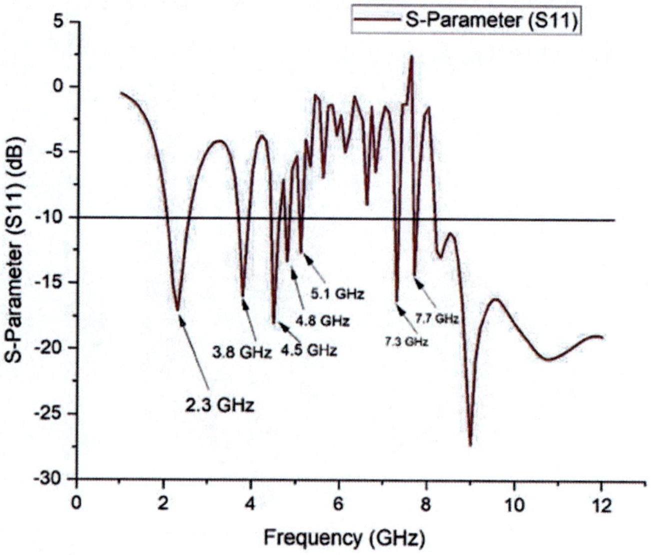


**Fig. 3** Multiple resonant frequencies with $S_{11}$

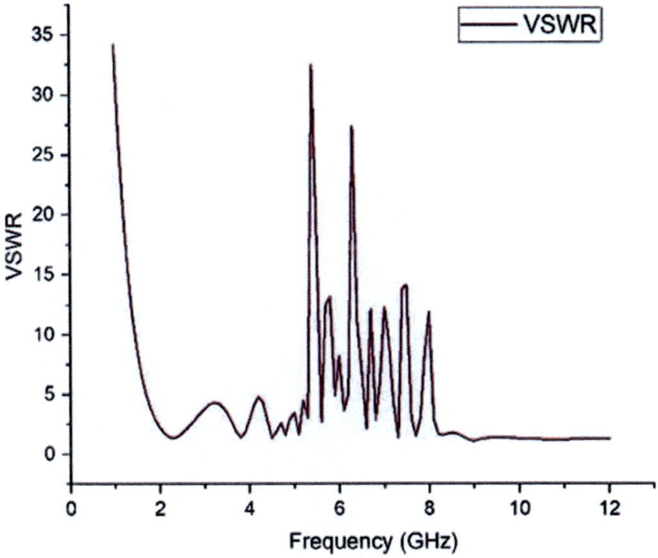


**Fig. 4** VSWR versus frequency graph

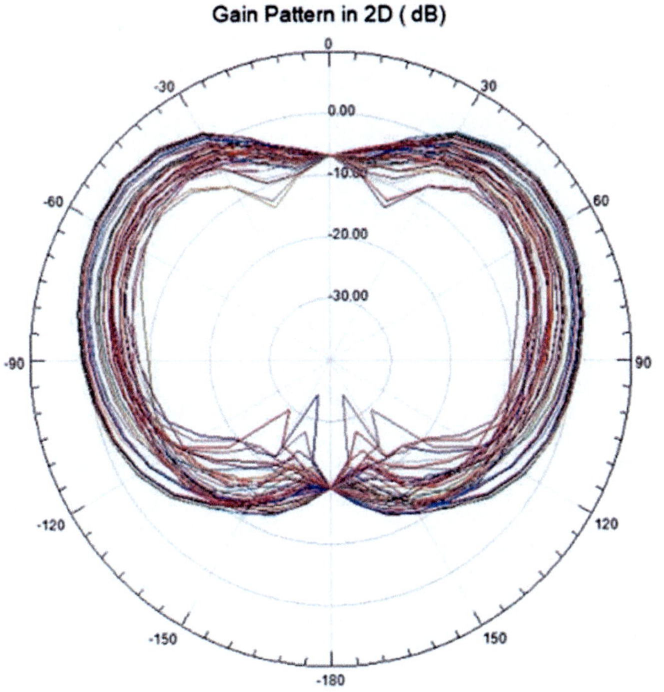


**Fig. 5** Gain pattern

**Fig. 6** Gain pattern in 3D

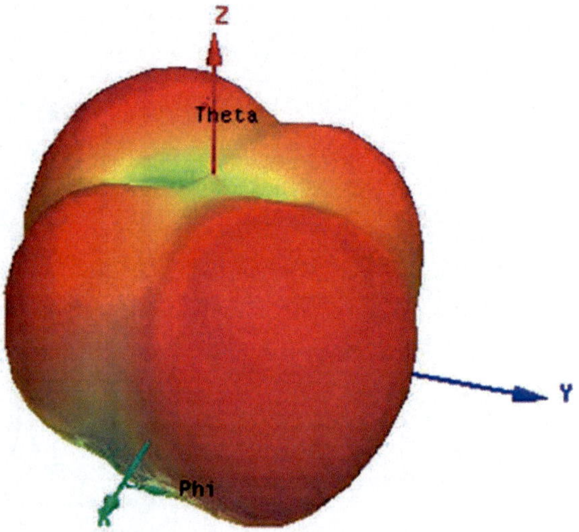


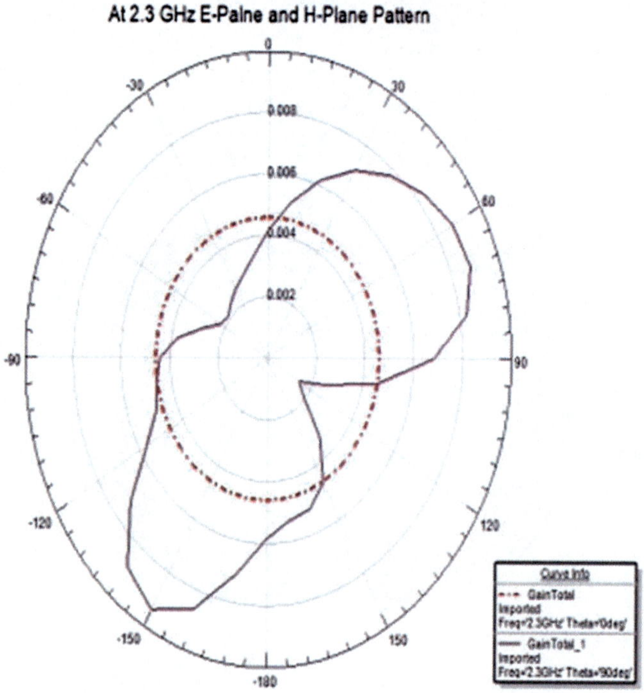


**Fig. 7** E and H field pattern for 2.3 GHz

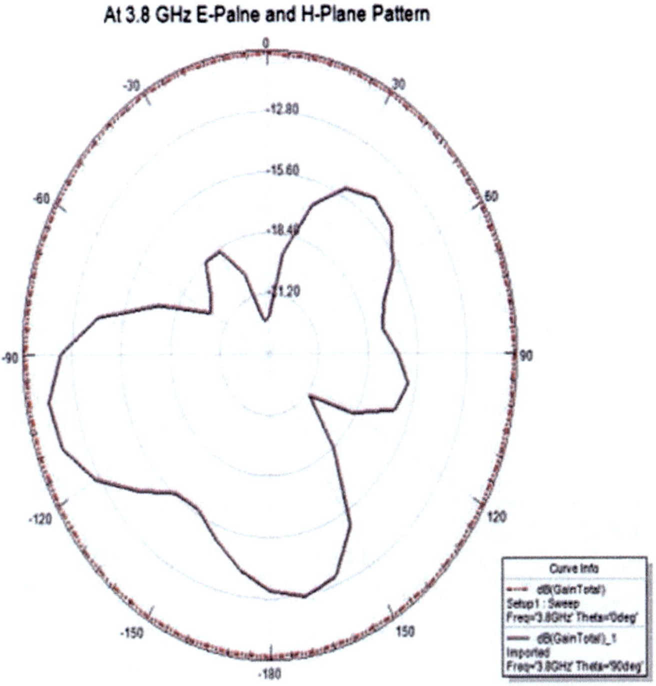


**Fig. 8** E and H field pattern for 3.8 GHz

**Table 2** Resonant frequencies with corresponding return loss and bandwidths

| S. No. | Resonant frequencies (GHz) | Return loss (dB) | Bandwidth (MHz) | Application |
|---|---|---|---|---|
| 1 | 2.3 | −17.05 | 490 | WBAN and WLAN |
| 2 | 3.8 | −15.95 | 240 | WLAN |
| 3 | 4.5 | −17.99 | 150 | Possible 5G |
| 4 | 4.8 | −13.27 | 100 | Possible 5G |
| 5 | 5.1 | −12.61 | 110 | C-Band |
| 6 | 7.3 | −16.28 | 100 | C-Band |
| 7 | 7.7 | −14.30 | 80 | C-Band |

**Table 3** Simulated parameters

| Parameters | Value |
|---|---|
| Peak antenna gain | 5.36 dB |
| Peak directivity | 6.6 dB |
| Radiation efficiency | 76% |

## 4 Conclusion

The implementation of perturbed slots in the circular patch leads multiple resonant frequencies for specific applications like WBAN and WLAN which includes 2.4 and 3.6 GHz frequencies. As several resonant frequencies has been detected in the range 4–8 GHz, the proposed design can also be used for C-Band application. The projected antenna provides a high gain of 5.36 dB.

## References

1. S. Lee, U. Kim, K. Kwon, W. Seo, J. Choi, Design of on-body antenna for wireless body area network, in *2012 14th International Conference on Advanced Communication Technology (ICACT)*, pp. 300–303 (2012)
2. S. Hussain, S. Hafeez, S. Ali Memon, N. Pirzada, Design of wearable patch antenna for wireless body area networks. Int. J. Adv. Comput. Sci. Appl. (ijacsa) **9**(9) (2018). https://doi.org/10.14569/IJACSA.2018.090920
3. S.M. Ali, V. Jeoti, T. Saeidi, W.P. Wen, Design of compact microstrip patch antenna for WBAN applications at ISM 2.4 GHz. Indonesian J. Electric. Eng. Comput. Sci. **15**(3), 1509–1516 (2019). ISSN: 2502-4752. https://doi.org/10.11591/ijeecs.v15.i3.pp1509-1516
4. K. Kwon, J. Ha, S. Lee, J. Choi, Design of a dual-band on-body antenna for a wireless body area network repeater system. Int. J. Anten. Propagat. (2012). Article ID 350797, 5p. https://doi.org/10.1155/2012/350797
5. M. Klemm, G. Tröster, Small patch antennas for UWB wireless body area network, in *Ultra-Wideband, Short-Pulse Electromagnetics*, vol. 7, ed. by F. Sabath, E.L. Mokole, U. Schenk, D. Nitsch (Springer, New York, 2007). https://doi.org/10.1007/978-0-387-37731-5_46
6. R.A. Panda, D. Mishra, Efficient design of bi-circular patch antenna for 5G communication with mathematical calculations for resonant frequencies. Wireless Pers Commun **112**, 717–727 (2020). https://doi.org/10.1007/s11277-020-07069-9
7. R.A. Panda, P. Kumari, J. Naik, P. Negi, D. Mishra, Flower shaped patch with circular defective ground structure for 15 GHz application, in *Innovations in Bio-Inspired Computing and Applications. IBICA 2019. Advances in Intelligent Systems and Computing*, vol 1180, ed. by A. Abraham, M. Panda, S. Pradhan, L. Garcia-Hernandez, K. Ma (Springer, Cham, 2021). https://doi.org/10.1007/978-3-030-49339-4_24
8. R.A. Panda, B. Hansdah, S. Misra, R.K. Singh, D. Mishra, Gauging trowel shaped patch including an optimized slot for 60 GHz WLAN, in *Innovative Data Communication Technologies and Application. ICIDCA 2019. Lecture Notes on Data Engineering and Communications Technologies*, vol. 46, ed. by J. Raj, A. Bashar, S. Ramson (Springer, Cham, 2020). https://doi.org/10.1007/978-3-030-38040-3_9
9. R.A. Panda, D. Mishra, E.P. Panda, N. Patnaik, Reshaped circular patch antenna with optimized circular slot for 5G application, in *Emerging Trends in Electrical, Communications, and Information Technologies. Lecture Notes in Electrical Engineering*, vol. 569, ed. by T. Hitendra Sarma, V. Sankar, R. Shaik (Springer, Singapore, 2020). https://doi.org/10.1007/978-981-13-8942-9_63
10. P.R. Meher, B.R. Behera, S.K. Mishra, Design and its state of the art of different shaped DRAs at millimeter wave frequency band. Int. J. RF Microwave Comput. Aided Eng. **30**(7), 1–15 (2020)
11. P.R. Meher, B.R. Behera, S.K. Mishra, A.A. Althuwayb, A chronological review of circularly polarized dielectric resonator antenna: design and developments. Int. J. RF Microwave Comput. Aided Eng. **31**(5), 1–23 (2021)

12. P.R. Meher, B.R. Behera, S.K. Mishra, A compact circularly polarized cubic DRA with unit-step feed for Bluetooth/ISM/Wi-Fi/Wi-MAX applications. AEU-Int. J. Electron. Commun. **128**, 1–8 (2020)
13. P.R. Meher, B.R. Behera, S.K. Mishra, Broadband circularly polarized edge feed rectangular dielectric resonator antenna using effective Glueless technique. Microw. Opt. Technol. Lett. **62**(10), 1–9 (2020)
14. P.R. Meher, B.R. Behera, S.K. Mishra, A.A. Althuwayb, Design and analysis of a compact circularly polarized DRA for off-body communications. AEU-Int. J. Electron. Commun. **138**, 1–7 (2021)
15. P.R. Meher, B.R. Behera, S.K. Mishra, Design of different shaped DRAs for 60 GHz millimeter-wave applications, in *Indian Conference on Antennas and Propagation*, Hyderabad, December 16–19, (2018)
16. P.R. Meher, B.R. Behera, S.K Mishra, Broadband circularly polarized cylindrical dielectric resonator antenna, in *TEQIP-III Sponsored International Conference on Microwave Integrated Circuits, Photonics and Wireless Networks*, Trichy, May 22–24 (2019)
17. W. Lee, J. Kim, C.S. Cho, Y.J. Yoon, Beam forming lens antenna on a high resistivity silicon wafer for 60 GHz WPAN. IEEE Trans. Antennas Propag. **58**, 706–713 (2010)

# A Secure Handshaking AODV Routing Protocol (SHS-AODV) with Reinforcement Authentication in MANET

**I. V. Ravi Kumar, G. Rajitha, and B. Nancharaiah**

**Abstract** One of the wireless networks is Mobile Ad-Hoc Network (MANET) in which the different nodes are communicated with the nodes even if they are not structured well. Therefore while transmission of data between two nodes or among number of nodes is too risky because of high chance to attacking threats. The Ad hoc On-Demand Distance Vector (AODV) routing protocol is intended to utilize through the mobile nodes in an ad hoc network. Security improvement and transmission execution up gradation are the two basic problems to be addressed in AODV's research field point of view. A secure handshaking AODV (SHS-AODV) routing protocol with reinforcement authentication in MANET is proposed in this paper. This new method of SHS-AODV Routing protocol (RP) uses a new authentication approach message digest 5 (MD5), hashing for MANET proposed SHS-AODV RP based on reinforcement learning. Moreover, the RSA (Rivest, Shamir, Adleman) based Symmetric Encryption algorithm AES is used to enhance the safety of such network that can be utilized with no need to a third party for distributing a secret key by detecting multiple black hole attack and prevent it. The network performance improvement of proposed SHS-AODV RP technique can be demonstrated in the results using throughput, energy and packet loss metrics. The result shows the average throughput and packet delivery ratio (PDR) are incremented and consequently increases the security of the path.

**Keywords** MANET · Routing protocol · AODV's · SHS-AODV

I. V. Ravi Kumar (✉) · G. Rajitha
Department of ECE, GIET University, Gunupur, Odisha, India
e-mail: iv.ravikumar@giet.edu

I. V. Ravi Kumar
Department of ECE, Swarnandhra College of Engineering and Technology, Narsapur, AP, India

B. Nancharaiah
Department of ECE, Usha Rama College of Engineering and Technology, Vijayawada, India

 
S. Patnaik et al. (eds.), *Smart and Sustainable Technologies: Rural and Tribal Development Using IoT and Cloud Computing*, Advances in Sustainability Science and Technology,
https://doi.org/10.1007/978-981-19-2277-0_9

## 1 Introduction

One of the infrastructures less network is MANET which is self-healing, self-configured and self-coordinated when the demand of communication arises. In military applications and disaster management systems MANETs are used because of its easy deployment [1]. Dynamic topology, multi-hop routing, healing and self-organizing are different characters of MANET. MANET routing protocols major challenging characteristics are mentioned as above [2]. Continuous mobility of nodes creates a routing problem in MANETs, this is because of topology changing which results in the breaking of routes frequently so re-establishment is required. Finding and selecting a path is very complex that can exist as long as possible in MANET reliable and optimal. A nontrivial mobile scenario is used for holding guaranteeing quality in terms of parameters as maximum end-to-end delay, requirement of minimum bandwidth and loss permit of maximum data packet. Most of the researchers are interested on MANET [3]. MANET Routing Protocols Caching Strategies are on demand protocols which are researched by many researchers.

The process of transmitting packets from one mobile node to other is used in the route discovery mechanism [4]. The route caching technique is used in order to eliminate the route finding mechanism for every packet transmission. Network flooding is decreased by the route caching and eliminated the process of route discovery operation which is performed too many times in the process. Delay is increased when the route discovery mechanism has frequent usage and also band width consumption increased which results to blocking of network with delay.

In the process of communication, MANET is moving in nature because of its infrastructure less network. Data packets are transmitted by all nodes which acts as router. Lack of fixed access point, network irregular connectivity and infrastructures less network are the challenges of the Ad hoc networks. Small changes in the networks are easily adopted by this network and communicate with nodes continuously. Different security problems are faced by the system if there are dynamic changes in characteristics of network. In AODV protocol usage, black hole attack is one security problem [5]. A malicious node is introduced in network with this black hole attack. While transmitting the packets from source to destination node then malicious nodes drops the data packets. Therefore, security is the challenging issue during the transmission of packets in network. Security can attain less importance in previous papers. But for secured data transmission, there is a need of improvement in security level. Limited resources, quality of service (QOS), scalability, sufficient admission control and confidentiality are the challenging parameters in the ad hoc networks. These challenges are overcome by designing secure solutions of routing protocol for communicating mobile nodes securely [6]. SHS-AODV secured routing protocol is proposed in this paper.

## 2 Ad Hoc on Demand Distance Vector

Ad-hoc networks operation is highly impacted with AODV algorithm. Network of every node acts as specialized router. The bandwidth requirement of the presented protocol is significantly less compared to other protocols of AODV and routing recurring advertisement is not used [7]. AODV is made as free from loop or cycle by using foregoing properties and router is connected. In the process of route discovering, broadcasting techniques is used by the AODV as Dynamic Source Routing (DSR), but source routing is not used at AODV intermediate node for entry into route table. Then dynamic routes are established [8]. Destination sequence numbers can be left by the AODV from DSDV and between the nodes recent routing information is maintained. The load of network and data traffic is minimized; efficient bandwidth is utilized in this DSR and DSDV combination [9].

### 2.1 *Path Discovery*

If there are no route entries in routing table then node starts to find a secure path for secure communication between the nodes, this process is called as path discovery. Broadcast ID and node sequence number are two different and specified counters for each and every node in the network. RREQ (Route REQuest) packet is sent by the source node to its neighbors for initiating the path discovery [10].

### 2.2 *Reverse-Path Setup*

RREQ is having the two sequence numbers along with broadcast_id and these numbers are the destination last sequence number, source sequence number in which both belongs to source. Any reverse route to the source new information is maintained in the sequence number of source and the source applied route freshness to the destination maintenance is the main of the destination sequence number. A reverse path is automatically setup by the all nodes to source node during the traveling of RREQ from source to different destinations. The neighboring node address is recorded or stored by each and every node in network and through the RREQ first copy is received which is used in reverse path designing. The reverse path is compulsory maintained by every node at least for the time duration of RREQ transmission from source to destination and reply is given to the sender [11].

### 2.3 Forward-Path Setup

RREQ is required to determine a recent route to destination. When there is availability of route entry then the freshness of route is calculated through utilizing sequence number of destination which is mentioned in RREQ own route entry table. If this obtained sequence number is greater than or equal to route value then a reply is obtained from intermediate nodes otherwise RREQ will be rebroadcasted. A RREP (route reply packet) is sent back by node from its neighbor if there are no recent routes to the destination and RREQ received [12].

### 2.4 Route Table Management

'Route request expiration timer' named timer and 'soft state' are some of the useful information which is stored in the route table. Reverse path routing entries are washed out by using timer which are from those nodes that don't lie between source and destination. Expiration time is affected by ad-hoc network size. Route-caching time out is also the main parameter in routing entries which is defined as time taken for considering or identifying the invalid route.

### 2.5 Path Maintenance

Route discovery process is restarted by source node in moving condition for finding the new route to the destination in an active transmission session. The routing path is not affected by any valid node which is moving on active path. In between the source node and other nodes or destination nodes a special RREP is sanded. Link failures detection and improving the symmetric links can be done by using periodic Hello messages.

### 2.6 Local Connectivity Management

The neighboring nodes are discovered by using two different ways. If there is a transmission of packets from its neighbors then connectivity information is updated along with its neighbor. A special unwanted RREP is also transmitted along with data packets transmission among the active neighbors and these contain the sequence number and identity.

## 3 Literature Survey

Feng et al. presented a technique named as improved routing protocol of Ad-AODV supported AODV [13] for enhancing the routing protocol of AODV which does not contemplate the residual energy. Hence nodes load condition selecting routes once, its potency sharply declines within high load and fast-paced speed case. For unraveling the on top of issues author has presented enhanced protocol Ad-AODV (Advanced-AODV) of routing formula of AODV supporting equalization and an energy model technique. Once the routing protocol of AODV is performed on route request then it can contemplate load condition and residual energy nodes. The Ad-AODV routing protocol simulation results enhances network throughput PDR and reduces the routing load, lessens average end-to-end delay.

Sridhara et al. [14] describes about the issues of final AODV routing such as time delay, quality, long route and several other while routing. Due to less energy within a node it cannot be in a position for complete routing. The QoS parameters such as delay area unit, PDR and turnout are directly affected. The projected EN-AODV (Energy based mostly AODV protocol) announced energy and supported nodes leads to receiving rates hence the info sizes to be sent and it satisfies either it's energy is attenuated or maintained. This can calculate the nodes energy levels before they are designated to routing path. Price of threshold is outlined and nodes area unit thought-about for routing providing its energy is on top of this threshold price. Decrease in delay and a rise in PDR is shown from simulation results and maintains turnout. Therefore presented EN-AODV produces extra reliable and consistent transfer knowledge compared to traditional AODV.

Liu et al. [15] presented a novel approach QoS-AODV (QoS-aware routing protocol supported AODV called QAODV). In the delay premises and out there information measure meet the demands of QoS. This protocol represents a new route metric with the count of hop and rate of load thus on selecting the simplest route consistent with it. From the results it is clear that compared to AODV presented QAODV performance is better end-to-end delay with small increment in control overhead and better throughput in each network.

## 4 SHS-AODV Routing Protocol

A framework of secure hand shaking-adaptive routing protocol for MANET is represented in Fig. 1.

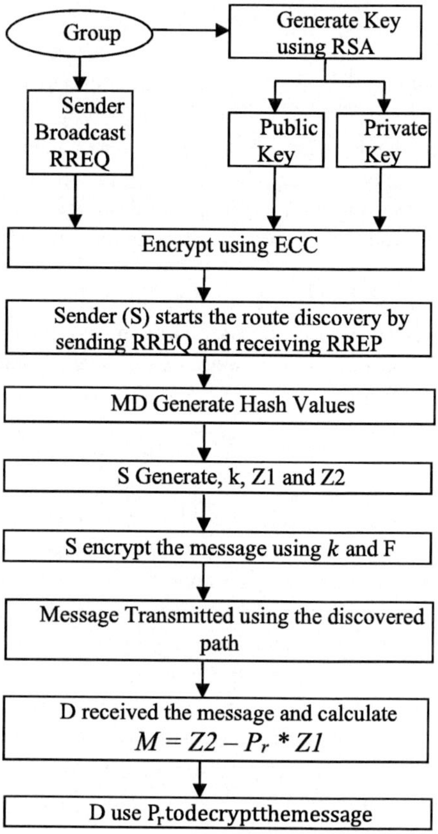


**Fig. 1** Framework of SHS-AODV Routing Protocol for MANET

## 4.1 Key Generation

Most important part is key generation, in which private key and public key are produced by the user. In light of number theory, an asymmetric (open key) cryptosystem is RSA and it is a block cipher device. Private keys and individuals are made by using two prime numbers with the help of this device. Encryption and decryption processers are uses of these special two keys. If there is a requirement of data transfer, then sender encrypt the data by using receiver's public key first and then by using its own private key, the receiver decrypt the information. Consider the '*k*' as particular range and number '*d*' within this range and general public key is designed as

$$P_u = d * F \quad (1)$$

where, public key is represented as '$P_u$' and private key as '$P_r$'. On a curve one of the factors is $F$ and within the range (1 to $k-1$) arbitrary quantity are to be choose.

**Message digest 5 (MD5)**: By the operation of some specified mathematical functions, results obtain the hashing values which are assigned values. When the data is transmitting on network routes then the security is provided by this operation. The receiver can only be identified by the message so packets will be safe in that condition. The data processing is specialized with this hashing algorithm including 512-bit blocks, total 16 words with each word of 32 bit is created. Message digest value with 128-bit as size is produced in this process. SHA-1 is slower than generating MD5. Therefore MD5 hashing is used in this proposed model because of its memory less and fast technique.

## *4.2 Elliptic Curve Cryptographic Algorithm (Ecc)*

The elliptic curve equation is given as,

$$j^2 = i^3 + ai + b \tag{2}$$

where, finite field elements are $a$ and $b$ with $pn$ elements, and a is most important element and it is a prime number. $(i, j)$ is ordered pairs which are selected on some points with $x$ and $y$ coordinates and the curve is given as in the form of equation as $j^2 = i^3 + ai + b$. Additional point with curve is said to be infinity.

**Encryption**: Sending data is considered as '$m$'. The corresponding values of data are represented on curve '$E$' with the factor '$M$'. '$k$' value is randomly decided from $[1 - (k - 1)]$. $Z1$ and $Z2$ are to cipher textual content

$$Z1 = k * F \tag{3}$$

$$Z2 = M + k * P_u \tag{4}$$

To other users Z1 and Z2 Ciphertext will be sent.

The sender private key is used in public key encryption and packet sensitive parts along with the application of ECC algorithm then RREQ packet are broadcasted. If RREQ packet is received by the destination node then it decrypts the data with public key. RREQ packet is now compared with public key and if there is matching found then RREP packet send through utilizing the public key of sender.

**Decryption**: The real message can be obtained by doing decryption operation with its own private key by the receiver node or destination node.

$$M = Z2 - P_r * Z1 \tag{5}$$

Essential handshaking task is performed by the AODV initially when malicious users attacking the network. Therefore a secure handshaking process is proposed in this process as SHS-AODV. The key pair is generated by the sender node with the help of RSA Algorithm. In that sender node is having one key as private key and destination node is having one key as public key. RREP packet is again received by sender node and with private key it decrypts the data. Therefore a secure communication is achieved with this secured hand shaking process. Then confidentiality, authenticity, integrity of the data packets is also improved.

A secured path is described in this process while transmitting the data from source to destination nodes without attacking of any malicious nodes. The path selection achieved in this process is fast while transferring the data.

## 4.3 Algorithm

| | | |
|---|---|---|
| **Step 1**: | Start | |
| **Step 2**: | Generate private and public key using RSA | |
| **Step 3**: | Assign has value to the key values using MD5 algorithm | |
| **Step 4**: | Start route discovery by source node *S* with propagation the RREQ packets to all its neighbor nodes | |
| | 1 | If neighbored nodes are having hash value changes nodes |
| | 2 | Kept these two neighbored as malicious and maintain this as blacklist |
| | 3 | Continue path discovery from the node where the cut was happened |
| | 4 | If 3 times use the path from black list as another node |
| | 5 | Treat it as normal node |
| | 6 | Delete it from the black list |
| **Step 5**: | Source *S* encrypts the message using ECC | |
| | 1 | *S* choose *k* |
| | 2 | *S* generate Z1 and Z2 |
| **Step 6**: | Transmit the packets from source to destination in discovered path | |
| **Step 7**: | *K* is used in decryption of message by destination | |
| | 1 | *D* uses $P_r$ |
| | 2 | D decrypts by calculating |
| | | $M = Z2 - P_r * Z1$ |

**Table 1** NS2 simulator for modified AODS, SHS-AODS routing protocols

| Parameter | Value |
|---|---|
| Simulator | NS2 (ns-allinone-2.35) |
| Network size | 1000 × 1000 m |
| Number of mobile nodes | 100 |
| No. of black holes nodes | 1–5 |
| Mobility speed | 0–15 m/s |
| Routing protocols | Modified AODV,SHS-AODV |
| Packet size | 512 B |

# 5 Results

## 5.1 Simulation Environment

On the NS-2 (Network Simulator) proposed system is simulated. Two languages are existed in Network simulator in that first one is c++ language and next one is object oriented extension of TCL (OTCL). Method of class calling is used in instance of class creation which is the first step of simulator script method for creation of different nodes. The IEEE 802.11 algorithm has been used in data link and physical layer. In network layer the system uses the SHS-AODV protocol. Consider the size of transmission packet as 2500 B and 1 Mbps as transmission rate. Choose the terrain area as 1000 × 1000 m with nodes 100 and 70 bps as speed. The user gives the inputs from source node and destination node. The movements of the nodes are in random manner in Mobile ad hoc network. In network, malicious nodes are placed or distributed at anywhere. All packets are absorbed by the malicious nodes which are transmitted toward it. Consider 50 m as transmission range for data packets. 10 spause time is taken by the system of random way point model (Table 1).

## 5.2 Performance Metrics

Network different performances are evaluated by comparing the protocols. For security evaluation, AODV is compared with SHS-AODV. Network performance is evaluated by considering different parameters and some are used in simulation results. Ad hoc network is highly impacted by attacking the malicious node in the network. Energy consumption, packet loss and Route Discovery Frequency are different parameters used in proposed method.

**Route Discovery Frequency**: It is defined as the generation of RREQ messages per second by all the sources.

The route discovery frequency is compared in between AODV and SHS-AODV and it is shown in above Fig. 2. The route discovery frequency value of SHS-AODV

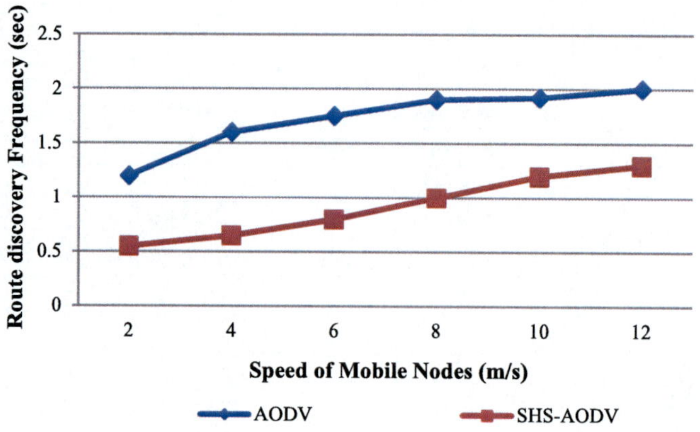


**Fig. 2** Route discovery frequency

is always lesser than 1 s than AODV at all speeds. Therefore the performance of SHS-AODV is better than AODV at all speeds. This is because of high transmission rate with discovering paths capability of SHS-AODV.

**Energy Consumption**: The total amount of consumed energy for transmission by all nodes in the network is known as energy consumption. If the presence of black hole nodes in network then, energy consumption is decreased, consequently packets are dropped in the network. Network energy consumption is increased during the packets transmission.

Figure 3 shows the throughput of SHS-AODV and AODV. High amount of energy is consumed by the AODV protocol than SHS-AODV. This is because of packets transmission at high rates is spending more time and its increased proportion then overall network efficiency is also increased.

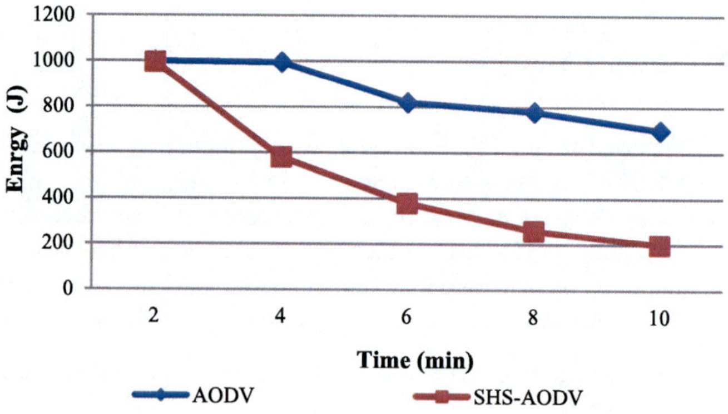


**Fig. 3** Energy consumption

**Average Delay**: Time taken by the packet for transmission from source to destination is called as average dely. Buffering delay such as at channel access, network layer, propagation delay and retransmission at the MAC are also included in this.

Average delay performance comparisons in between the SHS-AODV and AODV are represented in Fig. 4. The data packets transmission through links with better transmission rate and about 20% is close to HMMR (hybrid modular multilevel rectifier) performance than AODV.

The packet loss ratio of SHS-AODV is compared with AODV and it is less than the AODV which is represented in Fig. 5. From results it is observed that a secured transmission path is created for packets in SHS-AODV and the performance is highly improved than AODV.

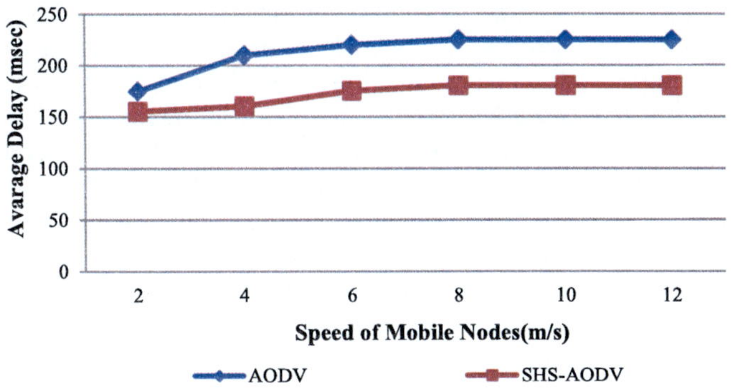


**Fig. 4** Average delay

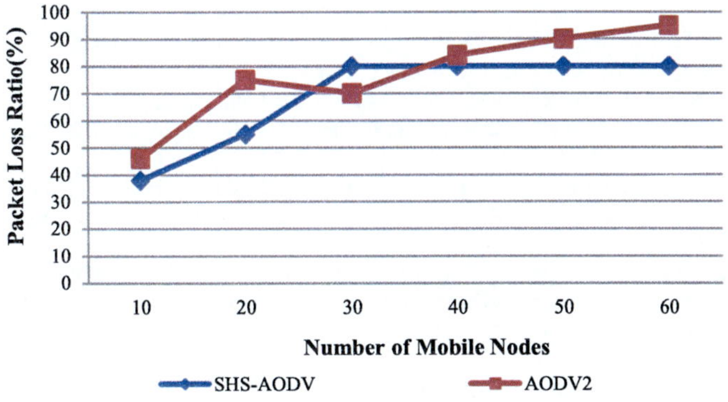


**Fig. 5** Packet loss ratio

# 6 Conclusion

A new protocol for MANET environment is proposed in this paper and it is called as SHS-AODV. There is no proper structure or framework for the MANET self- arranging mobile nodes aggregation. Therefore while transmission of data between two nodes or among number of nodes is too risky because high chance to attacking threats. A SHS-AODV routing protocol with reinforcement authentication in MANET is proposed in this paper. For designing a strong and secured RP for ad hoc networks number of researchers has been done. High-security as fundamental and delay-insensitive are the two conditions in which the SHS-AODV protocol is sensible. By suing proposed protocol, different undistinguishable attacks are also being detected and decreased. This SHS-AODV Routing protocol uses a new authentication approach hashing, MD5, for MANET presented SHS-AODV RP based on reinforcement learning. The security of the network is improved by using proposed Reinforcement technique through avoiding path attacks in the networks at nodes. In future there is also presented an approach for improving AODV performance through node caching. If node caching and route caching are combined and develop an AODV protocol based on the expectation of researcher then it will be a better performance than this SHS-AODV protocol. Doubly linked list is utilized by researchers as a structure for cache. The researchers have expects for more researchers over the structure and utilizes various data structures for route cache.

# References


1. J. Jang, B. Natarajan, Performance analysis of an enhanced cooperative MAC protocol in mobile ad hoc networks. EURASIP J. Wireless Commun. Netw. **2018**, 76 (2018)
2. A. Yasin, M.A. Zant, Detecting and isolating black-hole attacks in MANET using timer based baited technique. Wirel. Commun. Mobile Comput. **2018**, 1–10 (2018)
3. E.O. Ochola, L.F. Mejaele, M.M. Eloff, J.A. van der Poll, Manet reactive routing protocols node mobility variation effect in analysing the impact of black hole attack. SAIEE Afr. Res. J. **108**(2), 80–92 (2017)
4. A.K. Jain, A. Choorasiya, Security enhancement of AODV routing protocol in mobile ad hoc network, in *International Conference on Communication and Electronics Systems (ICCES)*, pp. 958–964 (2017)
5. S. Yadav, M. Chandra Trivedi, V.K. Singh, M.L. Kolhe, Securing AODV routing protocol against black hole attack in MANET using outlier detection scheme, in *2017 4th IEEE Uttar Pradesh Section International Conference on Electrical, Computer and Electronics (UPCON), 2017.*
6. K. Kosek-Szott, M. Natkaniec, A.R. Pach, A new busy signal-based MAC protocol supporting QoS for ad-hoc networks with hidden nodes. Wirel. Netw. **19**(2016), 1135–1153
7. R.C. Debarati, R. Leena, M. Nilesh, Implementing and improving the performance of AODV by receive reply method and securing it from black hole attack. Procedia Comput. Sci. **45**, 564–570 (2015)
8. E. Oketch Ochola, L. Mejaele, Analysing the impact of black hole attack on DSR-based MANET: the hidden network destructor, in *2015 Second International Conference on Information Security and Cyber Forensics (InfoSec)*, 2015

9. N. Choubey, A. Kushwaha, R. Chandrakar, Evaluation of routing protocols AODV, DSR and DSDV to enhance video transmission over Mobile Ad hoc Networks, in *2015 International Conference on Signal Processing and Communication Engineering Systems* (2015)
10. A. Venugopal Reddy, K.U. Rahman Khan, R.U. Zaman, Path load balanced adaptive gateway discovery in integrated internet-MANET, in *2014 Fourth International Conference on Communication Systems and Network Technologies* (2014)
11. D. Malik, K. Mahajan, D. Singh Karaulia, M.A. Rizvi, Event driven dynamic path optimization for AODV in MANET, in *2014 International Conference on Computational Intelligence and Communication Networks* (2014)
12. D. Lee, B. Shin, S. Yeong Han, An application-driven path discovery mechanism for MANET routing protocols, in *2014 IEEE International Conference on Communications (ICC)* (2014)
13. Z. Feng, L. Wang, X. Gao, An improved routing protocol Ad-AODV based on AODV, in *International Conference on Information Science and Computer Applications (ISCA 2013)*, Published by Atlantis Press (2013)
14. S. Sridhar, R. Baskaran, P. Chandraseka, Energy supported AODV (EN-AODV) for QoS routing in MANET, in *The 2nd International Conference on Integrated Information* (2013)
15. L. Liu, L. Zhu, L. Lin, Q. Wu, Improvement of AODV routing protocol with QoS support in wireless mesh networks, in *2012 International Conference on Solid State Devices and Materials Science* (2012)

# AI-Powered Smart Routers

**Gyana Ranjana Panigrahi, Nalini Kanta Barpanda, Sailesh Chandra Mohanty, and Ankit Das**

**Abstract** Complexity in network topologies and designs is rising, leading to the greater use of artificial intelligence and learning in the network control layer. The secret to efficient global activity for large-scale social constructivist platforms is deciding how to deploy intelligence. For deployment models (unified vs. ununified) to produce the best performance, we suggest a hybrid paradigm that incorporates network thought systems known as AI-powered smart routers. We put the control for high-oriented regulatory functions in a single place to guarantee good QoS. At the same time, for economic integration, we assign control to each connection to the AI-router for an IP hop-by-hop, giving more route path choices. The ununified/hybridized AI model involves distributed route management, using AI to provide successful centralized tunnelling of routing state and distributed control for path ingress and egress.

**Keywords** Machine automation · Network security · AI/ML · AI-powered routers · Smart network

## 1 Introduction

Recent research has shown that the network is changing drastically from Software-defined Networking, fifth-generation broadband services, and hyper-convergence [1–3]. Old systems are being eroded by the networking paradigms and built on open source. There are, however, new problems that arise with regard to the flexibility and

G. R. Panigrahi (✉) · N. K. Barpanda
Department of Electronics, SUIIT, Sambalpur University, Burla, Odisha, India
e-mail: gyana.ranjana.panigrahi@suiit.ac.in

N. K. Barpanda
e-mail: nkbarpanda@suniv.ac.in

S. Chandra Mohanty · A. Das
School of Forensic Sciences, CUTM, Bhubaneswar, Odisha, India
e-mail: 202105290002@cutm.ac.in

 
S. Patnaik et al. (eds.), *Smart and Sustainable Technologies: Rural and Tribal Development Using IoT and Cloud Computing*, Advances in Sustainability Science and Technology,
https://doi.org/10.1007/978-981-19-2277-0_10

scalability of the network. Specifically, with the introduction and advancement of the Internet of Things (IoT) and Augmented Reality (AR), traffic sizes and volumes are gaining ground, and QoS is in general demands network system management becomes even more complicated manual control schemes are unsuitable for scaling control of complex systems, especially. Indeed, more advanced methods of countering networking are urgently needed. More recently, with the outstanding progress of computer learning, the emphasis has shifted to Artificial Intelligence (AI/ML). It provides a general model and consistent learning methods for different network scenarios, tackles emotional challenges and high-dimensional scenarios that are a great help for deploying AI and advancing modern networking techniques [4, 5]. AI&ML is currently being applied in networking because of the network complexities and the quality of service (QoS) demands [5–7]. Networking with AI & ML was first discussed in "a communication layer for the world wide web." The most important thing is to learn, though, is just global optimization for a distributed network [8]. A fundamental failure that has stunted the growth of our intelligence is now possible with advances in network virtualization before proposing a knowledge-based networking model for networked AI where knowledge rules have been established in a single center. However, as the network increases, the clustered approach incurs coordination and computing costs due to real-time network activities like congestion routing. The high overhead is going to make AI-based algorithms do much worse [7, 8]. As previously said, both centralized paradigms are inherently flawed. Thus, in this article, we suggest synthetic AI-influenced control systems, which use dispersed "AI routers" to harness unified "network brains" to help sustain various network operations [9]. In detail, we discuss unified control for single-hop to multi-hop IPng version six and hop-by-hop IPv6 routing. Additionally, we use two types of machine learning algorithms to enhance traffic-control techniques to handle network constraints like router convergence congestion and performance quality measurement [10, 11]. The following section summarizes the paper's major contributions. We propose a hybrid deep learning architecture for PDU routing that combines distributed intelligence based on AI-driven routers with a unified information platform called the smart brainy network [12–14]. We address the effectiveness and superiority of centralized optimization for piggybacking routing (with a high-quality-of-service guarantee) and implement a path optimization strategy focused on deep reinforcement learning in the smart brainy network. To alleviate the overhead associated with unified, centralized management, we assign smart control to individual AI-based routers and use the smart brain to improve robust performance for hop-by-hop routing [15, 16]. The remainder of the paper has been classified appropriately. The second section discusses the related studies on AI-led network routing. The third section discusses the positioning of the smart controller and suggests a hybrid architecture of distinct operations [17, 18]. In the fourth section, we present a consolidated high-quality network infrastructure that is AI-dependent. We created a modular routing framework in the fifth Section to address the problem of congestion-related management. After that, it poses several issues and concerns that are self-evident.

## 2 Related Work

The big successes in machine learning have sparked considerable interest in AI and machine learning applications in networking in recent years. Though AI-driven networking is a hot topic of study at the moment, the notion of utilizing machine learning in traffic routing dates back to the 1990s. We are looking at work on AI-led network routing algorithms in this segment.

### *2.1 Ununified Routing (Loosely Coupled Mono-Cum-Multi-agent Evolution Strategies)*

Group of authors developed the Q-routing protocol to try to aid in packet-based routing control. Q-routing relies on the policies that have been configured for each router and the current connectivity details at the time of each connection. The experiment showed that Q-routing provided more efficiency on an under-optimized open shortest path (quasi) adaptive algorithm, which saw an even higher processing gain, particularly under heavy load [12]. The name of the algorithm suggested by researchers was predictive Q-routing, and it allowed the machine to preserve previous interactions to maximize the rate of learning and convergence. The backward expansion has suggested using DRQ, which allows the network to discover new knowledge and help the convergence process in pace. The principle of reinforcement learning (RL) was successfully applied in the case of the WANET, where the nodes that receive information from the sensors and send the data updates can self-adjust to the surroundings. However, single-user RL diverges in multi-systems have this unfortunate tendency to stop expanding; instead, collaboration among network nodes is more efficient [14]. A routing algorithm has been suggested by a team-partitioned Opaque Transition RL (TPOT-RL), enabling a network node team to learn how to accomplish a joint mission collaboratively [15]. In a sparse reinforcement learning algorithm was developed to change the behaviour of each network agent, called the Collective Intelligence algorithm [16]. The author, on the contrary, suggested a routing algorithm based on Collaborative RL (CRL) with no global status [18]. The CRL method for the late tolerant network routing has also been implemented with great results. However, state syncing is extremely difficult for all routers, especially with growing network size, speed, and load in an increasingly distributed environment [17].

### *2.2 Unified Routing (Close-Coupled Single-Cum-Multi-agent Evolution Strategies)*

Authors have suggested a deep evolution strategies (DES) algorithm for routing optimization in a centralized information space. The test results demonstrated very promising success as seen from a global control perspective [17]. The QoS-aware adaptive routing is used in clustering multi-layer software-defined networks employing the SARSA algorithm. Based on the QoS requirements, the controller configured the routing strategies of every stream and distributed the transmission table to every node along the path. AdaR, a Wireless Sensor Network (WSN) routing algorithm. The AdaR is used to improve the route routing duration, load balancing, and LSP Iteration Policy transmission rate (LSPI). The expense of unified AI administration is nevertheless substantial [18].

## 3 AI-Powered Network Routing

We begin by proposing a four-tier logical compatibility architecture for AI-led networking in this segment. After which, we analyze how far apart the automated-brain control layer and forwarding layer should position ("unified" or "ununified").

### *3.1 Regenerative Control Paradigm*

In traditional networks, forwarding can occur in both the control and data layers. However, the logic of an intelligent system cannot be fully explained by artificial intelligence (AI) and machine learning alone. In this article, we are drawing on the ability of the human brain's learning process to re-integrate itself, namely "observation–development–output–incorporation" to create a regenerative control network for AI-powered networking. We have used the four-dimensional model, as illustrated in Fig. 1, called the regenerative control paradigm with releasing level, the cognizance level, the brainy control level, and the synchronization level.

The releasing level takes packets from one network and sends them to another, and it relies entirely on the brainy control's layer guidance for its operation. The cognizance layer aims to log the network status and upload the results. However, we must be well-connected to use ML-based approaches to maintain synchronization between one and all. So, in order to provide overall network knowledge, we implement a new abstraction layer called the synchronization layer. The intellectual, intelligent control layer communicates control preferences to the releasing layer. At this stage, current and historical process data have been translated into monitoring programs through AI & ML-dependent algorithms. These four abstract layers form together with a regenerative framework for networking to carry out ML automation. The released layer is the "Operating Entity," the perceived layer is the "Observation-based Subject," and the brainy control layer is the "Subject of integration." AI & ML agents will actively learn and adapt network monitoring and control procedures focused on these four regenerative action layers through communication with the underlying network.

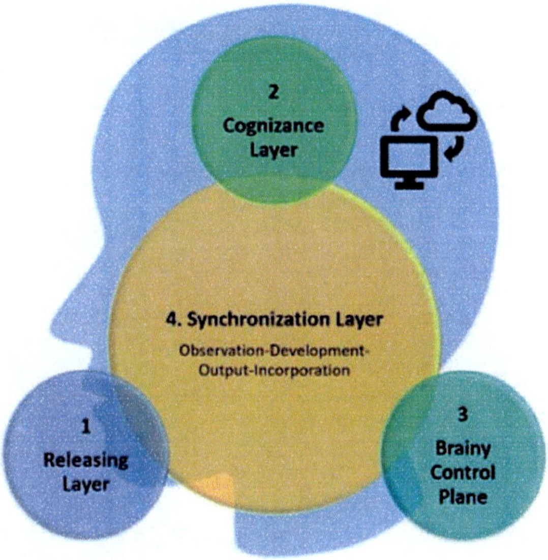


**Fig. 1** The regenerative control paradigm

## *3.2 Unified and Ununified AI-Powered Networking*

As previously mentioned, a four-tier distinct architecture has been proposed. Since this abstract mathematical theory has been applied in a real-world network, the position of the intelligent, brainy control layer (unified and ununified) is critical for the successful operation of AI-driven regulating architecture as in Fig. 2. Thus, the distance between the management layer and the information layer has long been a source of contention. The brainy control smart layer and releasing layer have tightly coupled in conventional ununified networking equipment. Each node has only had partial control over the network. In such a network, the optimization algorithm would have difficulty diverging. The smart control layer is, by comparison, separated from the network hardware in an SDN architecture and functions as a unified layer where the AI has access to the whole network to fine-tune a particular layer. However, the benefits of unified optimization are obvious, but the regenerative management overhead is strong. This overhead involves the overhead contact for processing and transmitting a vast volume of data and the computational overhead for the preparation and execution of AI operators. In the clustered paradigm, a single flow forwarding direction must be configured for all routers. Furthermore, the controller must report the forwarding rationale every time the network state changes. This extreme transmission friction and high computing stress are inappropriate for large-scale (in msec) networks with super high dynamics. As previously mentioned, the clustered and ununified paradigms are also flawed and have corresponding benefits and drawbacks.

Before we get into the specifics of our machine automation routing paradigm, let us take a look at the current state of routing protocol growth. Early on, the IP protocol succeeded over both routable and non-routable and source routing versus

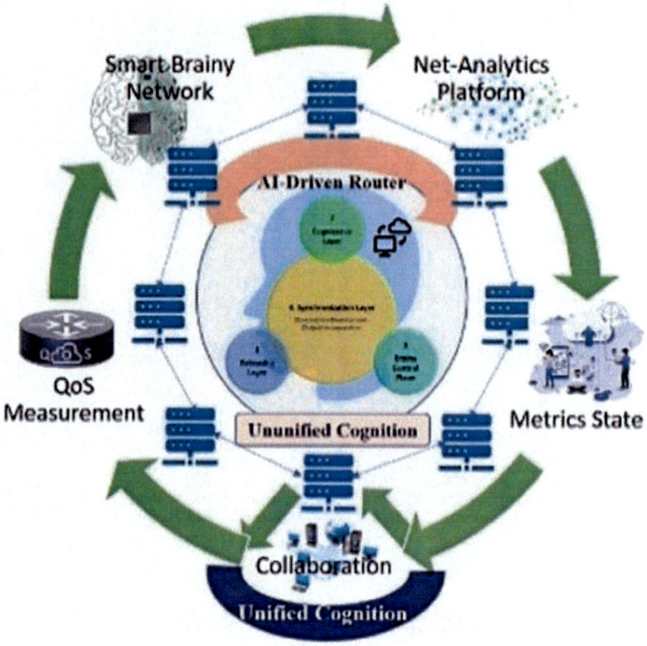


**Fig. 2** AI-powered regulating architecture

centralized unification. As in Fig. 3, each device in the end-to-end hoping protocol is based on its local data and communication.

This routing table contains the next-hop node and hops metrics. The "end-by-end" forwarding model reduces network scalability but increases network robustness. On the other hand, in the conventional sense, IP routing offers inadequate assistance with congestion engineering with QoS assurances because of the IP protocol's connectionless and dispersed characteristics. End-to-end source transmission systems are regaining popularity for accommodating high-QoS (high latency, delay-sensitive) networks. Software-defined WAN (SD-WAN), for example, uses end-to-end transmission layer switching for creating tunnels between senders and receivers on a temporary network, encapsulation-based standards as shown in Fig. 4. This present temporary tunnel routing technique makes QoS-guaranteeing faster and more effective for service providers.

## 4 Brainy (SMART/Intelligent) Network

High-quality service delivery is critical to the viability of several new market models for network implementations, e.g., video games and augmented/virtual realities,

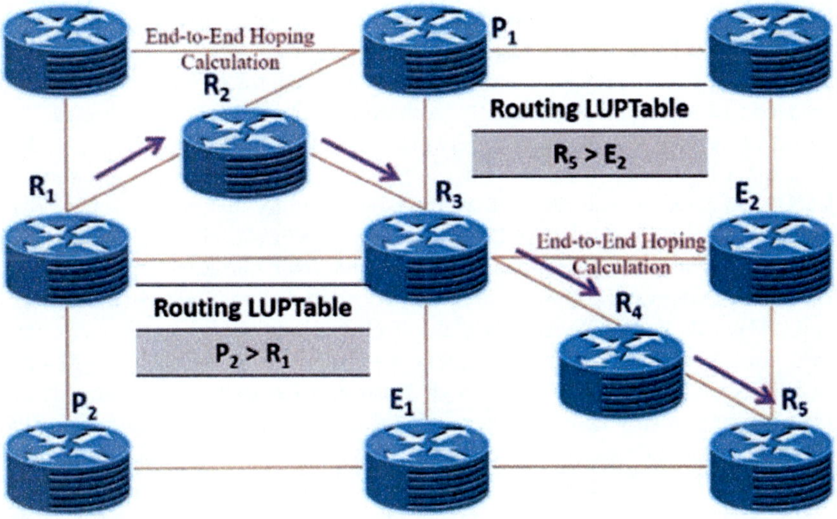


**Fig. 3** End-to-end hoping protocol

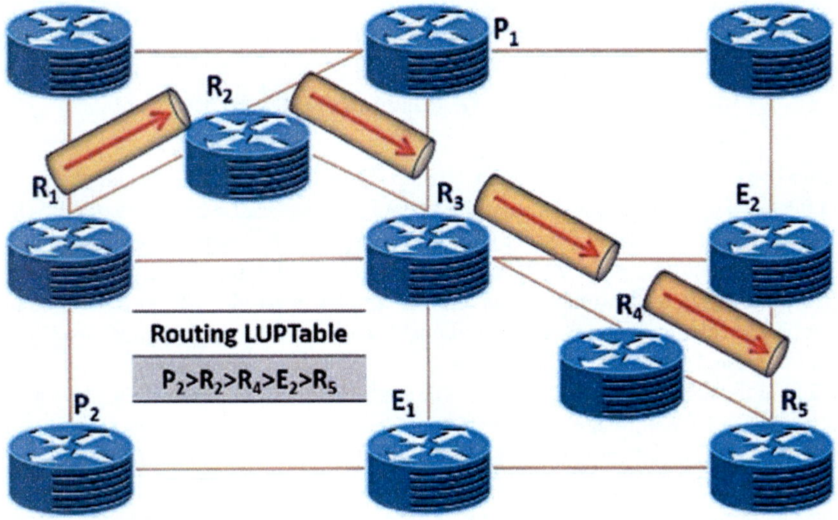


**Fig. 4** Encapsulation-based standard

which are resource-intensive and requires high bandwidths. For more than a decade, methods for ensuring QoS by tunnelling-based protocols were discussed and established; on the other hand, it would lead to a loss of stability and competition in network networks, leading to a lack of optimality, predictability, and slow integration. As a result, in our paradigm to ensure a good quality of operation for utilities that demand it, we unified the AI-powered smart management scheme using tunnelling-based routing. Thus, routing-based tunnelling protocols for secure service transmission coexist in today's networks with non-routable unified protocols. Intelligent regulation can be applied differently to these two types of routing systems. The upload connection makes use of a network monitoring protocol such as Cloudera Manager or Dynatrace to collect system states, traffic characteristics, configuration details, and service-level information; the download link makes use of a standard southbound application such as OpenNetMon or LogicMonitor to allow effective network control. The connections for upload and download establish an interface mechanism that gives the intellect of the network design standpoint and major hegemony. AI & ML algorithms are provided with closed-loop operations to generate and learn current and historical knowledge. However, it poses a major challenge to inject control policies from complex and high-dimensional system states. The network state is dimensionally expanded, particularly as the scale of the network and the granularity monitoring increase. Recent progress in adapting 3D games to various difficult decision-making domains indicates that this concept is not insurmountable. ES offers a paradigm for learning to create an effective behaviour policy by trial and error. DES may build and acquire information from raw high-dimensional data by using deep learning's representation learning capabilities. As a result, we use DES in this paper to generate efficient routing policies.

## 4.1 Unified and Ununified AI-Powered Networking

Black-Box Stochastic Optimization (BBSO) is a valuable statistical method for addressing similar questions in Reinforcement Learning (RL). The BBSO is a simplified formulation of the problem of learning by interaction in order to achieve precise control and automation goal. In our case, the clever, intelligent network and the distributed system environment function in concert to build a BBSO environment that produces control strategies continuously. The unified AI agent monitors $N_{st}$ at each point and takes a route determination based on the current proposal $\pi(r|N_{st})$. Which is the underlying network state. The controller then applies the relevant policy in the transmission directly to the network nodes. The network then switches to $N_{st}$ + 1, and the AI agent gets automatic environmental compensation C. More precisely, link conditions might express through intermediatory devices and its look-up table data, while forwarding direction may represent the behaviour. The incentive function quantifies the success of behaviour performed with the optimization objective (a latency prerequisite and bandwidth assurance). The work utilizes Exploration by Injecting Action Noise (EIAN) in the actions policy network architecture for policy creation. An EIAN agent has two components these are represented by two policy networks: a performer (deterministic policy network) $\zeta\left(St|\phi^{\zeta}\right)$ and a censor (also known as a critic network) $Z\left(St, \zeta|\phi^{\zeta}\right)$. The performer endeavours to do $\zeta\left(St|\phi^{\zeta}\right)$ the efficacy test of such policies with measures as $\phi^{\zeta}$. The EIAN agent applies an iterative control process that continually regulates both its policy (performers) and the policies that they have defined (censor). Also, during the training phase, the agent has selected a performer to proceed according to its usual stance given in Eq. 1:

$$P_t = \zeta\left(St|\phi^{\zeta}\right) + i_t \tag{1}$$

The performer then does the job $P_j$ and checks the grant $G_t$ and the new state of the underlying network is $N_{\mathrm{st}} + 1$. On data training, a repeat memory $M_r$ allows for reduced temporal correlation. The transitions $\left(N_{\mathrm{st}} + P_j + M_r + N_{\mathrm{st+1}}\right)$ which happen currently have saved to $M_r$ and then are chosen from $M_r$ using Black-Box Optimizer (BBO) with relatively short para-building periods and arbitrary mini-batches of $k$ transitions $\left(N_{st_K} + P_{j_k} + M_{r_k} + N_{st_k+1}\right)$ are applied to update the critic network (BBO). Additionally, the performer strategy has been revised with the goal of BBO the discounted total pay-out, which can state as in Eq. 2:

$$\Delta\phi^{\zeta} Q \simeq \frac{1}{m}\sum_i \Delta_P Z\left(St, \zeta|\phi^{\zeta}\right)|St = Nst_k P = \zeta(St_k)\Delta\phi^{\zeta}\zeta\left(St|\phi^{\zeta}\right)|Nst_k \tag{2}$$

ES has many benefits over conventional anomaly-based algorithms for network routing management. To begin, deep evolution strategies (DES) may produce information directly from spatial, complex, increased bandwidth structures without needing assumptions or simplifications. Second, as a BBO solution, DES enables the overhaul of incentive functions to accommodate multiple network goals without changing the algorithm model. Third, once qualified, an ES agent has the capacity of calculating a relatively single regenerative move. In addition, anomaly-based algorithms require many steps to develop into a new optimal solution, as the network state changes arbitrarily. The resulting spatial burden would have been considered most important, especially in broad, highly dynamic networks.

# 5 Simulation

To show our algorithm's viability and accuracy, we present simulation performance. The proposed model simulated a network comprised of 22 nodes and 30 full-duplex connections in our test along with 28 distinct traffic load levels to test the algorithm's success under various network overload scenarios. The traffic has developed using a Negative Binomial Distribution (NBD), and the parameter m has been used to specify the different traffic load intensities. We used a Sirius cognitive network with two hidden connected layers comprising 67 buffer components in the first layer and 57 buffer components in the second layer. We used OpenNMS to simulate network traffic and Torch and DeepPy to create EIAN agents in this experimentation [13]. Connection and node latency issues specify the network state, each operation is defined by collecting nodes that determine the forwarding path from the source to the destination, and the reward for forwarding is described by the total delay for routing from the sender to the recipient.

In Fig. 5, the EIAN agent's training process is depicted. EIAN and its agent eventually converge to the optimum approach as the number of training phases increases. Our algorithm has been compared to the Bellman-Ford process. As shown in Fig. 6, delivery on an average with low traffic load, the network is not congested. Thus, the Bellman-Ford routing algorithm outperforms the algorithm driven by artificial intelligence. However, as load volume increases, data traffic on the shortest routing route becomes excessive, and AI-powered routing outperforms the Bellman-Ford routing. As a result, we may infer that AI-based routing is efficient, even more so when network congestion exists.

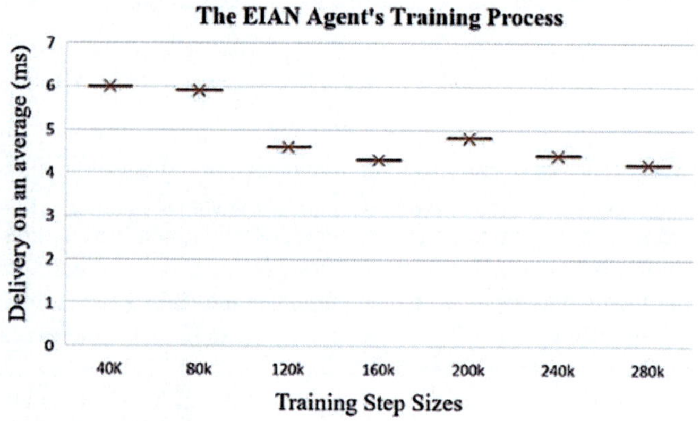


**Fig. 5** The EIAN agent's training process

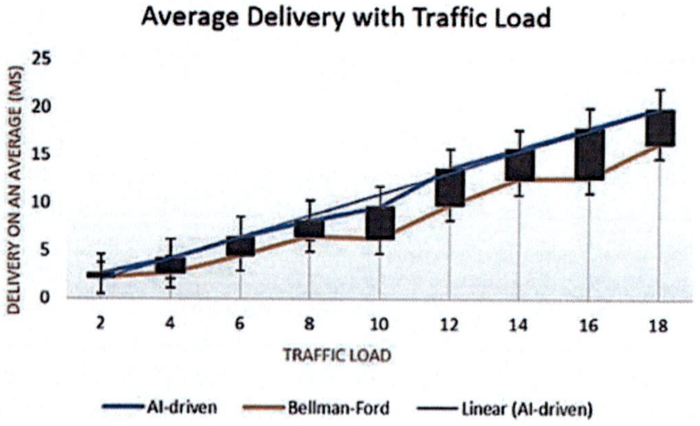


**Fig. 6** Delivery on an average with low traffic load

# 6 AI-Powered and Brainy Hardware Routers

Router tunnelling is better for traffic engineering and good quality of service (QoS) promise, but complete melting of the network causes organizational difficulty. So, work suggests an AI-powered (human-engineered) hybrid end-to-end transmission model through the domain authority, which reduces the administrative overhead while benefiting from the network's intelligence. We have given each router the ability to function as an individual master. Effort for each AI, engaging with its environment to maximize the predicted cumulative reward to one agent-only situation, the states of a multi-agent system to change together is that of individual router takes based on the local details, these choices affect each human transition. We implement a consolidated network brain to act as a hub of information for global experience sharing via unified control, and the minds can view global information and communicate it with others. Cooperative information exchange is better than peer-to-to-peer sharing. Moreover, it is critical to decide whether the knowledge is more relevant; the sooner it is transferred, the quicker the convergence can be. As described in depth below, we have applied a "change reward" to boost the quality of the AI-powered routers' collective performers.

## 6.1 Framework Design Formulation

Multiple agent coordination may be expressed mathematically using quasi-unified probabilistic-Black-Box Stochastic Optimization (P-BBSO). The notation denotes this Pro-BBSO $\{R_k, S_t, P_r, L_r, G_t\}$ where $R_k$ denotes the collection of representatives, $S_t$ denotes the collection of network states, $P_r$ denotes the collection of performers, $L_r$ denotes the collection of localized remarks of the individual agent, and $G_t$ denotes the collection of grants. In our example, each AI-powered router selects a performer $P_r$ based on its localized remarks $L_r$ including the latest rule $r(P_r, L_r)$. The control system then provides an automatic local reward $Gt_l$ And the network state $N_{\text{st}}$ shifting to the current state $(N_{\text{st}} + 1)$. DES-based AI-enabled brainy routers must collaborate and fight for the network's limited resources. In this article, we describe various router resources such as throughput, buffer, and computing capacity. For router j the remark $R_j$ is equal to $\{Wn_1, Sz_1, Mc_1, Srs_1, t\}, \ldots, \{Wn_l, Sz_l, Mc_l, Srs_l, t\}$, where $Wn_l$ is the weightiness of the resources $l$, $Sz_l$ is the size of that resources $l$, $Mc_l$ is the metrics count of the resources $l$ and $Srs_l$ t is the size of that resource spent. The next-hop router symbolizes the performer $P_j$, and the immediate local incentive $L_r$ is as follows given in Eq. 3:

$$L_r\left(R_j, t\right) = f\left(R_j\right) \sum_{l \in j} W_l \, e \frac{-Sz_l}{Src_l, t} \tag{3}$$

However, since this localized granted signal aims to maximize accumulated reinforcement, it promotes just despicable performances. To improve coordination and other AI-driven routers, we use a differential reinforcement to change the cumulative reward by eliminating the majority of the noise introduced by edge routers. The following section contains a definition of the variance remark:

$$\nu_j(N_{\text{st}}, P) = \mathbb{U}(N_{\text{st}}, P) - \mathbb{U}\big(N_{\text{st}}, P_{-j}\big) \tag{4}$$

Here, $\mathbb{U}(N_{\text{st}}, P)$ is a universal comment, which represents the universal. The whole scheme's utility depends on the coordinated activities carried out by a slew of AI-driven routers. The universal remark is described as the total of all localized remarks as in Eq. 5:

$$\mathbb{U}(t) = \sum_{R_j \in J} L_r(R_j, t) \tag{5}$$

As a result, the following can be used to rewrite the modification of the Q-value in an AI-driven router in Eq. 6:

$$Q_j(R_t, P_t) \leftarrow \nu_j(N_{\text{st}}, P) + \eta Q_j(R_{t+1}, P_{t+1}) \tag{6}$$

The unified brainy network will constantly modify the approach of each router resulting from variations by rewards of individual agents. The transmission of packets cognition may be anticipated to react appropriately to network state improvements, circumventing the need to centralize AI-driven response, recalculation, and up-gradation.

## 6.2 *Simulation Results*

Our simulated network has seven nodes and nine inertial links; this created our testing setting with a pay-out of 514 PDU packets. Each packet originated from the same source and ended up at the same destination. Each packet was sent out in a different route. As shown in Fig. 7, we applied a probabilistic routing approach, a single-agent DES strategy for the paradoxical approach. All packets followed the same direction, and then there would be serious congestion on the network; thus, this approach will be very unhelpful. RL will respond to the network congestion. However, the learning method for single-agent RL has an extraordinary pre-convergence due to the non-static environment of the DES. The future looks bright for artificial intelligence-powered networking, but there are several issues yet to be resolved.

It is hard to apply AI algorithms on such large data sets that the new routers cannot supply AI & ML deployment. Even in the AI age, recently, cross GPUs have become the backbone. Some experiments have shown that GPUs enhance packet processing.

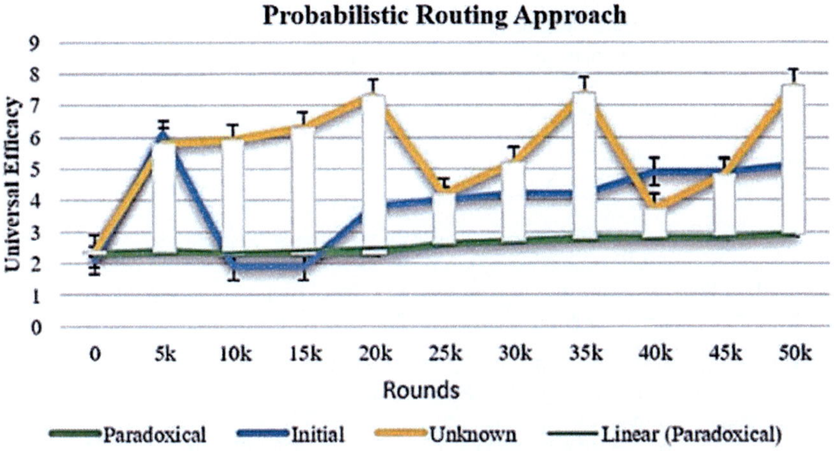


**Fig. 7** Probabilistic routing approach for universal efficacy

However, there is a vast distance between introducing universal AI and their future-use possibilities, owing to the need for fast processing speeds and short response times. Computation of the network data is becoming more difficult to deal with as traffic increases by many-fold and the number of computers in use increases by 82%. Meanwhile, the global dimension of networking is increasing the complexity of platform rollout. Logs, metrics, telemetry, and remote nodes all increase problems emerge as the volume and number of flows grow to the millions per-of-per-millisecond. End-to-to-end implementations, which use various technologies like Apache Spark and Hadoop, are laborious and time-intensive. Thus, an efficient, scalable, large network data analytics framework is needed. Another significant enabler for machine learning-based networking is program libraries. Plan, practice, and validate ML algorithms on ML frameworks. Present machine learning systems, such as Azure and RapidMiner, are not optimized for networking use for high-speed, low-level, easy, lightweight applications.

## 7 Conclusion

We began our exploration of two alternative deployment models by explaining their differences: a concept known as a brainy control layer and followed by a summary of their advantages and their problems and shortcomings. After coming up with a hybrid packet-routing system, we turned to apply unified networking to various network functions, with packet routers working together with an ununified network mind. The ununified/hybridized AI model involves distributed route management, using AI to provide centralized tunnelling of routing state and distributed control for path ingress and egress. Additionally, we use two routing techniques: exponential and exhaustive search to simplify our routing strategies.

## References


1. S. Ramisetty, S. Varma, The amalgamative sharp wireless sensor networks routing and with enhanced machine learning. J. Comput. Theor. Nanosci. **16**(9), 3766–3769 (2019)
2. Y. Zhao, Y. Li, X. Zhang, G. Geng, W. Zhang, Y. Sun, A survey of networking applications applying the software-defined networking concept based on machine learning. IEEE Access **7**, 95397–95417 (2019)
3. C. Benzaid, T. Taleb, AI-driven zero-touch network and service management in 5G and beyond: challenges and research directions. IEEE Network **34**(2), 186–194 (2020)
4. Z. Ghaffar, A. Alshahrani, M. Fayaz, A.M. Alghamdi, J. Gwak, A topical review on machine learning, software defined networking, internet of things applications: research limitations and challenges. Electronics **10**(8), 880 (2021)
5. D.M. Casas-Velasco, O.M.C. Rendon, N.L. da Fonseca, Intelligent routing based on reinforcement learning for software-defined networking. IEEE Trans. Netw. Serv. Manage. (2020)
6. M.A. Ridwan, N.A.M. Radzi, F. Abdullah, Y.E. Jalil, Applications of machine learning in networking: a survey of current issues and future challenges. IEEE Access **9**, 52523 (2021)
7. Y. Li, *Deep Reinforcement Learning: An Overview* (2017). arXiv preprint arXiv:1701.07274
8. C. Chen, Z. Wang, Q. Pei, C. He, Z. Dou, Distributed computation offloading using deep reinforcement learning in internet of vehicles, in *2020 IEEE/CIC International Conference on Communications in China (ICCC)*, pp. 823–828 (2020). IEEE
9. J. Liu, D. Jiang, Y. Luo, S. Qiu, Y. Huang, Minimally buffered deflection router for spiking neural network hardware implementations. Neural Comput. Appl. **33**, 1–12 (2021)
10. S. Arivarasan, An energy-efficient Qos routing protocol based on red deer algorithm in MANET. Turkish J. Comput. Math. Educ. (TURCOMAT) **12**(5), 1461–1471 (2021)
11. V.R. Verma, D.P. Sharma, C.S. Lamba, QoS improvement in MANET routing by route optimization through convergence of mobile agent, in *2018 5th IEEE Uttar Pradesh Section International Conference on Electrical, Electronics and Computer Engineering (UPCON)*, pp. 1–6 (2018). IEEE
12. R.M. Desai, B. Patil, Dual reinforcement q routing for ad hoc networks. Indonesian J. Electr. Eng. Comput. Sci. **7**(3), 786–794 (2017)
13. R. Dudukovich, G. Clark, J. Briones, A. Hylton, Microservice architecture for cognitive networks, in *2020 IEEE International Conference on Wireless for Space and Extreme Environments (WiSEE)*, pp. 39–44 (2020). IEEE
14. A. Serhani, N. Naja, A. Jamali, AQ-routing: mobility-, stability-aware adaptive routing protocol for data routing in MANET–IoT systems. Clust. Comput. **23**(1), 13–27 (2020)
15. G.M. Jinarajadasa, S.R. Liyange, A survey on applying machine learning to enhance trust in mobile ad-hoc networks, in *2020 International Research Conference on Smart Computing and Systems Engineering (SCSE)*, pp. 195–201 (2020). IEEE
16. Y. Weng, H. Chu, Z. Shi, An intelligent offloading system based on multiagent reinforcement learning. Secur. Commun. Netw. (2021)
17. Y. Ge, Y. Nan, X. Guo, Maximizing network throughput by cooperative reinforcement learning in clustered solar-powered wireless sensor networks. Int. J. Distrib. Sens. Netw. **17**(4), 15501477211007412 (2021)
18. C. Xu, W. Zhuang, H. Zhang, A deep-reinforcement learning approach for SDN routing optimization, in *Proceedings of the 4th International Conference on Computer Science and Application Engineering*, pp. 1–5 (2020)

# Modified Sierpinski Gasket Monopole Fractal Antenna for Sub 6 GHz 5G Applications

**S. Malathi, Subrajith Pradhan, and K. Srinivasa Naik**

**Abstract** A compact wideband monopole fractal antenna with a modified Sierpinski Gasket structure to provide broader bandwidth and coplanar waveguide (CPW) feed for sub-6 GHz 5G applications is presented. It is achieved by modifying the Sierpinski Gasket fractal structure with sectoral shape and circle inscribed in it. The antenna dimensions are $35 \times 27 \times 1.6$ mm$^3$, and it can be operated within the frequency range of 2.2–6.8 GHz with more than 3 dB gain, 4.6 GHz bandwidth, and good return loss, which is highly suitable for sub 6 GHz 5G applications.

**Keywords** Monopole antenna · Sierpinski Gasket antenna · Sub-6 GHz 5G applications

## 1 Introduction

The rapid growth in Wireless Technologies leads to an increase in demand for future-generation 5G technologies. Nowadays, for wireless devices and end terminals, the antenna plays an essential role. The 5G network is divided across two bands: sub-6 GHz and millimeter wave. The spectrum below 6 GHz is referred to as the sub 6 GHz band. This spectrum offers less depth of penetration and can travel faster than the speed of light which is more useful for 5G applications [1]. The Flame Retardant 4 (FR4) substrate with a defected ground structure (DGS) is proposed for low cost and high insulating qualities. It becomes an appropriate material for substrate because of its properties such as 4.4 relative permittivity, 1 permeability, and 0.026 loss tangent [2]. At a low cost, this material offers high electrical conductivity of 59.7 $\times 10^6$ S/m and improves gain with minimum reflection coefficient. The license-free UWB (Ultra-wide Band) spectrum occupies the frequency range of 3.1–10.6 GHz

S. Malathi (✉) · S. Pradhan
Gandhi Institute of Engineering & Technology University, Gunupur, Odisha, India
e-mail: s.malathi@giet.edu

S. Malathi · K. Srinivasa Naik
Vignan's Institute of Engineering for Women, Visakhapatnam, AP, India



S. Patnaik et al. (eds.), *Smart and Sustainable Technologies: Rural and Tribal Development Using IoT and Cloud Computing*, Advances in Sustainability Science and Technology,
https://doi.org/10.1007/978-981-19-2277-0_11

band [3, 4]. This UWB unlicensed spectrum has many useful wireless applications, which makes researchers develop a low-profile antenna for broadband applications [5, 6]. Because of their uniplanar construction, ease of production, and circuit integration, They are more suitable for miniature wireless communication devices using a coplanar waveguide (CPW) feed [7]. A meandering radiating structure can create compact antennas, and various antennas [8, 9] have been reported utilizing this technique. Fractal antennas [10] have several advantages, including their compact size and multiband capabilities. Most fractal geometry offers self-similarity structures with varying scaling factors [11, 12]. The fractal form was created using the multiple reduction copy machine technique by applying a finite number of iterations [13]. The design procedures and simulation results have been examined in this paper.

## 2 Evaluation of Antenna Design

The design approach for the modified fractal structure of triangular shape is shown in Fig. 1. The construction starts with the essential triangular patch height of h, and the angle $\alpha = 60°$ is converted into a sectoral shape. A modified fractal structure with a diameter of '*d*' is etched from the sectoral form.

The proposed fractal antenna-designed steps in HFSS are shown in Fig. 2. A triangular patch antenna with $35 \times 27 \times 1.6$ mm$^3$ was initially designed on FR 4 substrate, as shown in Fig. 2a. To resonant at desired frequency defected ground structure (DGS) with a small semi-circular slit considered on ground structure as shown in Fig. 2b. This triangular antenna with DGS operates at 3.6 GHz with a less return loss of −37.04 dB and an operating bandwidth of 1.27 GHz.

To operate in the required frequency band with a reduction in size for 5G applications, Fractal technology of Sierpinski fractal geometry is used. The overall schematic

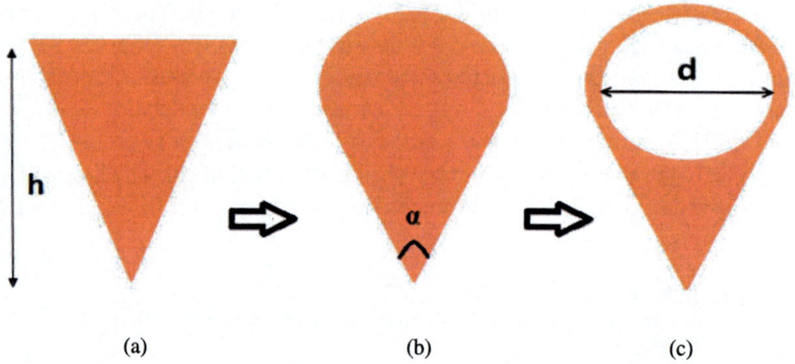


**Fig. 1** Iterative steps in modified fractal structure **a** Basic triangular, **b** Sectorial structure, **c** Circle inscribed structure

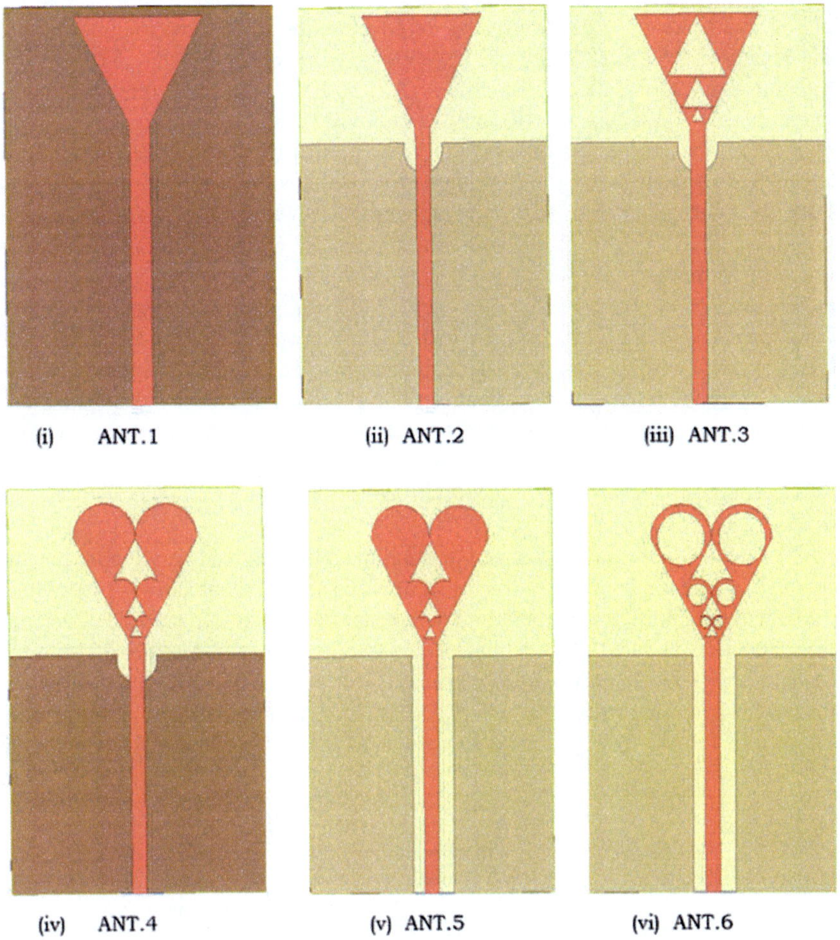


**Fig. 2** Geometry of modified fractal monopole Antenna. **a** Basic Triangular Antenna (TA), **b** TA with DGS, **c** 3rd Iteration Fractal TA with DGS, **d** Sectorial FTA with DGS, **e** CPW feed SFTA, **f** Circle inscribed SFTA

illustration of the recursive technique for the modified proposed shape is shown in Fig. 2c. A simple triangle with a height of h and an angle of 60° is transformed into a sectoral shape. A circle with a diameter of d is removed from the geometry. It has already been shown that the operational bands do not require the whole sub-gap structure created with each fractal iteration. The four primary gaps placed along the central vertical axis of the structure are principally responsible for the Sierpinski Gasket's periodic multiband activity. And with that in mind, Sierpinski Gasket's next design has been created.

The Sierpinski Gasket's first iteration geometry is created by dropping a triangle of 19.3 mm in length. The second iteration structure is created by dropping a triangle of 9.65 mm in length. The 1st iteration of construction loses three triangles with a side length of 9.65 mm. Sierpinski Carpet's first iteration geometry is created by dropping a rectangle with the dimensions (8.12 × 10.28) mm. And three rectangles with dimensions of (2.04 × 4.19) mm are removed from the first iteration structure to obtain the second iteration structure. The Sierpinski Gasket modified structure is designed in the same way. Still, only single triangles are removed with each iteration shown in Fig. 2c as a Fractal triangular antenna (FTA) with DGS. The recommended design and simulations are carried out using simulation software of a High-Frequency Structural Simulator (HFSS).

The semi-circle patch attached to each of the triangular form sectorial shapes represented in Fig. 2d Sectorial FTA with DGS was used to shift the frequency of operation to the desired frequency with circular polarization. CPW feed gives wider bandwidth (BW) with optimum characteristic parameters shown in Fig. 2e CPW feed sectorial FTA (SFTA). The proposed design consists of an inner circle with a radius d in a sectorial shape. The antenna geometry becomes simpler and operated in sub-6Ghz 5G band with the reduction in size as shown in Fig. 2f Circle inscribed SFTA.

The antenna's size can be changed depending on the desired operating frequency and the $\varepsilon_r$ dielectric constant of the substrate. Stub structures of various types yield a variety of outcomes. It is mainly used to increase bandwidth and provide high radiation properties [14–18]. With a relative permittivity of $\varepsilon_r$ and a height of 1.6 mm, the geometry is developed on the FR-4 substrate ($W \times L$). The geometry and dimensions of the proposed structure are depicted in Fig. 3, and measuring parameters are tabulated in Table 1. The antenna is fed with a 50 Ω feed line ($W_f \times L_f$) and which

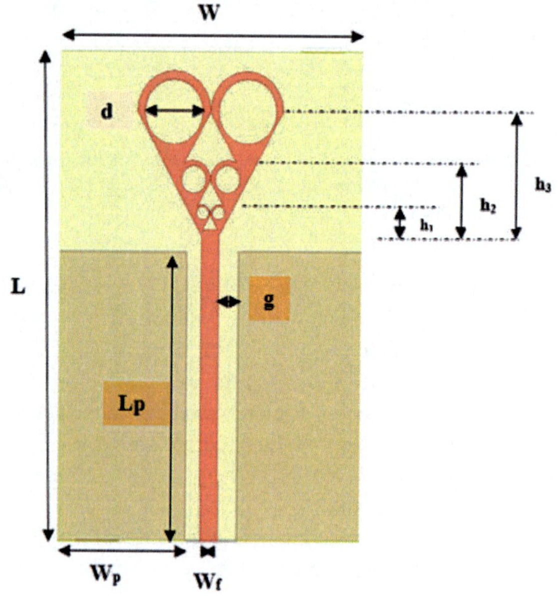


**Fig. 3** Geometry of modified Sierpinski Gasket monopole fractal antenna

**Table 1** Parameters and values of the proposed antenna (all measurements are in millimeters)

| Parameter | Value | Parameter | Value |
|---|---|---|---|
| $L$ | 35 | $L_p$ | 20.84 |
| $W$ | 27 | $D$ | 4.26 |
| $h_1$ | 4 | $W_p$ | 12.47 |
| $h_2$ | 7.12 | $G$ | 0.16 |
| $h_3$ | 13.36 | $W_f$ | 1.5 |

is computed with Eqs. (1), (2), and (3).

$$W_f = \frac{c}{2f_r}\sqrt{\frac{2}{\varepsilon_r + 1}} \tag{1}$$

$$L_f = L_{\text{reff}} - 2\Delta L \tag{2}$$

$$\varepsilon_{\text{reff}} = \frac{\varepsilon_r + 1}{2} + \frac{\varepsilon_r - 1}{2}\left[\frac{1}{\sqrt{1 + \frac{12h}{w}}}\right] \tag{3}$$

## 3 Performance Analysis of Proposed Antenna

The proposed antenna reflection coefficients $S_{11}$ curves for the standard triangular antenna (TA), Fractal triangular antenna (FTA), and the CPW feed modified sectoral fractal triangular antenna (SFTA) are shown in Fig. 4. It is worth noting that the return loss of the Sierpinski and triangle antennas is −10 dB. On the other hand, the proposed SFTA covers a bandwidth of 3.2–6.5 GHz (S11 <− 10 dB). The suggested antenna's simulated return loss for various iteration levels was observed. With the increase of iterations (maximum up to two iterations) the frequency of resonance moves towards the lower frequency band. As the number of iterations rises, the operating frequency moves towards the lowest band and also enhances the impedance bandwidth with proper impedance matching. However, the operational frequency cannot be attained after the second iteration because the antenna gets too convoluted.

The VSWR plot for various structures of the introduced antenna is shown in Fig. 5. As seen in the graphs above, the introduced gives a VSWR of 2 over the frequency band 2.2–6.8 GHz. The highest BW and return loss of less than −10 dB can be reached by setting the feed gap $g = 0.16$ mm.

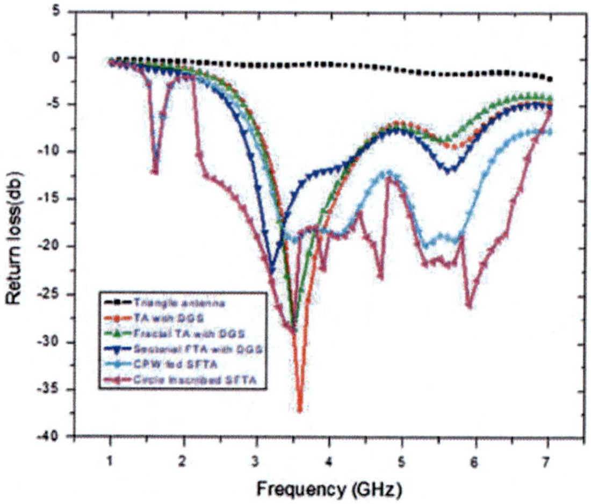


**Fig. 4** S parameters of Basic Triangular Antenna (TA), TA with DGS, 3rd Iteration Fractal TA with DGS, Sectorial FTA with DGS, CPW feed SFTA, Circle inscribed SFTA

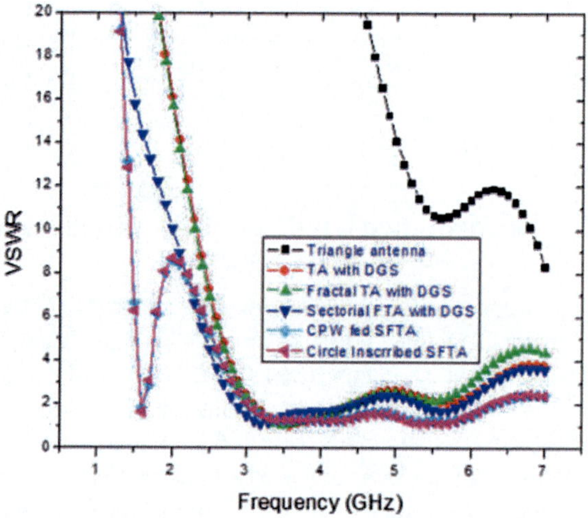


**Fig. 5** Simulated VSWR for different structures

# 4 Performance Analysis of Proposed Antenna

Figure 6 illustrates the simulated surface current distribution of the designed antenna at different operating frequencies. Surface current is primarily spread on the radiator's conductor edges and rectangular ground in this case, indicating that antenna shape impacts resonant frequency. The distributed current on the surface at 3.5, 4.2, and 5.5 GHz frequencies are substantially identical. The antenna behaves like a dipole, with an oscillating current and observed a standing wave pattern due to more distribution at the edges of the patch. Their doughnut-shaped radiation patterns

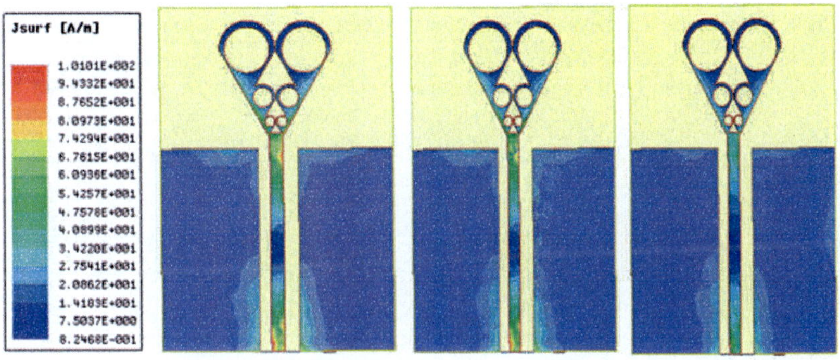


(a) 3.50GHz (b) 4.2 GHz (c) 5.50GHz

**Fig. 6** Surface current distribution of proposed SFTA at various resonant frequencies

attest to this. The distribution shows required operating modes at 3.50 and 5.5 GHz frequencies.

The current travels through the feed line and in the ground plane but oscillates at the edges of patch, which leads to an omnidirectional radiation pattern at 3.5 and 4.2 GHz and nearly omnidirectional at 5.5 GHz in the *XY* plane. These findings indicate that the gain of the SFTA antenna is greater than 2 dB. The omnidirectional 3D radiation patterns are observed in Fig. 7 and are nearly omnidirectional at higher frequencies. Many of the commercial bands, including Wi-Fi 802.11y (3.6–3.7 GHz), WiMAX (3.4–3.6 GHz and 3.7–4.2 GHz), and WLAN 802.11 (3.6–3.7 GHz), have bandwidth requirements that are met by the observed bandwidth (5.31–6.32 GHz). The maximum peak gain of 3.68 dB was achieved within the frequency range of 3.2–6.2 GHz. Throughout the band, the entire simulated antenna efficiency, including all

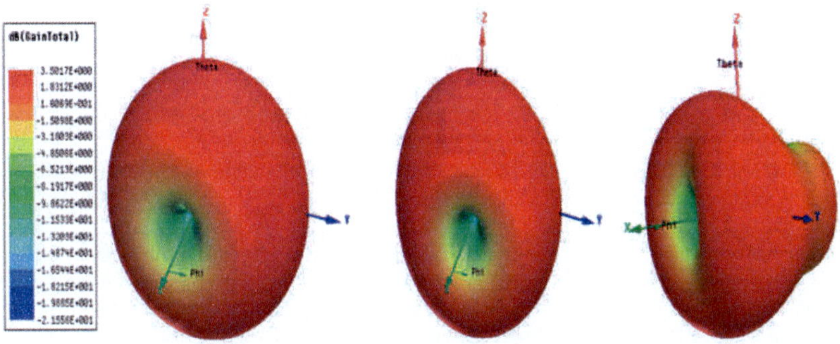


(a) 3.50GHz (b) 4.2GHz (c) 5.50GHz

**Fig. 7** 3D radiation patterns at resonant frequencies

**Table 2** Performance parameters of the various structures of monopole antenna

| Performance parameter | Triangle antenna (TA) | Triangle antenna with DGS | Fractal triangle antenna with DGS | Sectorial triangle antenna with DGS | CPW Fed SFTA | CPW fed inscribed circle SFTA |
|---|---|---|---|---|---|---|
| Gain (dB) | −2.45 | 3.165 | 3.125 | 3.018 | 3.526 | 3.68 |
| Return loss | −1.48 | −37.04 | −28.45 | −22.34 | −20.12 | −28.85 |
| VSWR | 42 | 0.612 | 0.981 | 1.16 | < 2 | < 2 |
| Bandwidth (GHz) | – | 1.27 | 1.28 | 1.49 | 3.27 | 4.6 |

losses, is greater than 80%. Table 2 shows the performance parameters of the different structures of an antenna in different stages at a 3.5 GHz frequency.

## 5 Conclusion

A sectoral fractal concept is used to develop monopole antenna for more comprehensive bandwidth sub 6 GHz 5G applications. The suggested antenna was designed for handheld terminals and built on a FR 4 substrate with a volume of $35 \times 27 \times 1.6\ \text{mm}^3$. The proposed SFTA is suitable for fine adjustment of the resonant frequency within the desired band. The suggested antenna has an enhanced broad bandwidth of 4.6 GHz (2.2–6.8 GHz) based on the measured return loss. At the lower resonant band, the suggested antenna has a gain of 2.14 dB, while at the upper resonant band, it has a gain of 3.68 dB. As a result, the suggested antenna can be used as a low-profile wideband antenna to serve a variety of wireless communication services below the 6 GHz band and for 5G applications.

## References

1. A.K. Saurabh, M.K. Meshram, Compact sub-6 GHz 5G-multiple-input-multiple-output antenna system with enhanced isolation. Int. J. RW Microw. Comput. Aided Eng. **30**(8), e22246 (2020)
2. G. Kaur, A. Singh, D. Mittal, A. Prince, Kaur, P. Panday and E. Sidhu, Performance analysis of conductive patch materials for the design and fabrication of microstrip patch antennas, in *Progress in Electromagnetics Research Symposium-Spring (PIERS)*, pp. 502–508 (2017)
3. I.K. Kim, J. Ghimire, J. Maharjan, I. Nadeem, S. Kim, D. Choi, Ultra Wideband (UWB) microstrip patch antenna with adjustable notch frequencies, in *IEEE International Conference on Industry 4.0, Artificial Intelligence, and Communications Technology (IAICT)*, pp. 70–73 (2019)
4. N.M. Awad, M.K. Abdelazeez, Multislotmicrostrip antenna for ultra-wide band applications. J. King Saud Univ. Eng. Sci. **30**(1), 38–45 (2018)
5. Z. Li, X. Zhu, C. Yin, CPW-fed ultra-wideband slot antenna with broadband dual circular polarization. AEU-Int. J. Electron. Commun. **98**, 191–198 (2019)

6. P. Khanna, A. Sharma, A.K. Singh, A. Kumar, A CPW-Fed octagonal ring shaped ultra wide band antenna for wireless applications. Adv. Electromagn. **7**(3), 87–92 (2018)
7. H. Chen, H. Chen, A cpw fed dual- frequency monopole antenna. IEEE Trans. Antennas Propag. **52**, 978–982 (2004)
8. H.-D. Chen, Compact cpw-fed dual frequency monopole antenna, IEE Electron. Lett. **38**, 1622–1624 (2002)
9. W.C. Liu, W.R. Chen, Cpw-fed compact meandered patch antenna for dual-band operation. IEE Electron. Lett. **40**, 1094–1095 (2004)
10. M.R. Werner, *Frontier of Electromagnetic* (1999)
11. H. Jones, *Fractals and Chaos* (1990)
12. B. Mandelbort, *The Fractal Geometry of Nature* (1983)
13. H. Peitgeo, Chaos and fractals (1990); M.C. Lagarias, A. Wilks, Compact frequency notched ultrawide band fractal printed slot antenna. IEEE Microwave Wirel. Compon. Lett. **40**, 224–227 (2004)
14. M.G. Sorwar Hossain, Y. Ohashi, Stub coupled monopole antenna for a broadband application, in *Asia-Pacific Microw Conference*, pp. 1965–1968 (2011)
15. K. Xu, Z. Zhu, H. Li, J. Huangfu, C. Li, L. Ran, A printed single-layer UWB monopole antenna with extended ground plane stubs. IEEE Antennas Wireless Propag Lett. **12**, 237–240
16. O.M. Haraz, M. Ashraf, S. Alshebili, M.R. Al Shareef, H.M. Behairy, UWB monopole antenna chipless RFID tags using 8-bit open-circuit stub resonators, in *21st International Conference on Microwave, Radar and Wireless Communications (MIKON)* (2016)
17. N. Takemura, D. Kaneko, J. Suzuki, T. Takeda, T. Hiraguri, Design of UWB printed monopole antenna with short stub, in *Proceedings International Symposium on Electromagnetic Compatibility–EMC Europe 2016*, Wroclaw, Poland, pp. 566–569, 5–9 Sept 2016
18. C. Yanni, Q. Zhou, X. Zhang, D. Shen, Y. Zhou, A novel multiple-stub ultra-wide band antenna, in *2017 International Workshop on Antenna Technology: Small Antennas, Innovative Structures, and Applications (iWAT)*, Athens, pp. 229–232 (2017)

# Design of a Dual-Ring Resonator Antenna for WBAN Applications

Y. E. Vasanth Kumar, G. Rajita, and K. P. Vinay

**Abstract** A novel dual ring resonator working in the ISM band was anticipated. The radiating element with denim as substrate designed for WBAN. The proposed antenna has a dimension of 80 mm × 70 mm × 0.7 mm. Two parasitic elements are introduced inside rings for better impedance matching. For On/Off body applications with −32.78 dB as reflection coefficient, gain as 3.6 dB, 3.71 directivity, and wide bandwidth of 4.49 GHz was achieved with the aid of defected ground structure. All the results are simulated by the CST Studio Tool.

**Keywords** Defected ground structure · ISM band · MPA · Ring resonator · WBAN

## 1 Introduction

Nowadays on-body and off-body communication plays a vital role in wearable electronic technology or devices. The gadgets which are designed are for live usage for extracting the information from different sources. Wearable technology came into existence in defense machinery with world-class quality. The devices became a day-to-day practice in the area of health care monitoring to get back the information at any instant. The role of wearable technology focuses mainly on health and medicine [1, 2]. The frequency bands which come with WBAN are 402–405 MHz Medical Implant Communication System, 420–450 MHz and 863–870 MHz Wireless Medical Telemetry Systems, 902–928 MHz Industrial, Scientific and Medical

Y. E. Vasanth Kumar (✉) · G. Rajita
Department of ECE, GIET University, Gunupur, Rayagada, Odisha 765022, India
e-mail: y.eswaravasanth@giet.edu

G. Rajita
e-mail: g.rajita@giet.edu

Y. E. Vasanth Kumar · K. P. Vinay
Department of ECE, Raghu Engineering College (A), Visakhapatnam, Andhra Pradesh 531162, India

 
S. Patnaik et al. (eds.), *Smart and Sustainable Technologies: Rural and Tribal Development Using IoT and Cloud Computing*, Advances in Sustainability Science and Technology,
https://doi.org/10.1007/978-981-19-2277-0_12

Radio Band, 2400–2450 MHz worldwide ISM, 3100–10,600 MHz Ultra Wide Band to enhance the performance of the wearable devices for multiple applications [3].

Here every section is discussed such as Sect. 2 briefs about the structure of dual ring resonator, in Sect. 3 discussion of results, and in Sect. 4 final conclusion.

## 2 Implementation of Dual-Ring Resonator

With the help of microstrip transmission model analysis, the antenna is calculated and simulated using CST tool for the wideband frequency with denim fabric as a substrate having $\varepsilon_r = 1.6$, height $h = 0.7$ mm, tan $\delta = 0.002$ for the dual ring resonator [1–6]. A novel ring structure with an outer radius of 20 and 15 mm inner radius is depicted in Fig. 1a with defected ground structure in Fig. 1b. Additional two rectangular slots are attached inside with dimensions of 6.5 mm × 9.2 mm to the ring resonator to develop gain and reduce return loss. The input is fed through the transmission line feeding technique for the dual ring resonator antenna [5].

To find the actual radius '$a$' of the patch we have [7]

$$a = F/\{1 + 2h/\Pi\varepsilon r F[\ln(\Pi F/2h) + 1.7726\}^{1/2}$$

where $F = 8.791 \times 10^9/\text{fr}\sqrt{\varepsilon\text{r}}$.

**Fig. 1** Front view of dual ring resonator

**Fig. 2** Defected ground structure

The proposed design as shown in Figs. 1 and 2 are simulated with computer simulation technology tool with two rectangular slots designed to enhance the bandwidth of the projected antenna. The toil is carried out with defected ground structure having denim as a substrate for off-body and on-body applications [8]. The versions of the real-time simulated design for the dual ring resonator and the particulars of dimensions are tabulated in Table 1 [9, 10].

## 3 Results

The below figures are the simulated values for proposed antenna at 5 GHz with − 32.78 dB as return loss in Fig. 3 satisfying that it is below the threshold value for better gain [2, 3, 11]. The radiation pattern is shown in Figs. 5, 6 for which is 3.6 dB gain and 3.71 dB as directivity. Here, in Fig. 4, achieves 4.49 GHz wide bandwidth. The proposed antenna also produces another band with a reflection coefficient of − 23.46 dB at 3 GHz [12].

Table 2 shows the antenna parameters values for the proposed dual ring resonator antenna.

**Table 1** Description of antenna components

| S. No. | Component | Parameter | Dimensions (mm) |
|---|---|---|---|
| 1 | Substrate | Length | 80 |
| | | Width | 70 |
| | | Height | 0.7 |
| 2 | Ground | Length | 80 |
| | | Width | 70 |
| | | Height | 0.035 |
| 3 | Patch | | |
| | Left ring | Inner radius | 15 |
| | | outer radius | 20 |
| | Right ring | Inner radius | 15 |
| | | outer radius | 20 |
| 4 | Feed line | Length | 32 |
| | | Width | 3.3 |
| | | Height | 0.035 |
| 5 | Rectangular slot 1 and rectangular slot 2 | Length | 6.5 |
| | | Width | 9.2 |

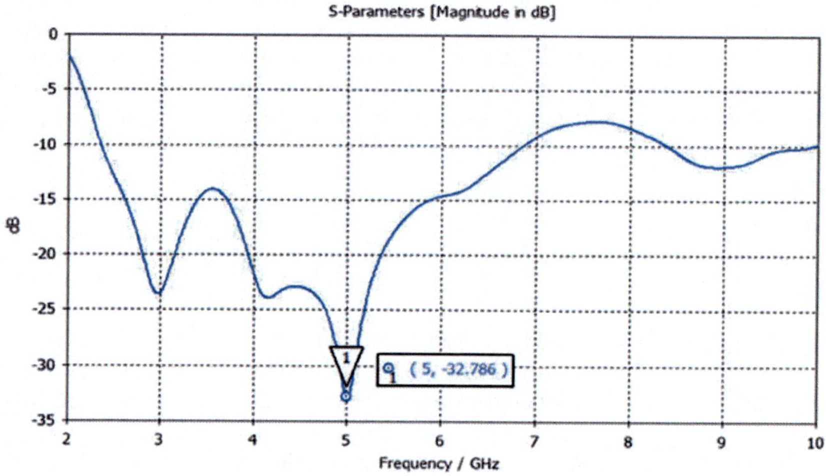


**Fig. 3** −32.78 dB reflection coefficient at 5 GHz

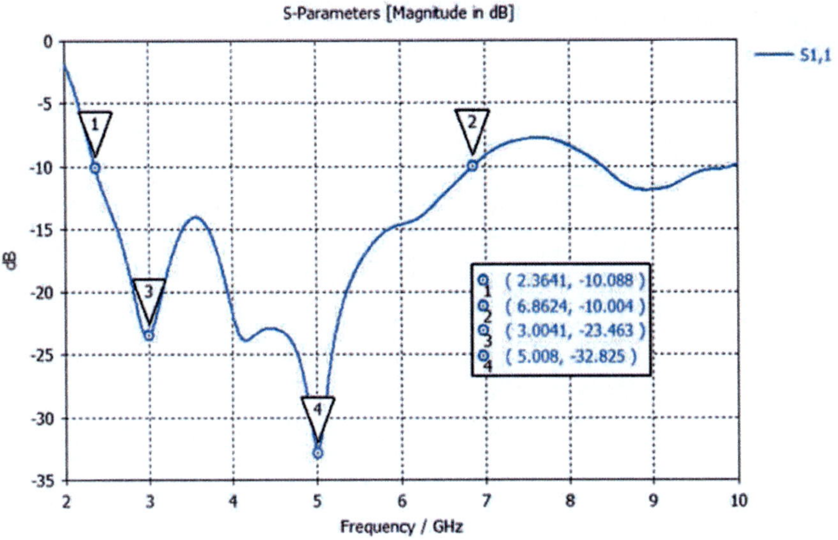


**Fig. 4** Wide bandwidth ranging from 2.36 to 6.86 GHz

**Fig. 5** 3.6 dB gain at 5 GHz

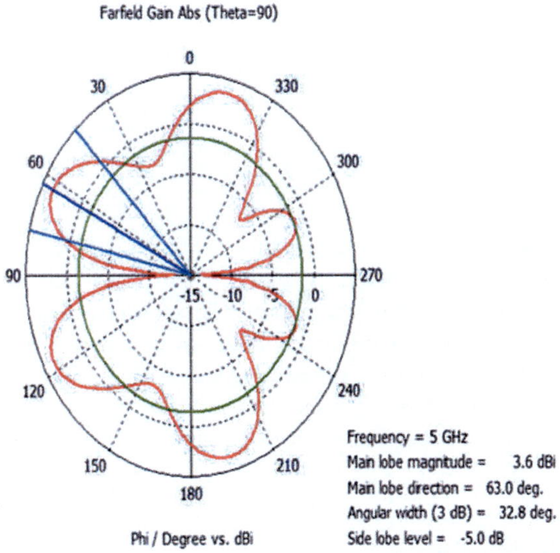


## 4 Conclusion

In this communication, a novel dual ring resonator structure was designed to operate at 5 GHz Wi-Fi frequency for WBAN applications. The antenna was made with copper for conducting portion and denim material as a substrate. The proposed

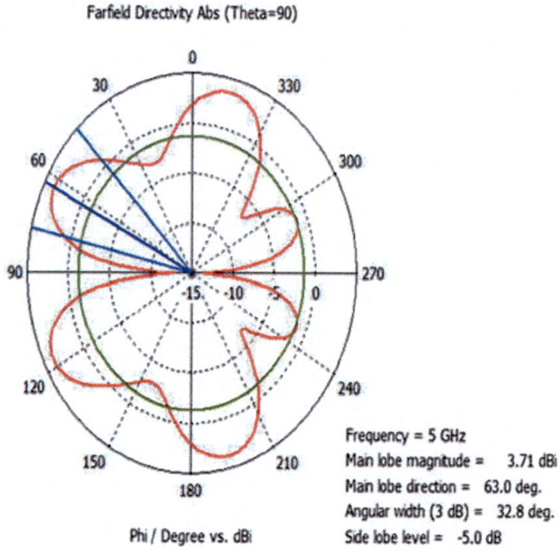


**Fig. 6** 3.71 dB directivity at 5 GHz

**Table 2** Antenna parameter for proposed design

| S. No. | Antenna parameters | Simulated values |
|---|---|---|
| 1 | Operating frequency (GHz) | 5 |
| 2 | $S_{11}$ (dB) | −32.78 |
| 3 | Gain (dB) | 3.6 |
| 4 | Directivity (dB) | 3.71 |
| 5 | Bandwidth (GHz) | 4.49 |

antenna with transmission line technique provides a bidirectional radiation pattern with gain 3.6 dB. The two rectangular slots inside the ring resonator with defected ground structure help to improve the impedance matching; it also provides wide bandwidth of about 4.49 GHz ranging from 2.36 to 6.86 GHz with a center frequency of 5 GHz where the return loss is −32.78 dB. The proposed antenna also produces another band with a reflection coefficient of −23.46 dB at 3 GHz. The results show that the proposed antenna can be used for ISM band between (5.7–5.8 GHz).

## References

1. P. Manish Potey, K. Tuckley, Design of wearable textile antenna for low back radiation. J. Electromagn. Waves Appl. (2019)
2. A.Y.I. Ashyap, Z. Zainal Abidin, S.H. Dahlan, H.A. Majid, W.A., Kamarudin, J.M. Noras, et al, Inverted e-shaped wearable textile antenna for medical applications. IEEE Access (2018)
3. Y.J. Li, Z.Y. Lu, L.S. Yang, CPW-fed slot antenna for medical wearable applications. IEEE Access (2019)

4. R. Aprilliyani, P.A. Dzagbletey, J.H. Lee, M.J. Jang, J.-H. So, J.-Y. Chung, Effects of textile weaving and finishing processes on textile-based wearable patch antennas. IEEE Access (2020)
5. R.B.V.B. Simorangkir, Y. Yang, L. Matekovits, K.P. Esselle, Dual-band dual-mode textile antenna on PDMS substrate for body-centric communications. IEEE Antennas Wirel. Propagat. Lett. **16**, 677–680 (2017)
6. S. Sankaralingam, B. Gupta, Determination of dielectric constant of fabric materials and their use as substrates for design and development of antennas for wearable applications. IEEE Trans. Instrum. Measur. **59**(12), 3122–3130 (2010)
7. C.A. Balanis, *Antenna Theory Analysis and Design*, 3rd ed. (Wiley Interscience, Hoboken, 2005)
8. D. Gopi, A.R. Vadaboyina, J.R.K.K. Dabbakuti, DGS based monopole circular-shaped patch antenna for UWB applications. SN Appl. Sci. **3**(2) (2021)
9. G. Dattatreya, K.K. Naik, A low volume flexible CPW fed elliptical ring with split triangular patch dual band antenna. RF Microwave Comput. Aided **29**(8) (2019)
10. K.P. Vinay, M. Meena Kumari, Y.E. Vasanth Kumar, Design and analysis of LOL-P textile antenna, in *10th International Conference on Innovations in Bio-Inspired Computing and Applications* (Springer, 2019)
11. D. Gopi, P. Vinay Kokilagadda, S. Gupta, V.R. Koti Reddy Dodda, *Asymmetric Coplanar Strip-Fed Textile-Based Wearable Antenna for MBAN and ISM Band Applications*, First published: 23 June 2021
12. L. Prasad, B. Ramesh, K.S.R. Kumar, K.P. Vinay, Design and implementation of multiband microstrip patch antenna for wireless applications. Adv. Electromagn. (2018)

# Performance Analysis of MIMO Antenna for Isolation Improvement

**K. Srinivasa Naik, Pasumarthi Suneetha, Pachiyannan Muthusamy, and Pechetti Priyanka**

**Abstract** The four-element orthogonal MIMO antenna simulated and fabricated at a frequency of 5.26 GHz for WLAN applications is presented, where each antenna is designed using a rectangular monopole in addition to a circle to bring out the desired frequency. The overall dimension of the four-element antenna and the orthogonal four-element antenna is $48 \times 50$ mm$^2$ and $48 \times 44$ mm$^2$, respectively. First, a single radiating element has been simulated and then fabricated. The size of the single antenna is $25 \times 25$ mm$^2$. The gain for the single antenna is 3.23 dB and the VSWR is less than 2. After that, the isolation improvement analysis is carried out for side-by-side and orthogonal MIMO structures. Using orthogonal MIMO 7 dB isolation is improved. The diversity characteristic mode analysis of both MIMO antennas is examined with the help of the Envelope Correlation Coefficient (ECC) and Diversity Gain (DG) by using CSTMW 2018. According to observations, the orthogonally arranged MIMO antenna meets the wireless communication link's diversity performance characteristics. The S-parameter of the proposed MIMO antenna was measured using the MITS eleven lab and the Rohde & Schwarz ZVL vector network analyzer, and then the prototype MIMO antenna was fabricated.

**Keywords** DG · ECC · MIMO antenna · Mutual coupling

## 1 Introduction

MIMO technology is one of the performance significant elements for wireless communication systems. MIMO antenna systems are designed to enhance and increase the link capacity and data throughput without increasing transmit power as compared to conventional systems. However, the MIMO antennas face the problem

K. Srinivasa Naik (✉) · P. Suneetha · P. Priyanka
Vignan's Institute of Information Technology (A), Visakhapatnam, AP, India
e-mail: nivas97033205@gmail.com

P. Muthusamy
Vignan's Foundation for Science, Technology, and Research, Vadlamudi, Andhra Pradesh, India



S. Patnaik et al. (eds.), *Smart and Sustainable Technologies: Rural and Tribal Development Using IoT and Cloud Computing*, Advances in Sustainability Science and Technology,
https://doi.org/10.1007/978-981-19-2277-0_13

of signal interference with adjacent antennas called mutual coupling interference. For this purpose, several designers are existing for good isolation between antenna elements such as polarization diversity antenna orientation and defected grounded structures (DGS) using metamaterial structures. Kaimi [1] presents the use of MIMO technology and methods to increase the quality of the communication link. In [2], Balanis described isolation for MIMO systems, two symmetrical F-shaped MIMO system is reported in [3]; the isolation has provided a T-junction, which is used to provide good isolation. Another study reports an FSS-based multiband MIMO is reported in [4], high isolation achieved meandered line based defected ground plane respectively. By utilizing circular polarization, the proposed MIMO offers high insulation of 5.10–5.85 GHz across the frequency spectrum in [5]. Another study found that by interrupting the I-shaped ground plane slot with [6] tri-band MIMO antenna and preventing decoupling of standard techniques like DGS and EBG structures, [7] triple-band MIMO antennas can achieve low interference. For four elements [8–15], several antennas for WLAN and Wi-Max services have been developed.

In the proposed design, the overall size is $130 \times 130\ \text{mm}^2$ and isolation value is $< 15$ dB whereas ECC and DG values are not analyzed [14, 15]. The overall size of the substrate is $40 \times 40\ \text{mm}^2$ which is designed with good isolation at the desired frequency band, in addition to this performance is also analyzed with ECC and DG parameters [12]. With a tiny dimension of $30 \times 30 \times 1.6\ \text{mm}^3$, a CPW antenna with one port feed achieved acceptable dual-band polarization. Dual-band MIMO for Wi-MAX/WLAN applications was designed in [13]. T-shaped junctions, which function as a short stub-loaded resonator (SSLR) are utilized for high isolation. In [14] a lattice of $2 \times 5$ unit cells of metamaterial design is used to improve the gain, bandwidth, and also for low mutual coupling of the antenna. In [15] a structural metamaterial decoupling is designed for the improvement of the MIMO separation array of two antenna patch components, operating at 5.3 GHz. In [16] A coplanar waveguide (CPW) antenna with a single port feed achieved good dual-band polarization with a compact size $30 \times 30 \times 1.6\ \text{mm}^3$. In [17] isolation is improved by extruding the T-shape reflecting stub on the ground plane and also the T-shape parasitic element is placed in between the two radiating elements. A circularly polarized $4 \times 4$ pentagonal microstrip patch MIMO antenna and The CP elements are placed orthogonally with the same CP diagonal element to minimize Interconnectivity between components [18]. In [19] four/eight MIMO antenna was designed for 2.4, 5.2 and 5.8 GHz for WLAN applications, In comparison to the upper-lower configuration, the overall size of the eight-element is $102 \times 52\ \text{mm}^2$ and the isolation is improved by side-by-side antennas. A 4 element MIMO/diversity antenna was designed for WLAN among four two elements that resonate at 2.4 Ghz and another two operate at 5.8 GHz and the isolation is improved with antenna orthogonality method was presented in [20]. In [21], a new metamaterial superstrate structure is placed above the two-element MIMO antenna by a 9 mm distance to improve the isolation. A two-element and four-element MIMO antenna was designed for WLAN and isolation is $> 30$ dB using antenna Orthogality presented in [22].

In this work, a rectangular patch with a circle is used as a basic antenna to construct a four-element orthogonal MIMO antenna to resonate at 5.2 GHz. In order to improve the isolation between the four antenna elements, orthogonality is used. Furthermore, the characteristics of the MIMO antenna are examined using ECC and DG. Each section discusses and deliberates on the design details and results. All the simulated results were done using CSTMW software and Rohde & Schwarz ZVL vector network analyzer and then the prototype MIMO was fabricated.

## 2 Single Antenna Design Calculations

The basic parameters for designing the microstrip antenna are width (W), length (L), resonant frequency ($f_o$), and substrate height (h) for obtaining the desired frequency. In this work rectangular microstrip antenna is designed for 5.26 GHz frequency. In this proposed design FR4 substrate was used as a substrate. In a microstrip patch antenna the signal is propagated through the width of the patch only, so to achieve good radiation the value of width and the length can be calculated by the following formula.

$$W = \frac{C}{2fr}\sqrt{\frac{2}{(\in r + 1)}}$$

$$\varepsilon_{\mathrm{reff}} = \frac{\varepsilon_r + 1}{2} + \frac{\varepsilon_r - 1}{2}\left[1 + \frac{12h}{w}\right]^{-1/2}$$

$$L = \frac{1}{2fr\sqrt{\varepsilon_{\mathrm{reff}}}\sqrt{\mu_o \varepsilon_0}} - 2\Delta L$$

The novelty of this present work is achieved by embedding the triangular over the circular to achieve the desired frequency of 5.26 GHz for WLAN applications. In this work, a single radiating patch is designed, simulated results show good return loss, and then it is fabricated using an FR4 substrate having a 1.6 mm height with a dielectric material of 4.3. The evolution of the single-stage antenna and the desired single antenna are shown in Fig. 1a, b, and Table 1 shows the design parameters of the basic single radiating element. Figure 2 shows a fabricated desired antenna. The measured setup with S-parameter results is represented in Fig. 3.

The measured and simulated S-parameter values at the desired operating frequency are shown in Fig. 4a, b. The simulated antenna operates at 5.26 GHz with a low S-parameter value of –29.76 dB and the fabricated desired antenna resonates at 5.30 GHz with an S-parameter of –18.33 dB, which covers the minimum required value. As compared to the practically calculated resonant frequency and simulated values of the designed antenna, Error in the measured value due to the fabrication process Fig. 5 represents the v current distribution of the desired antenna when port.1 is excited, and we observe that the current is highly concentrated on the radiating patch so low return loss is achieved. The 1-D E-plane and H-plane radiation plots for

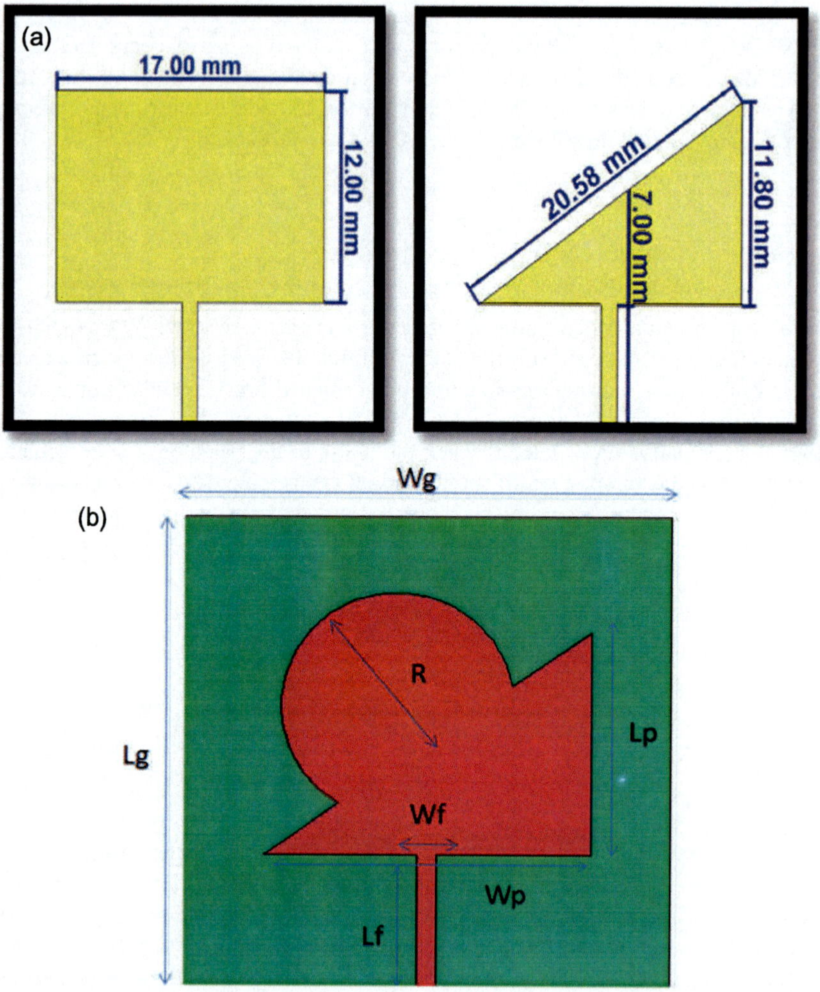


**Fig. 1** **a** Stages of single antenna, **b** proposed single antenna

**Table 1** Design parameters of the antenna

| Parameters | Dimensions in mm |
|---|---|
| Lg—ground length | 25 |
| Wg—ground width | 25 |
| Lp—Rectangular patch length | 12 |
| Wp—Rectangular patch width | 17 |
| Lf—Microstrip feed length | 7 |
| Wf—Microstrip feed width | 1 |
| R—circle radius | 6 |

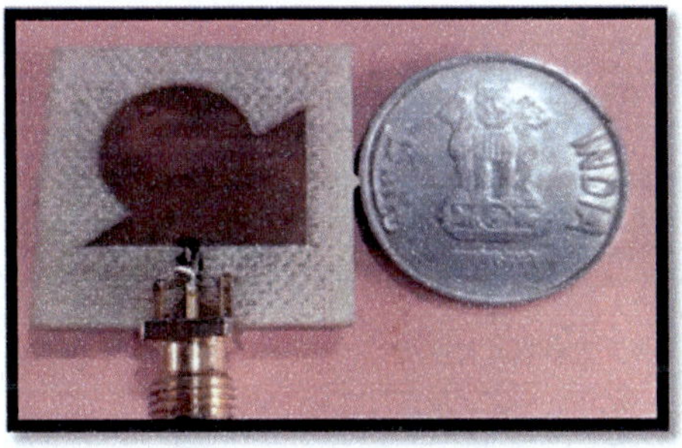

**Fig. 2** Fabricated prototype single antenna structure

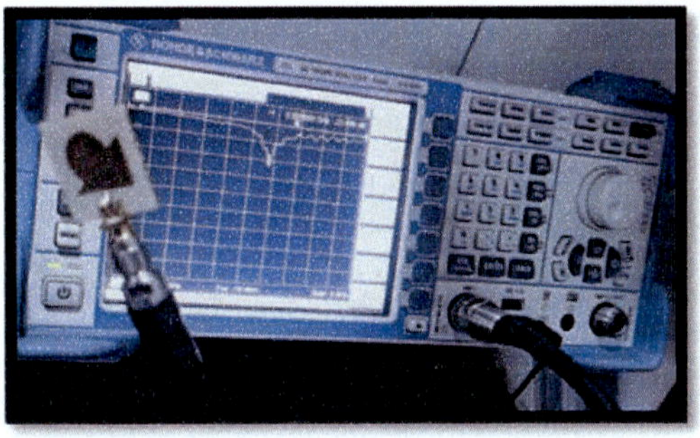

**Fig. 3** Measured $S_{11}$-parameter plot displayed in network analyzer

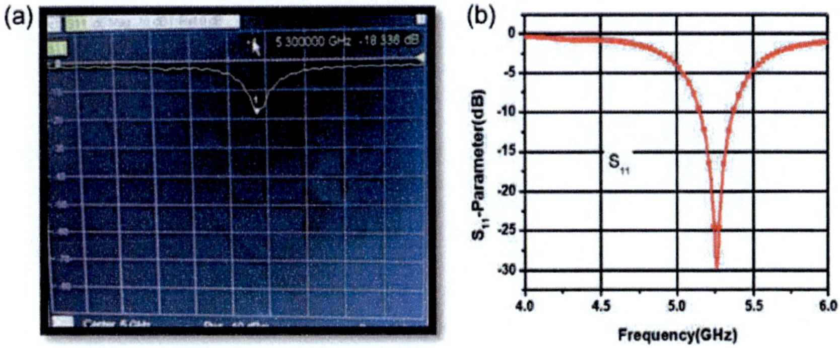


**Fig. 4** **a** Measured S-parameter result. **b** Simulated S-parameter result

the designed single antenna are represented in Fig. 6. From that notice, for the XZ-plane, the highest radiation is at $-15°$ and for the YZ plane the maximum radiation pattern is $-20°$ respectively. The simulated directivity and gain plots are depicted in Fig. 7a, b. The simulated VSWR plot is depicted in Fig. 8.

Fig. 5 Port-1 surface current distribution

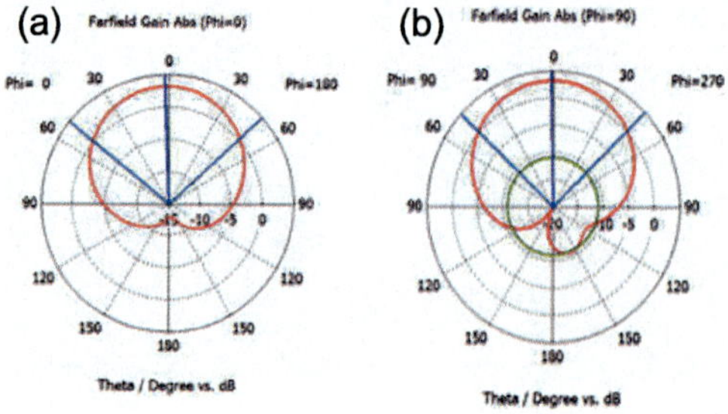


**Fig. 6** 1-D plot for **a** E-plane, **b** H-plane

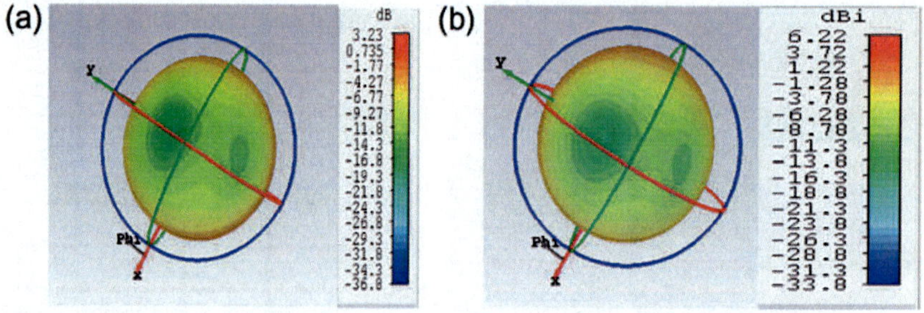


**Fig. 7** **a** Gain plot. **b** Simulated 3-D plot for directivity

# 3 Design of MIMO Antenna

An analysis was carried out in this section for the purpose of improving isolation. For this, two MIMO antennas were designed, one using space diversity and the other using orthogonality, and the results were compared based on return losses, gain and

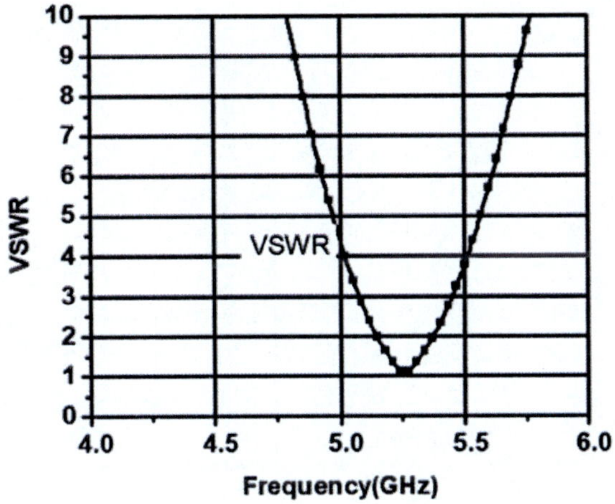


**Fig. 8** Simulated VSWR plot

directivity, ECC and DG, and most notably, compactness dimensions. The isolation provided by this orthogonal MIMO antenna is excellent.

**ECC**

The MIMO system's performance is assessed using ECC (envelope correlation coefficient). The ECC value is a measure of the antennas' close vicinity. Significant separation is possible because of the low ECC. This can be measured using the reference Eq. 1 as described in [6].

$$\rho_{eij} = \frac{\left|S_{ii}{}^{*}S_{ij} + S_{ji}{}^{*}S_{jj}\right|^{2}}{(1 - |S_{ii}|^{2} - S_{ij}{}^{2})(1 - \left|S_{ji}\right|^{2} - S_{jj}{}^{2})} \tag{1}$$

**DG**

The Diversity Gain (DG) measures the transmission power loss in MIMO systems. Can be evaluated by the following Eq. 2 in reference [10].

$$DG = 10\sqrt{1 - \left|\rho_{eij}\right|^{2}} \tag{2}$$

## 4 Analysis of the Space Diversity of MIMO Antenna

This session discusses the space diversity MIMO antenna. Figure 9 shows the overall dimensions of the space diversity MIMO antenna, which is $50 \times 48 \times 1.6\ \text{mm}^3$. The antennas are separated by a tenth of a wavelength. The simulated S-parameter values

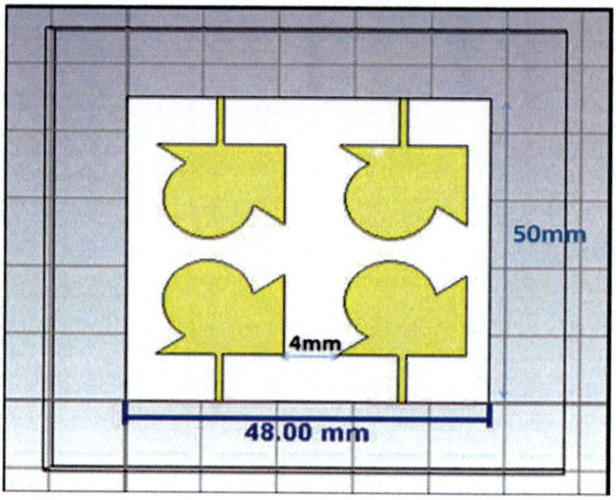


**Fig. 9** MIMO antenna structure

are shown in Fig. 10. When port-1 is excited as a result of the observation, some current is distributed to the adjacent radiating element, resulting in mutual coupling between the adjacent radiating elements. The VSWR results of the desired MIMO antenna are represented in Fig. 12. The envelope correlation coefficient (ECC) for desired space diversity MIMO is less than 0.01, as shown in Fig. 13a, and the Diversity Gain (DG) is 9.99 dB, as shown in Fig. 13b. The 1-D E-plane and H-plane radiation plots for the designed MIMO antenna are represented in Fig. 14. From that noticed, for the *XZ*-plane, the highest radiation is at −20° and for the *YZ* plane the maximum radiation pattern is −20°, respectively (Fig. 11).

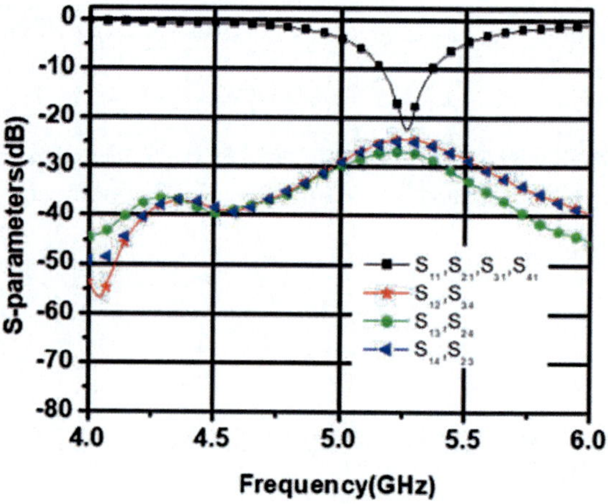


**Fig. 10** Simulated MIMO S-parameters plot

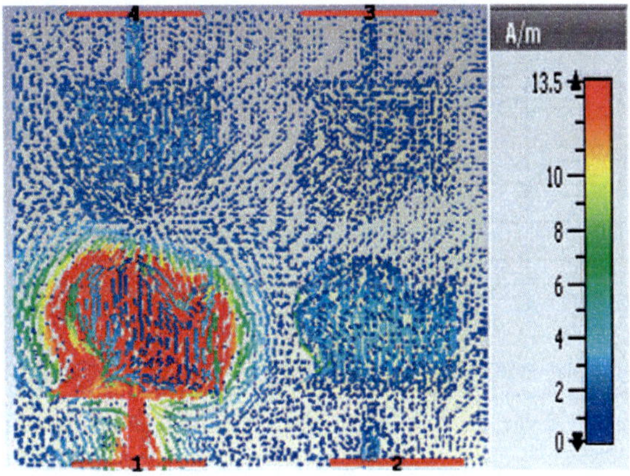


**Fig. 11** Current distribution at 5.2 GHz

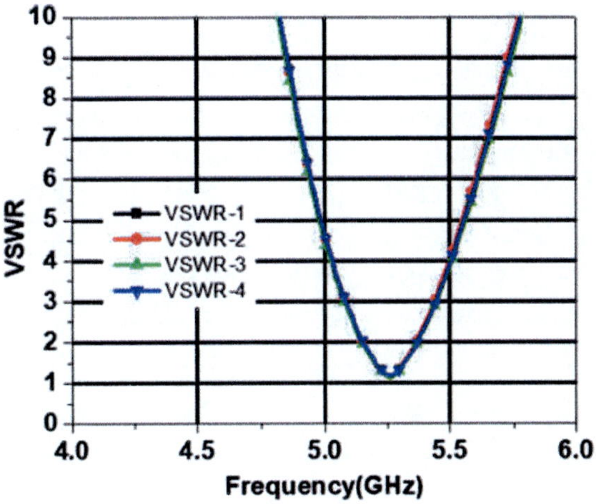


**Fig. 12** Proposed antenna VSWR

## 5 Analysis of Prototype Orthogonal MIMO

The prototype of the four-element MIMO is constructed on FR-4 substrate and practically measured using MITS eleven laboratory shown in Fig. 15. The measured setup and prototype are represented in Figs. 16 and 17. Figure 18 shows the surface current distribution of the proposed MIMO due to the orthogonality low mutual coupling is achieved. The designed MIMO's voltage standing wave ratio (VSWR) simulated values are shown in Fig. 19. The simulated S-parameter results $S_{11}$, $S_{21}$, $S_{31}$, $S_{41}$, as well as between different ports, are shown in Fig. 20, and the diversity parameter values are shown in Fig. 21a, b. This proposed four-element antenna obtained the

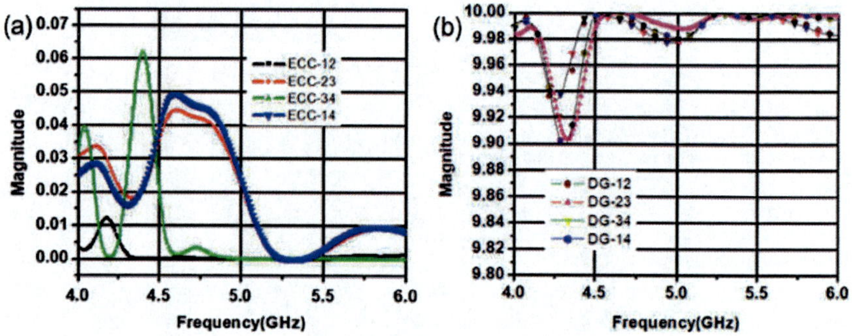


**Fig. 13** **a** Simulated ECC plot for Space diversity MIMO antenna. **b** Simulated DG plot for space diversity MIMO antenna

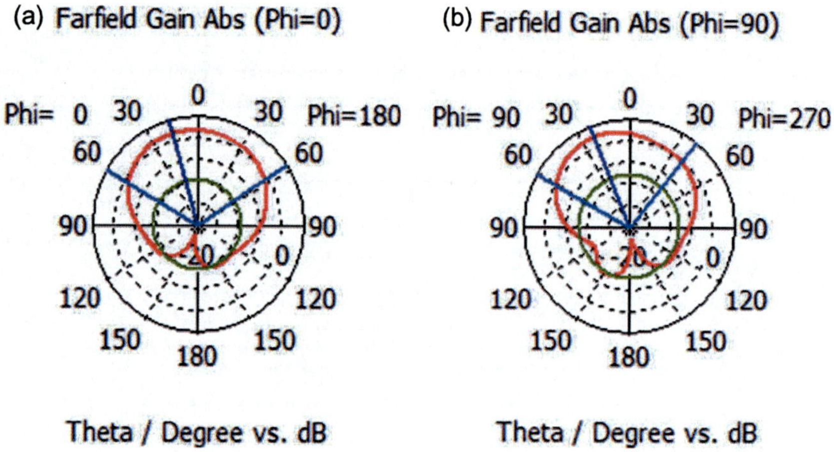


**Fig. 14** 1-D plot for **a** E-plane, **b** H-Plane

required results in terms of diversity performance, with ECC > 0.001 and DG > 9.8. From Table 2, which provided the results of a simulated and measured comparison between the single, four-element MIMO and orthogonal four-element MIMO antenna, from Table 2 it was found that the isolation is improved by 7 dB using the orthogonal MIMO structure. The 1-D E-plane and H-plane radiation plots for the designed single antenna are represented in Fig. 22. From that noticed, for the *XZ*-plane the highest radiation is at −20° and for *YZ* plane the maximum radiation pattern is −30° respectively Table 3 depicts a comparison of the proposed work with existing work. From the literature survey, it is evident that the orthogonal-based antenna has a compact size, low return loss, and good ECC and DG with existing works.

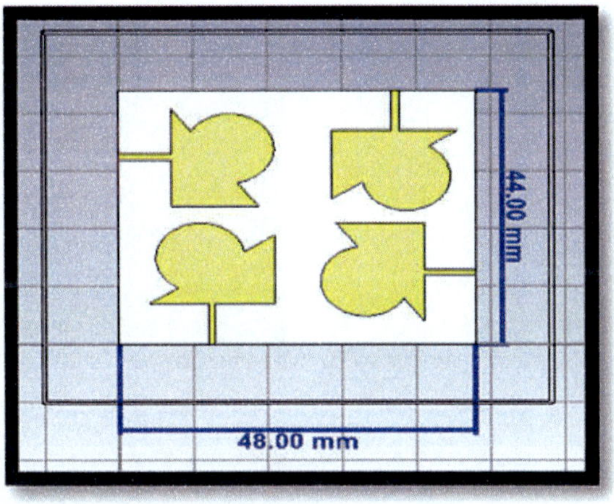


**Fig. 15** Orthogonal four-element MIMO antenna

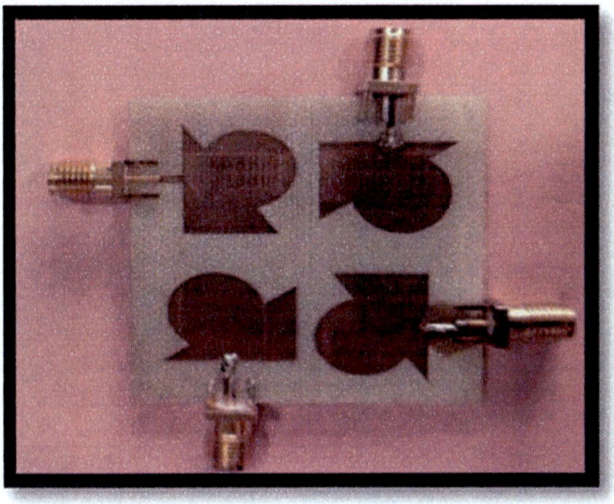

**Fig. 16** Fabricated prototype orthogonal MIMO antenna

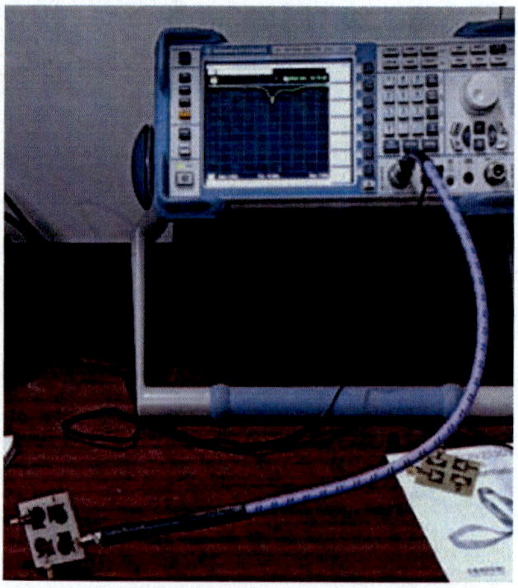

**Fig. 17** Setup for measurement of orthogonal MIMO antenna

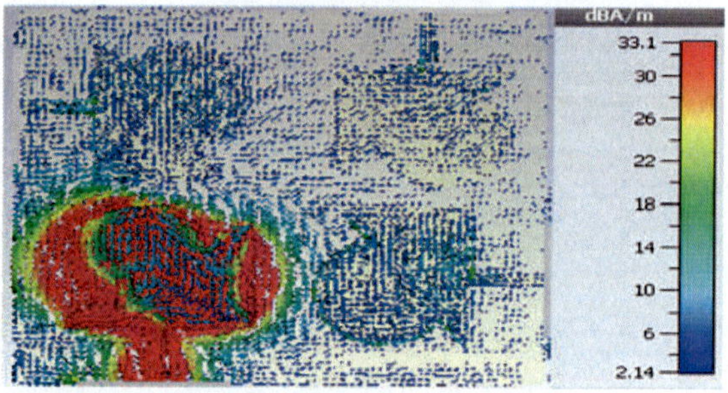


**Fig. 18** Surface current distribution at 5.2 GHz

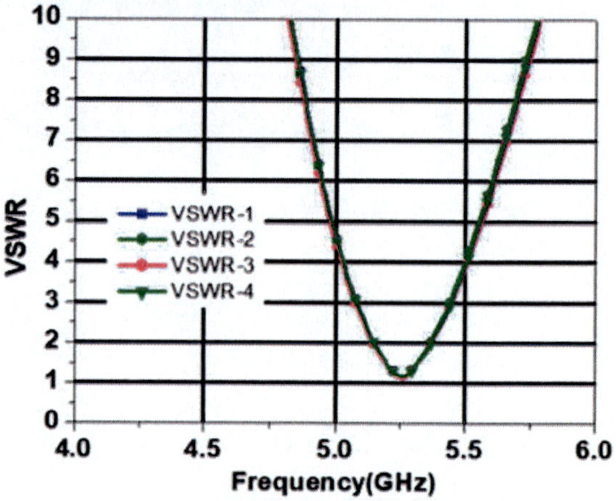


**Fig. 19** VSWR of the proposed antenna

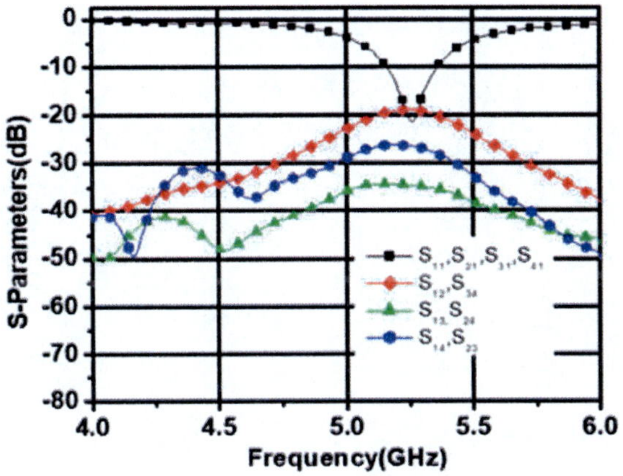


**Fig. 20** Simulated S-parameter results of orthogonl MIMO

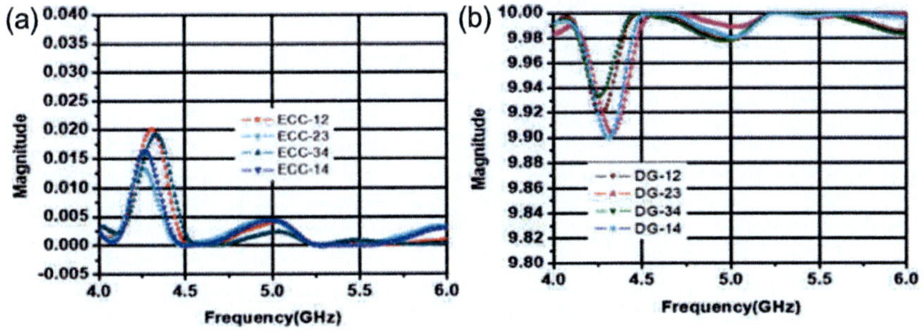


**Fig. 21** **a** ECC plots for orthogonal. **b** DG plots for orthogonal MIMO

**Table 2** Comparison of simulated and measured S-Parameters results

| | Class of the antenna | $S_{11}$(dB) | $S_{21}$(dB) | $S_{12}$(dB) | $S_{22}$(dB) |
|---|---|---|---|---|---|
| Simulated | Single element | 29.76 | | | |
| Measured | Single element | 18.33 | | | |
| Simulated | 4 element antenna | 21.51 | 18.75 | 34.74 | 26.59 |
| | 4 element orthogonal antenna | 22.25 | −25.06 | −27.01 | −23.93 |
| Measured | 4 element orthogonal antenna | 20.12 | 23.34 | 25.49 | 21.29 |

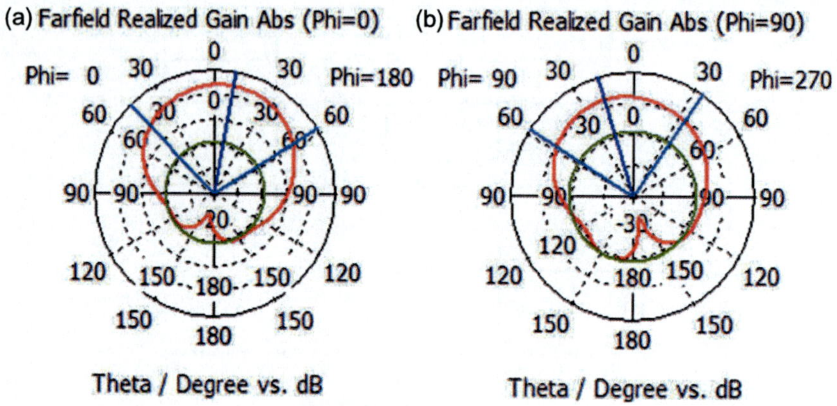


**Fig. 22** 1-D plot for **a** E-Plane, **b** H-plane

**Table 3** Comparison table for current work with previous work

| Ref. antenna | Frequency | Dimensions | MIMO elements | Isolation (dB) | ECC | DG |
|---|---|---|---|---|---|---|
| [4] | 2.4, 2.95, 3.2, 3.6, 5.5 | 38 × 36 mm$^2$ | 2 | 18 | 0.001 | > 9.8 |
| [13] | 2.5/5.8 GHz | 38 × 36 × 1.6 mm$^3$ | 2 | < 20 | < 0.16 | 9.81/9.98 |
| [14] | 5.8 GHz | 137 × 77 × 3.048 mm$^3$ | 4 | 18 | < 0.02 | Not reported |
| [15] | 5.3 GHz | 40 × 30 mm$^2$ | 2 | 12.5 | Not reported | Not reported |
| [17] | 3.63/5.64 GHz | 18 × 36 mm$^2$ | 2 | < 20 | 0.022 | Not reported |
| [18] | 2.4 GHz | 186 × 188 mm$^2$ | 4 | > 30 | < 0.5 | 9.99 |
| [19] | 5.2/5.8 GHz | 102 × 52 mm$^2$ | 4 | > 40 | < 0.5 | > 9.0 |
| [20] | 2.4/5.8 GHz | 20 × 47 mm$^2$ and 20.5 × 20 mm$^2$ | 4 | < 20 each two element | < 0.05 | > 9.95 |
| [22] | 4.5/5/5.5 GHz | 60 × 60 mm$^2$ | 4 | > 30 | 0.003 | > 9.98 |
| Proposed work | 5.26 GHz | 48 × 44 mm$^2$ | 4 | > 30 | 0.00012 | > 9.999 |

## 6 Conclusion

In this research, a four-element orthogonal MIMO antenna is simulated and fabricated for WLAN applications. The analysis was carried out for the improvement of isolation. Compared to the side-by-side MIMO elements, orthogonally placed MIMO structure isolation is improved by 7 dB. The performance of the MIMO parameters is also observed through simulation. The orthogonal structure MIMO ECC is less than 0.0012 and DG is greater than 9.999 dB, respectively. The proposed MIMO operates at a 5.17–5.37 GHz frequency. The prototype four-element antenna has been fabricated, and the results show good performance at the desired frequency band.

**Acknowledgements** This work is funded by the DST Science and Engineering Research Board (SERB) under grant number EEQ/2016/000391.

## References


1. F.M. Kaimi, Antenna design challenges for 4G. IEEE Wirel. Commun. **18**, 4–5 (2011)
2. C.A. Balanis, *Fundamental Parameters of Antennas, Antenna Theory, Analysis and Design*, 3rd ed. (JWS, Hoboken, 2005), Chapter 1, Section 2.2.3, pp. 34–36
3. A. Dkiouak, A. Zakrit, M. El Ouahabi, A. Mchbal. *Design of a Two Symetrical F-Shaped MIMO Antenna for Wi-MAX and WLAN Applications* (2019). IEEE-978-1-5386-7850-3/19/$31.00
4. R. Saleem, M. Bilal, H. Tariq Chattha, An FSS based multiband MIMO system incorporating 3D antennas for WLAN/WiMAX/5G cellular and 5G Wi-Fi applications. IEEE Access (2019). https://doi.org/10.1109/ACCESS.2019.2945810
5. H.H. Tran, N. Hussain, T.T. Le, Low-profile wideband circularly polarized MIMO antenna with polarization diversity for WLAN applications. AEU-Int. J. Electron. C. **108**, 172–180 (2019)
6. N. Kumar, U.K. Kommuri, MIMO antenna mutual coupling reduction for WLAN using spiro meander line UC-EBG. Progr. Electromagn. Res. C **80**, 65–77 (2018)
7. A.K. Gangwar, M.S. Alam, *A Compact Size Tri-band MIMO Antenna with Reduced Mutual Coupling for WLAN and Wi-MAX Applications* (2017). IEEE-978-1-5090-6674-2/17
8. X. Zhang, X. Zhong, B. Li, Y. Yu, A dual polarized MIMO antenna with EBG for 5.8ghz WLAN application. Progr. Electromagn. Res. Lett. **51**, 15–20 (2015)
9. X. Liu, Y. Wu, Z. Zhuang, W. Wang, Y. Liu, A dual-band patch antenna for pattern diversity application. IEEE Access **6**, 51986–51993 (2018)
10. M.M. Hasan, M.R.I. Faruque, M.T. Islam, 'Dual band metamaterial antenna for LTE/Bluetooth/WiMAX system. Sci. Rep. **8**, Art. no. 1240 (2018)
11. Y. Xia, S. Luo, Y. Li (2018) *MIMO Antenna Array Decoupling Based on a Metamaterial Structure* (2018). 978-1-5386-5648-8/18/$31.00 c IEEE. https://doi.org/10.1109/APCAP.2018.8538027
12. K. Kavitha, S.P. Rajan, *Single Band MIMO Antenna for WLAN/Wi-MAX Application*, vol. 04, no. 1, pp. 1686–1689 (2017)
13. A. Dkiouak, A. Zakriti, M. El Ouahabi, A. Mchbal, *Design of Two Element Wi-MAX/WLAN MIMO Antenna with Improved Isolation Using Short Stub Loaded Reasonator (SSLR)* (2019). https://doi.org/10.1080/09205071.2020.17569927
14. N. Ngoc Lan, V. Van Yem, Gain enhancement for MIMO antenna using metamaterial structure. Int. J. Microwave Wirel. Technol. (2019)

15. H. Zhang, Z. Wang, J. Yu, J. Huang, A compact MIMO antenna for wireless communication. IEEE Antennas Propag. Mag. **50**, 104–107 (2008)
16. K.C.R. Madaka, P. Muthusam, Mode investigation of parasitic annular ring loaded dual band coplanar waveguide antenna with polarization diversity characteristics. RF Microwave Comput. Aided Eng. 30(4), (2019). https://doi.org/10.1002/mmce.22119
17. P.R. Kurekar, S.S. Khade, Design and implementation of MIMO antenna for WLAN application. IEEE (2017). https://doi.org/10.1109/ICCSP.2017.8286403
18. R. Subhanrao Bhadade, S.P. Mahajan, Circularly polarized 4 × 4 MIMO antenna for WLAN applications. Electromagnetics **39**(5), 325342 (2019). https://doi.org/10.1080/02726343.2019.1619227
19. S. Roy, S. Ghosh, U. Chakarborty, Compact dual wide-band four/eight elements MIMO antenna for WLAN applications. RF Microwave Comput. Aided Eng. https://doi.org/10.1002/mmce.21749
20. A. Birwala, S. Singhb, B.K. Kanaujiac, S. Kumard, Low-profile 2.4/5.8 GHz MIMO/diversity antenna for WLAN applications. J. Electromagn. Wave Appl. (2020) https://doi.org/10.1080/09205071.2020.1757516
21. R. Mark, N. Rajak, K. Mandal, S. Das, Metamaterial based superstrate towards the isolation and gain enhancement of MIMO antenna for WLAN application. Int. J. Electron. Commun. (AEÜ) (2019). https://doi.org/10.1016/j.aeue.2019.01.011
22. A.W. Mohammad Saadh, K. Ashwath, P. Ramaswamy, T. Ali, J. Anguera, A uniquely shaped MIMO antenna on FR4 material to enhance isolation and bandwidth for wireless applications. Int. J. Electron. Commun. (AEÜ) (2020) https://doi.org/10.1016/j.aeue.2020.153316

# Low Power and High-Speed Full Adder with Complemented Logic and Complemented XOR Gate

**Bhaskara Rao Doddi, Rajita Gullapalli, and Leela Rani Vanapalli**

**Abstract** Low power and high-speed adder designs are most popularly used in high-speed computations. Complemented logic-based innovative circuits for sum and carry are proposed in this paper. The proposed designs are efficient in power and performance. Complemented logic is used to design sum and carry outputs that require less number of transistors. A novel XNOR cell has been designed for generating the sum and carry outputs. T-spice simulations are carried out to compute the efficiency of the proposed and conventional designs. Simulations are based on the 0.25um CMOS process. Simulation results conclude that the proposed design has a considerable reduction in power and delay metrics of about 10–33%.

**Keywords** Full adder · Power-consuming event · Critical path · Complemented XOR gate · Complemented logic

## 1 Introduction

Low power and high-speed adder designs are most popularly used in high-speed computations. Hybrid Full adder presented in [1] has achieved low power consumption at the cost of more transistors. Adder has low activity, but the use of transmission gates for producing carry output can degrade the speed and it cannot drive for larger word lengths [2]. Multiplexer-based adder [3] is designed by using XNOR/XOR and AND/OR as the basic building blocks. This structure needs more number of

B. R. Doddi (✉) · R. Gullapalli
Department of ECE, GIET University, Gunupur, Rayagada, Odisha 765022, India
e-mail: bhaskararao.doddi@giet.edu

R. Gullapalli
e-mail: g.rajita@giet.edu

L. R. Vanapalli
Department of ECE, GVP College of Engineering (A), Visakhapatnam, Andhra Pradesh 530048, India

 
S. Patnaik et al. (eds.), *Smart and Sustainable Technologies: Rural and Tribal Development Using IoT and Cloud Computing*, Advances in Sustainability Science and Technology,
https://doi.org/10.1007/978-981-19-2277-0_14

internal nodes which can be power hungry. Design proposed in [4] is also multiplexer-based with XNOR/XOR and NAND as building blocks. This structure can also be power hungry. Novel XNOR/XOR cell is designed [5] with three transistors can cause problem of lack of full swing when technology is scaled down. Carry logic is designed with static CMOS to overcome the drawback of transmission gates and sum logic is designed with pass transistor logic [6]. This structure has taken more number of transistors. Carry logic is designed with static CMOS and sum logic is designed with transmission gate and pass transistor logic [7]. This XNOR/XOR structure has a cross-coupled connection which increases the delay. Adder is designed with minimal number of transistors with 3 T XOR can suffer from lack of full swing, when technology is scaled down [8]. With the review of the above papers, we can conclude that carry logic is to be designed with static CMOS such that to avoid driving problems for larger word lengths. Sum logic can be designed using transmission gates or pass transistors since sum output is not going to drive any input in a ripple carry structures.

## 2 Review of Full Adders

Full adder with hybrid logic is presented in this section. Design of full adder is analyzed for estimating the delay and power consumption.

### 2.1 *Hybrid Full Adder [9]*

Hybrid logic is a mix of logic styles that are intended to achieve high-speed and low power consumption. As a designer, those things should be addressed properly. XNOR/XOR block designed using hybrid logic [2] requires 12 transistors with two transistors in the critical path, where the critical path has taken two transistors individually for both XNOR as well as XOR. All other paths consist of a single transistor except the critical path.

## 3 Proposed Full Adder

### 3.1 *Sum Output Generation Using XNOR Cell*

XNOR cell is designed using four transistors and one not gate to generate the four possible paths of the outputs as shown in Fig. 1. To produce the solid '0', the logic of XNOR is when $A$ = '1', then XNOR = $B$, and when $B$ = '1', then XNOR = $A$. To produce the solid '1', when $A$ = '0', then XNOR = $B'$ and when $B'$ = '0', then XNOR = $A$. Two PMOS transistors ensure to give solid '1' and two NMOS

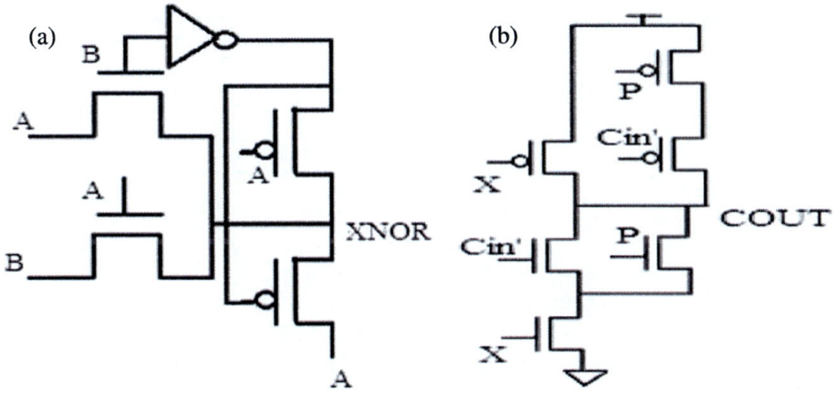


**Fig. 1** Schematic of XNOR and OAI (OR AND INVERT)

transistors ensure to give solid '0'. Two XNOR cells are required in the design of adder and to produce the sum output, twelve transistors are required. Expression for sum output in the proposed full adder is given in Eq. (1).

$$\text{SUM} = A \text{ XNOR } B \text{ XNOR } C_{\text{in}}. \tag{1}$$

## 3.2 *Carry Output Using XNOR Cell*

Carry output can be designed using one nand gate and one OR-AND-INVERT (OAI) cell. Nand gate produces complemented output when carry needs to be generated. OAI cell includes both carry generate and propagate part in the pull-up network. Carry propagate logic can be implemented using complemented logic with an XNOR cell to produce carry output. Carry output requires 10 transistors to produce full swing outputs. Carry output logic can be expressed with the proposed equations as shown below in Eqs. (2)–(5).

$$C_{i+1} = (X.(P + Q))'. \tag{2}$$

$$X = (A.B)'. \tag{3}$$

$$P = A \text{ XNOR } B. \tag{4}$$

$$Q = (C_{\text{in}})'. \tag{5}$$

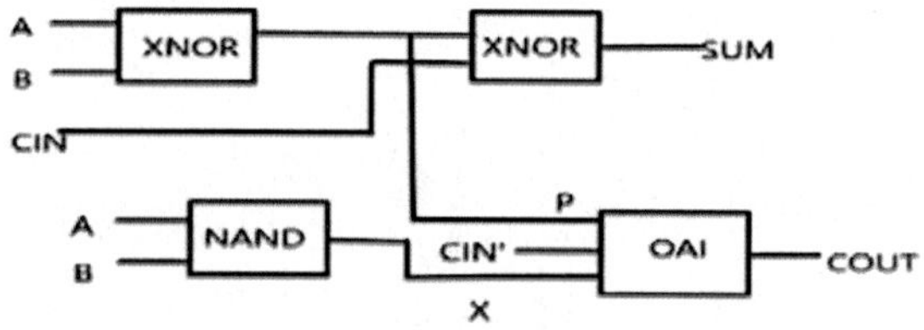


**Fig. 2** Schematic of full adder

Schematic of full adder presented in Fig. 2 requires 22 transistors which require two levels of logic. Sum output can be generated with XNOR in the first level as well as in the second level. Carry output can be generated with NAND and XNOR in the first level and OAI21 cell in the second level. Different cells required for full adder are XNOR, NAND, OAI cells, respectively.

## 4 Delay Analysis

To estimate the delay of 1-bit full adder, all possible output events are to be evaluated. To activate events, 32 test vectors have been applied to know the propagation delay from low to high and high to low. Delay is measured by considering the maximum value. *X*–*Y* input sequence means *X* is the previous and *Y* is the current input.

Table 1 shows the delays for the sample possible events of the 1-bit adder carry. The proposed adder critical path delay is 185 ps for "011–001" input test vector when previous input is "011" and the present input is "001". Existing 1-bit adder critical path delay is 186 ps for "110–100" input test vector, when previous input is "110" and present input is "100". Critical path delay is the maximum delay that is highlighted.

Table 2 shows the delays for the sample possible events of the 1-bit adder Sum. Proposed adder critical path delay is 143 ps for "010–000" input test vector. Existing 1-bit adder critical path delay is 177 ps for "010–110" and "100–110" input test vector.

**Table 1** Delay for 1-bit carry output

| S. No. | Input sequence (*X*–*Y*) | Delay (ps) | Delay (ps) |
|---|---|---|---|
| | | Existing full adder [9] | Proposed full adder |
| 1 | 011–000 | 110 | 100 |
| 2 | 011–001 | 142 | **185** |
| 3 | 011–010 | 122 | 103 |
| 4 | 011–100 | 140 | 092 |
| 5 | 101–000 | 140 | 094 |
| 6 | 101–001 | 162 | 168 |
| 7 | 110–010 | 149 | 123 |
| 8 | 110–100 | **186** | 090 |

**Table 2** Delay for 1-bit sum output

| S. No. | Input sequence | Delay (ps) | Delay (ps) |
|---|---|---|---|
| | | Existing full adder [9] | Proposed full adder |
| 1 | 010–000 | 145 | **143** |
| 2 | 100–000 | 152 | 130 |
| 3 | 111–000 | 130 | 047 |
| 4 | 111–101 | 118 | 093 |
| 5 | 001–110 | 142 | 048 |
| 6 | 010–110 | **177** | 064 |
| 7 | 100–110 | **177** | 073 |
| 8 | 111–110 | 120 | 050 |

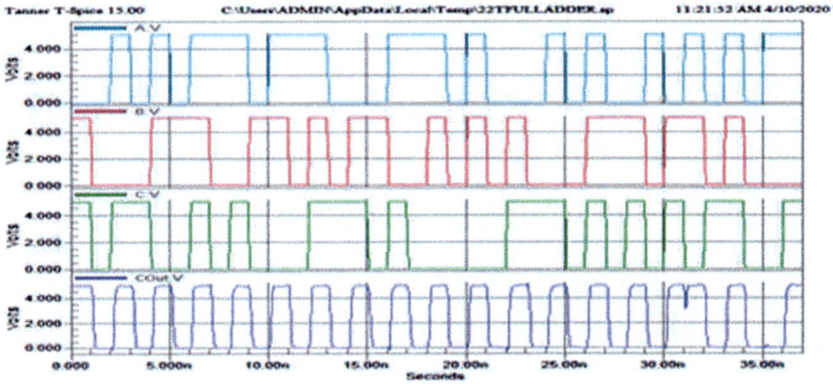


**Fig. 3** Delay measurement for carry output in proposed design

Figure 3 shows the possible events for the carry outputs in the proposed full adder design.

Figure 4 shows the possible events for the carry output in the existing full adder design.

## 5 4-Bit Adder

One bit adder can be cascaded to form a 4-bit ripple carry structure. To evaluate the delay, bits of A and B at each and every position are taken such that carry output and sum output depends on the initial carry input. Carry input is taken as ‘1’, such that its value will be propagated to the carry output and it will result in a change of value to measure the delay as in Fig. 5. Similarly, carry input is taken as ‘0’ to measure the delay for the carry output.

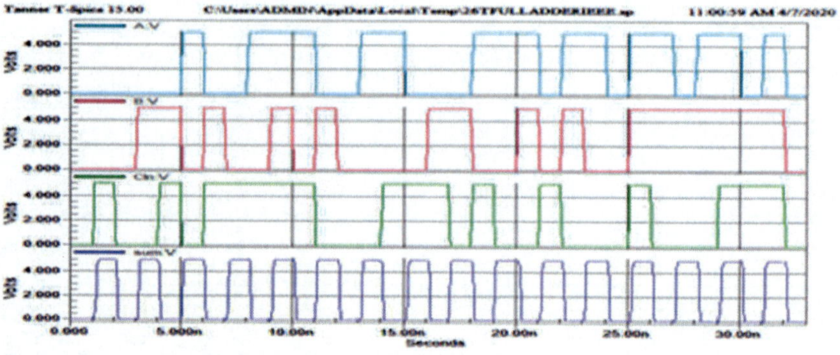


**Fig. 4** Delay measurement for carry Output in existing design

Figure 5 shows delay measurements in the proposed adder with the applied test vectors as $A =$ "0000", $B =$ "1111", $C_{\text{in}} = 1$.

Table 3 shows the delay values for the proposed and existing adder circuits. To generate 4-bit full adder carry output, the proposed design takes the delay for 62 ps and the existing one needs 82 ps. Delay measurements are based on Fig. 5.

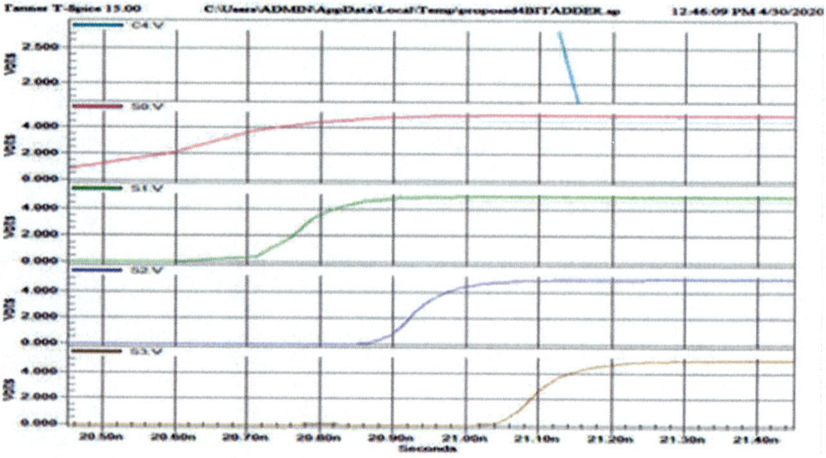


**Fig. 5** Delay measurement of the 4-bit proposed adder

**Table 3** Delay measurements for 4-bit adder with initial carry input as '1'

| S. No. | Design | Delay (S0) | Delay (S1) | Delay (S2) | Delay (S3) | Delay (C4) |
|---|---|---|---|---|---|---|
| 1 | Proposed adder (ns) | 0.12 | 0.27 | 0.42 | 0.60 | 0.62 |
| 2 | Existing adder (ns) | 0.01 | 0.265 | 0.50 | 0.71 | 0.82 |

**Table 4** Delay measurements for 4-bit adder with initial carry input as '0'

| S. No. | Design | Delay (S0) | Delay (S1) | Delay (S2) | Delay (S3) | Delay (C4) |
|---|---|---|---|---|---|---|
| 1 | Proposed (ns) | 0.06 | 0.32 | 0.50 | 0.71 | 0.80 |
| 2 | Existing (ns) | 0.13 | 0.42 | 0.62 | 0.82 | 0.90 |

Table 4 shows the delay values for the proposed and existing adder circuits. To generate 4-bit full adder carry output, the proposed design takes the delay for 80 ps and the existing one needs 90 ps. All sum outputs in four-bit proposed adder take less delay compared to the existing adder. Delay measurements are based on the $A$ = "0000", $B$ = "1111", $C_{in}$ = 1'b0.

## 6 Power Analysis

Power is consumed in the circuit with number of node transitions in a circuit from '0' to '1' or '1' to '0'. Power consumption can be decreased by reducing number of transitions in the circuit. To activate the required transition, necessary test vectors can be applied to control the node transitions. Table 5 shows the average power consumption measured for different vectors applied to proposed and existing adders.

Figure 6 shows the power-consuming events and for the first clock cycle all the inputs were '0' to force all the outputs to '0'. Test vectors generated during the second pulse make all the outputs to a high state. In the second pulse, test vector is generated such that all outputs go high so that we can have power-consuming events. To activate several portions of the circuit, test vector is generated to activate events on the outputs. In a similar way, test vectors were generated for eight cycles. Table 5 shows that the proposed adder consumes less power compared to the existing design.

**Table 5** Power consumption with input test vectors

| S. No. | Input | Test vector | Proposed adder (mw) | Existing full adder (mw) [9] |
|---|---|---|---|---|
| 1 | A3 | 01,010,101 | 2.260500 | 2.998898 |
| 2 | A2 | 01,010,110 | | |
| 3 | A1 | 01,001,100 | | |
| 4 | A0 | 01,001,111 | | |
| 5 | B3 | 01,010,101 | | |
| 6 | B2 | 01,001,001 | | |
| 7 | B1 | 01,010,011 | | |
| 8 | B0 | 01,010,000 | | |
| 9 | $C_{in}$ | 01,001,010 | | |

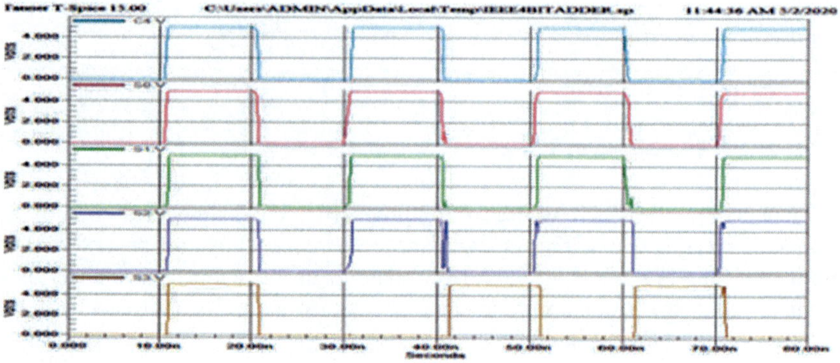

**Fig. 6** Power-consuming events for 4-bit full adder

# 7 Conclusion

4-bit adder has been evaluated for power and delay. The proposed adder optimizes different metrics like power and delay. From the simulation results, it can be concluded that the proposed design could achieve 10% reduction in the delay, a 25% reduction in the power, and 33% reduction in the PDP (Power delay product) compared to the existing adder design. With the complemented logic used in the proposed design, it requires 88 transistors to implement the 4-bit adder design. Whereas the existing 4-bit adder requires 104 transistors to implement the same. Hence it can be concluded that the proposed design is better compared to the existing design in terms of power, speed, and area than the conventional one. The proposed design can be extended to N-bit ripple carry adder, since the use of static CMOS logic for designing the carry part of the logic.

# References


1. A. Vesterbacka, 14-transistor CMOS full adder with full swing nodes, in *Proceedings of IEEE Workshop on Signal Process Systems*, Taipei, Taiwan, pp. 22–22 (1999)
2. P. Bhattacharyya, B. Kundu, S. Ghosh, V. Kumar, A. Dandapat, Performance analysis of a low-power high-speed hybrid 1-bit full adder circuit. IEEE Trans. Very Large Scale Integr (VLSI) Syst **23**, 2001–2008 (2015)
3. M. Aguirre-Hernandez, M. Linares-Aranda, CMOS full-adders for energy-efficient arithmetic applications. IEEE Trans. Very Large Scale Integrat. (VLSI) Syst. **19**, 718–721 (2010)
4. P. Kumar, R.K. Sharma, Low voltage high performance hybrid full adder. Eng. Sci. Technol. Int. J. **19**, 559–565 (2016)
5. S. Wairya, R.K. Nagaria, S. Tiwari, New design methodologies for high-speed low-voltage 1 bit CMOS Full Adder circuits. Int. J. Comput. Technol. Appl **2**, 190–198 (2011)
6. I. Hassoune, D. Flandre, I. O'Connor, J.D. Legat, ULPFA: a new efficient design of a power-aware full adder. IEEE Trans. Circ. Syst. I, Reg. Pap. **57**, 2066–2074 (2008)

7. C.H. Chang, J. Gu, M. Zhang, A review of 0.18-μm full adder performances for tree structured arithmetic circuits. IEEE Trans. Very Large Scale Integrat. (VLSI) Syst. **13** (2005)
8. S.R. Chowdhury, A. Banerjee, A. Roy, H. Saha, A high speed 8 transistor full adder design using novel 3 transistor XOR gates. Int. J. Electron. Circuit. Syst. **2**, 217–223 (2008)
9. N. Hamed, T. Somayeh, Low-power and fast full adder by exploring new XOR and XNOR gates. IEEE Trans. Very Large Scale Integrat. (VLSI) Syst. **26**, 1481–1493 (2018)

# Determining the Number of Bit Encryption That Is Optimum for Image Steganography in 8 Bit Images

**Amisha Agarwal and Avinash Tandle**

**Abstract** The most challenging task is to hide an image into another without the viewer getting an indication of the presence of secret data while looking at it. The message could be in any form of digital media like text, codes, or images. The paper proposes image steganography using LSB (Least Significant Bit) encryption. This technique can be applied to install secret messages in an 8-digit, 24-bit, or grayscale structure. In our research, 8-bit format has been used. The transmitter encrypts the secret image with respect to cover image to form stego image. Receiver decrypts the encrypted image by dividing 8 bit of secret image into '$x$' bit of the cover image and '8-$x$' bit of secret image. The paper has also evaluated the PSNR (Peak Signal to Noise Ratio) of retrieved secret and cover image with respect to its original secret and original cover image. As a result, this paper explains 4-bit encryption is best for the purpose of steganography as PSNR of retrieved images (cover and secret both) is highest here.

**Keywords** Steganography · Image · LSB · Encryption · Decryption · PSNR

## 1 Introduction

The significance of steganography is to make a message unnoticeable to the third party so he/she doesn't try to decrypt it. But once its existence is known it can be decrypted by a third party. This technique has to be done with a precision that the attacker doesn't know about the presence of such data. It merely covers or hides the existing data with the intention that no one knows that there is some data however steganography doesn't secure the data. Cryptography on the other hand is related to

A. Agarwal (✉) · A. Tandle
Department of Electronics and Telecommunication, Mukesh Patel School of Technology, Management & Engineering, Mumbai, India
e-mail: agarwal.amisha02@nmims.edu.in

A. Tandle
e-mail: Avinash.Tandle@nmims.edu

 
S. Patnaik et al. (eds.), *Smart and Sustainable Technologies: Rural and Tribal Development Using IoT and Cloud Computing*, Advances in Sustainability Science and Technology,
https://doi.org/10.1007/978-981-19-2277-0_15

the security of the data [1]. In cryptography, message is in form of secret code or cyphered using a certain algorithm and then sent to receiver [2].

Information break is a significant network protection issue that has caused enormous monetary misfortunes and compromised numerous peoples' security. The Privacy Rights Clearinghouse (PRC) reports 9015 information leak in 2005 to 2019, representing 11,690,762,146 records. The Identity Theft Resource Center and Cyber Scout reports 1244 information leak episodes in 2018, uncovering 446,515,334 records, which is a lot higher (or a 126% leap) from the 197,612,748 records uncovered in 2017. As indicated by NetDiligence, for little to-medium ventures (i.e., under $2 billion in yearly income), the mean breach cost from 2014 to 2018 is $178 K [3]

Nowadays, network has an important role in transferring data accurately and fast from the receiver to the transmitter. The data and the channel cannot be considered secure enough to transmit confidential information. The security of data has now become one of the principle challenges of resource sharing with data communication over computer networks. An image has become a widely used feature in multimedia now and in an individual's day-to-day life. Steganography is a hidden communication methodology that means: "covered writing" [4, 5].

Steganography is the technique for concealing privileged information inside any type of advanced media. When an individual views the article wherein the data is covered up inside, the person will have no sign that there is any secret data [6]. So the individual won't attempt to decode the data. When we talk about image (a form of signal) steganography, the aim is that since images are made up of pixels and we play with the bits of that particular pixel. In the greyscale images, the pixel values vary from 0 to 255, 0 being black and 255 being white [7]. Imperceptibility or in simpler terms transparency or anti-detection performance is another important term used in steganography. Imperceptibility is improved when secret and cover information has good matching relationship [1].

A ton of procedures are utilized for concealing the data in pictures. Least Significant Bit (LSB) is the traditional strategy used to conceal data within a picture [8]. Since the Steganographic techniques have progressed, the LSB Steganography strategies in coloured pictures have become conceivable which were at first restricted distinctly to grayscale pictures [9].

## 2 Various Techniques of Image Steganography

### 2.1 LSB Method

For concealing a picture into a picture, LSB-based calculation replaces the last '$x$' bits of the cover picture pixels with each up to '$x$' maximum significant bits of secret picture bits. As we increasingly hide more information into a picture by utilizing LSB, it actuates more commotion and subsequently covers picture's goal (Quality and appearance of picture) is changed. The security of data then becomes less as it gets simpler to distinguish the message by a person, and can be done by taking out the LSBs of the pixels [10]. LSB strategy is a helpful method to insert data in pictures, however, the information can be effectively decoded. This strategy can be used for both GIF (Graphics Interchange Format) and PNG (Portable Network Graphics) formats [11].In our experiment, we have used PNG format images. Literature suggests that JPEG (Joint Photographic Experts Group) format performs best especially as cover image since distortion is minimum as compared to other image formats. This is true for both LSB and MSB (Maximum Significant Bit) steganography. Substitution of 1 LSB showed a difference of 0.081 in the green channel [12].

This maximum number of bits that a pixel can accommodate is known as capacity per pixel (Bits per Pixel) [13]. The technologies nowadays have gone too advanced, hence instead of hiding a secret message in a grayscale image, hiding it in a colour image is much preferable. The reason behind it is because colour image provides more spaces, or we call it bits, which allows the existence of a message almost undetectable as it only causes a minor change in the original image [14].

## 2.2 *Palette-Based Technique*

Picture which is based on palette techniques comprises of colour indexes and colour palette. In an image, the list of colour pixels is contained by colour palette while colour palettes have pointers which are known as colour indexes. This tells the RGB (Red Green Blue) colour in the image. The secret data in this technique can be hidden in the palette's bits or data of the image. This method arranges the colour in the image palette in a particular way along with secret data. With this technique, HVS (Human Visual System) attack could be prevented. If the chosen cover image is not very similar to the neighbouring palette pixel then this technique can result in a completely different colour pixel. Therefore it is necessary to choose the right cover image [15].

## 2.3 *Secure Cover Selection*

It is a complicated approach in which the hackers compare the blocks of the carrier picture to the blocks of their particular malware. If an image is having similar blocks like malware, it is selected as the candidate to take the malware with itself. The same malware blocks are then precisely put into the carrier photo. As a result, the picture is identical to the original but this photo is not marked as a danger by the detection software and programs.

## 3 Terms Used

For the purpose of this proposed approach, we will come across a few terms like cover image, secret image, stego image, retrieved cover image and retrieved secret image [9]. Cover image is the original image inside which the information is to be concealed which is precisely our secret image. The combination of these two images where only the cover image is visible to the naked eyes is known as stego image [16]. This 8-bit stego image is then broken down into x bits of MSB (Maximum Significant Bits) and '8-x' bits of Least Significant Bit which are known as retrieved cover image and retrieved secret image respectively.

## 4 Proposed Method

In our paper, we are using LSB technique. We know that LSB technique is one of the classic methods for steganography but this research addresses the gap as to how many bit encryption should be done to get optimum results. With optimum result, we mean that the secret image should be so inconspicuous. If the third party is not getting suspicious about the data having secret content they will not try to decrypt it. So its execution is not so complicated but execution in the right way is.

The method is segregated into two parts: 'encryption' and 'decryption'. We start with encryption, i.e. combining the cover image with the secret images that are to be hidden. For the same, the images of the same size are required. The reason behind this is that an image is a form of matrix and if the dimensions are not appropriate then matrix calculations are impossible and code will yield an error.

Into a new image, the MSBs of the cover image and MSBs of secret image are taken. On merging these bits we get our stego image. The receiver receives the stego image. This stego image is then broken down into two images by dividing its 8 bits into $x$ and 8-$x$ bits. The '$x$' maximum significant bits become the retrieved cover image whereas the '8-$x$' least significant bits become the retrieved secret image. The reason behind LSB being the most used method for steganography is because of its ease of use. If we change one of the MSB of a pixel, the change in information is very high and apparent. The change is even visible to the naked eyes. But if we change one of the LSB of a pixel, the change comes out to be minor and unnoticeable [10].

Algorithm to embed secret images using LSB is:

1. Input the cover and secret image
2. Retrieve first 4 MSB of each pixels of secret image
3. Retrieve first 4 MSB of each pixels of cover image
4. Replace LSB of cover image with retrieved MSB of secret image
5. Output the stego image.

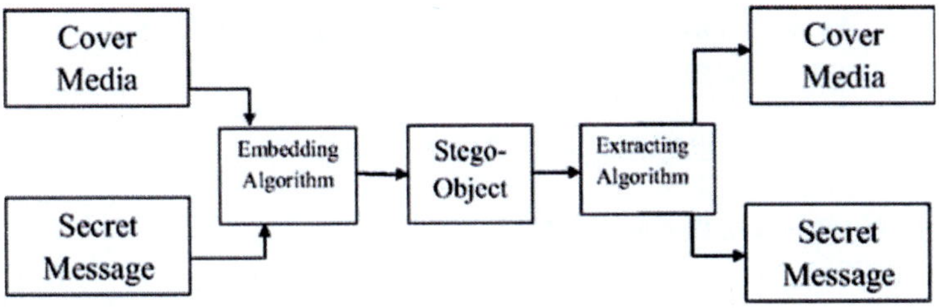


**Fig. 1** Block Diagram for steganography [5]. *Source* https://link.springer.com/article/10.1007/s11042-019-7507-6

Steps to decrypt secret image are:

1. Input the stego image
2. Extract initial 4 MSB of every pixel of stego picture
3. Extract initial 4 LSB of every pixel of stego picture
4. Separate them as individual images
5. Output the retrieved secret and cover image.

In Fig. 2; 2 bits of the cover image and 6 bits of secret image have been added to form 8 bit stego image. After decryption 4–4 bits have been separated to get 'Retrieved Cover Image' and 'Retrieved Secret Image'.

In Fig. 3; 4 bits of the cover image and 4 bits of secret image have been added to form 8 bit stego image. After decryption 4–4 bits have been separated to get 'Retrieved Cover Image' and 'Retrieved Secret Image'.

In Fig. 5; 6 bits of the cover image and 2 bits of secret image have been added to form 8 bit stego image. After decryption 4–4 bits have been separated to get 'Retrieved Cover Image' and 'Retrieved Secret Image'.

**PSNR**

Peak signal to noise ratio is the ratio of highest power possible and noise that impacts the originality of the picture. For PSNR, MSE (Mean Square Error) has been calculated using the following formula [17].

$$\text{MSE} = \frac{1}{n}\sum_{i=1}^{n}(y_i - \widetilde{y}_i)^2$$

MSE is the mean of the squared difference between original images and compressed images.

This has been further used to find PSNR and the following formula has been used in the python program.

$$\text{PSNR} = 20\log 10(\text{MAX}/(\text{MSE})^{(1/2)})$$

MAX is the maximum value a pixel can have which is 255 (8-bit unsigned int). Hence, the formula becomes: -

$$\text{PSNR} = 20 \log 10(255/(\text{MSE})^{(1/2)})$$

[Note: If MSE = 0, PSNR = 100].

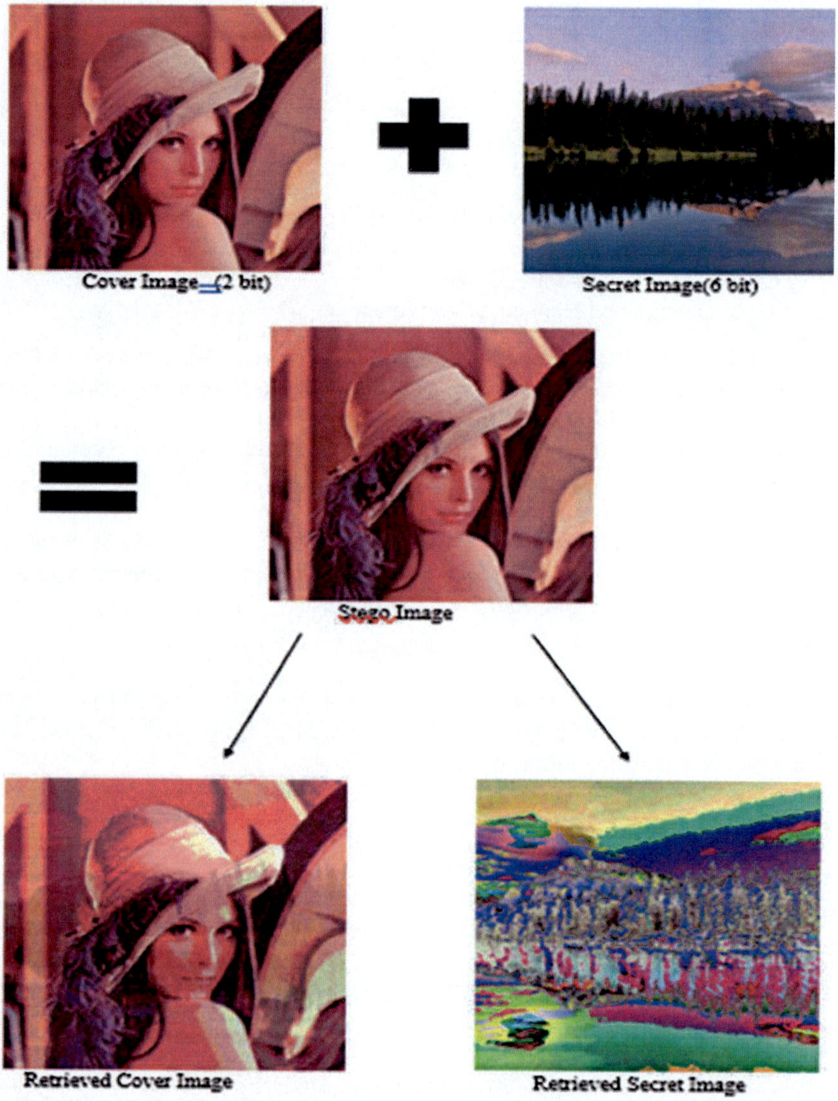


**Fig. 2** 2–6 bit encryption

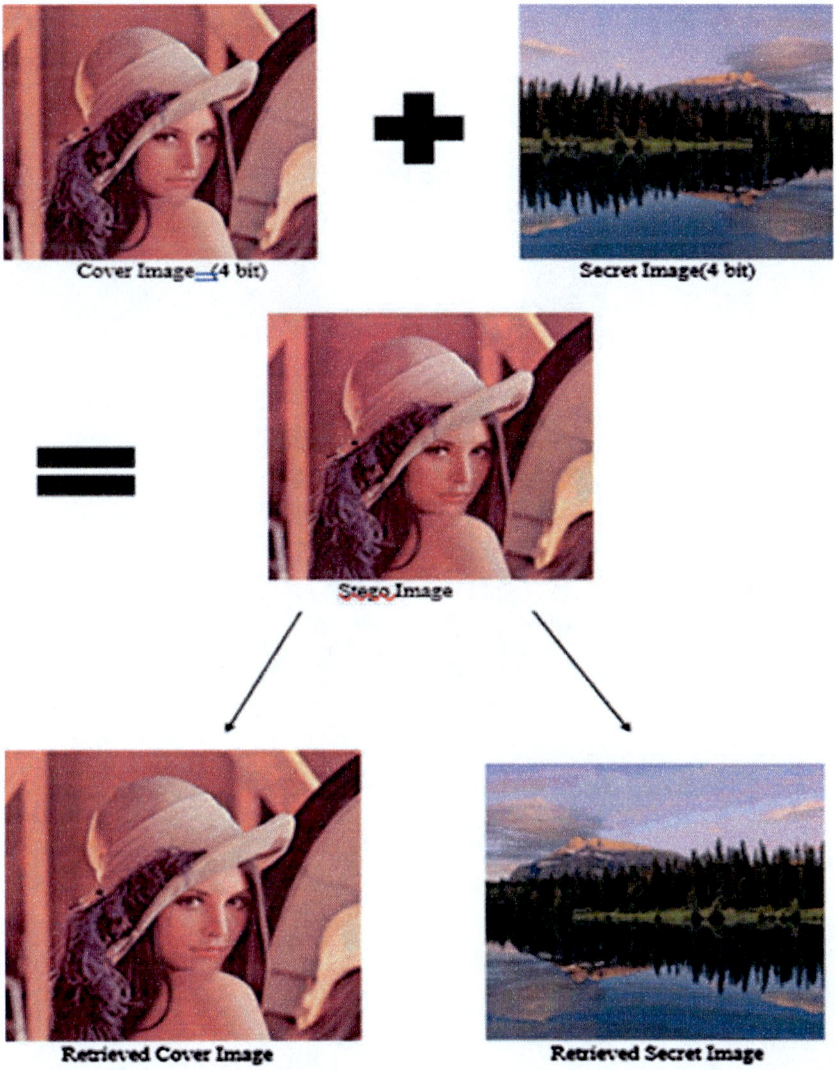


**Fig. 3** 4–4 bit encryption

## 5 Result

In the initial results of stego image, we notice that traces of secret image is visible even though it is in the Least Significant Bit. This is because an excess number of bits have been used from the secret image. In the last stages of stego images, we notice that it is very much similar to the cover image and that is because excess number of bits have been used from the cover image (Fig. 7).

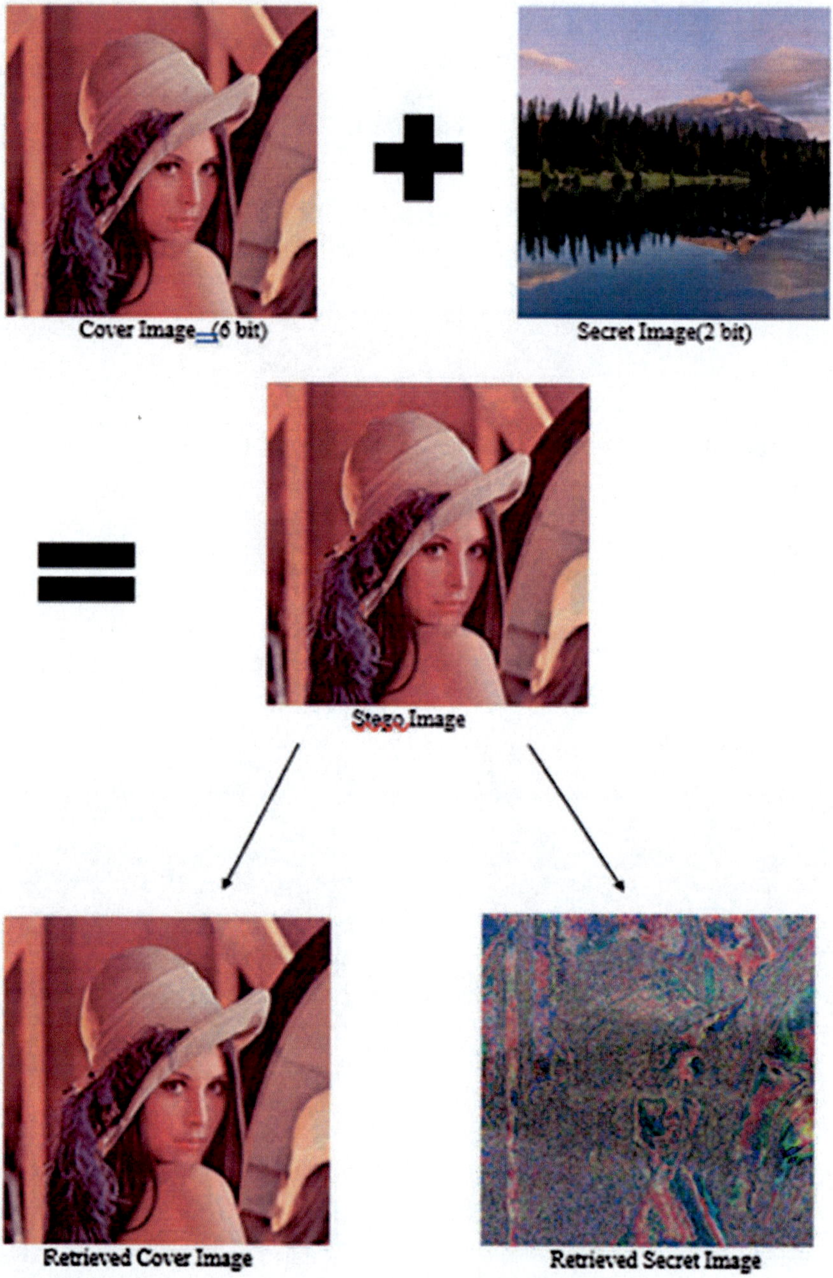


**Fig. 4** 6–2 bit encryption

**Fig. 5** Retrieved Cover images for 1–8 bits

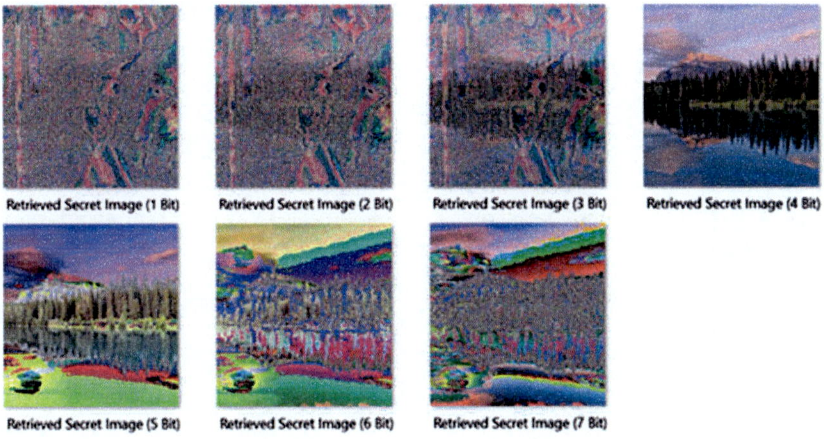


**Fig. 6** Retrieved Secret images for 1–8 bits

We have now calculated the PSNR of the retrieved image with respect to its original image and plotted it for 1–7 bits (Figs. 8 and 9; Tables 1 and 2).

## 6 Discussion

The results reflect that the above-explained technique is efficient to hide data without too much distortion. It is not easy for an unauthorized user to detect such minor differences in the stego image, especially with naked eyes [16].

**Fig. 7** Stego images for various bit combination

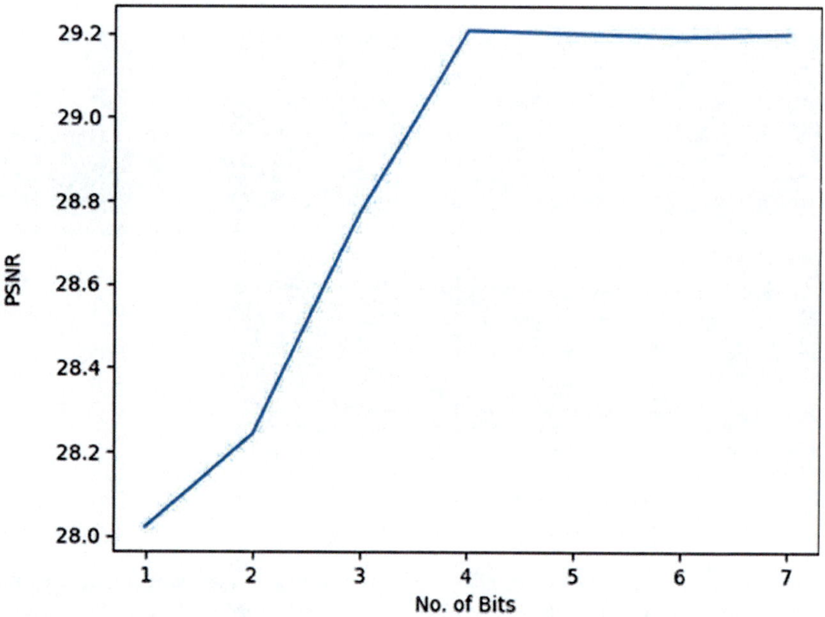


**Fig. 8** PSNR of 'Original Cover Image' with respect to 'Retrieved Cover image' plotted against number of bits

After extracting the hidden image the resolution is affected. The nature of the recovered cover picture and recovered secret picture is comparatively distorted [9]. This happens because to form the stego image we are using x and '8-x' MSBs of cover and secret image which are also originally of 8 bits. This means that we are

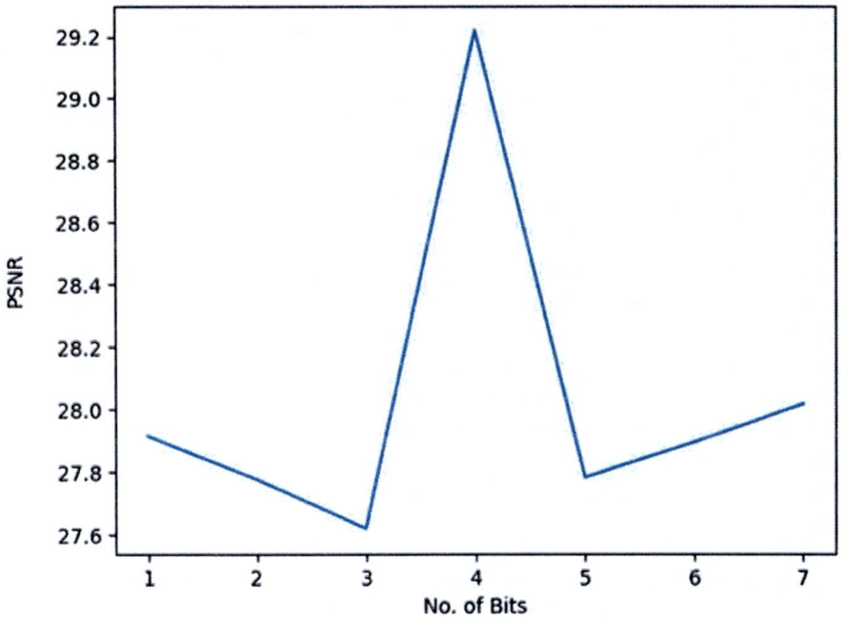


**Fig. 9** PSNR of 'Original Secret Image' with respect to 'Retrieved Secret Image' plotted against number of bits

**Table 1** PSNR of 'Original Cover Image' with respect to 'Retrieved Cover image'

| Retrieved cover image bit ($x$) | PSNR |
|---|---|
| 1 | 28.022 |
| 2 | 28.243 |
| 3 | 28.774 |
| 4 | 29.207 |
| 5 | 29.200 |
| 6 | 29.193 |
| 7 | 29.199 |

**Table 2** PSNR of 'Original Secret Image' with respect to 'Retrieved Secret Image'

| Retrieved secret image bit ($x$) | PSNR |
|---|---|
| 1 | 27.913 |
| 2 | 27.774 |
| 3 | 27.617 |
| 4 | 29.219 |
| 5 | 27.780 |
| 6 | 27.894 |
| 7 | 28.015 |

losing out some LSBs of both the images to form stego image and hence the minor details are lost. This further affects the quality of the picture but not so much that it cannot be identified.

From the results above refer to Fig. 5, we notice that the retrieved cover images that the quality improve gradually. This happens because the number of bits in stego image is increasing leading to its greater impact on stego image. When stego image is decrypted, images having a greater impact on cover image have clearer retrieved cover image. Similarly in Fig. 6, the main point of focus is that 4-bit encryption has the best result for retrieved secret images. From 1 to 3 bits we see traces of cover picture in the retrieved secret image. This is because the impact of cover images is on higher side on the stego image and while decrypting 4–4 bits, some traces of cover images have gone in secret image also. Let us now notice Fig. 7, it shows stego images for various bits combination. The main intention of stego image is to look as similar as a cover image. That is better fulfilled when bits of cover images are maximum hence the last (c7_s1) looks best. But doing this is leading to a compromise on secret image and secret image no longer remains itself when bits of cover image is more in stego image.

From the graphs, we get that for cover image PSNR value goes up till 4 bits and then becomes constant. For secret image PSNR goes up till 4 bits and then comes down. Hence in both images PSNR is highest for 4 bits.

## 7 Conclusion

This work defines and simulates a steganography system that can be used practically for sending image type data securely. The proposed strategy tells us how to conceal the picture and extract (decrypt) it with the most minimal expense as far as loss in the data is concerned. Experiment was done to find out number of bits best suited for steganography. Encryption was done using different number of bits for cover and secret image. Decryption was done for 4–4 bits of cover and secret image. PSNR value was calculated for cover and secret images and graphs are plotted.

From all the discussion above done using PSNR value, we can say that 4-bit encryption is the best (for 8-bit pixel image) when it comes to LSB as it helps to maintain a balance between quality of the retrieved images along with hiding it in the Least Significant Bits.

## References

1. K. Tiwari, S.J. Gangurde, LSB steganography using pixel locator sequence with AES, in *2021 Second International Conference on Secure Cyber Computing and Communication (ICSCCC)* (2021)

2. C. Patsakis, N.G. Aroukatos, LSB and DCT steganographic detection using compressive sensing. J. Inf. Hiding Multimedia Sig. Proces. **5**(1) (2014). ISSN 2073-4212
3. Z. Fang, M. Xu, S. Xu, Senior Member, IEEE, T. Hu, A framework for predicting data breach risk: leveraging dependence to cope with sparsity. IEEE Trans. Inf. For. Secur. **16** (2021)
4. M. Sharma, Steganography is the art of hiding data. IJARSE 4 (2015)
5. G. Murugan, R. Uthandipalayam Subramaniyam, Performance analysis of image steganography using wavelet transform for safe and secured transaction. Multimed Tools Appl. **79**, 9101–9115 (2020)
6. S. Channalli, A. Jadhav, Steganography an art of hiding data. Int. J. Comput. Sci. Eng. **1**(3), 137–141 (2009)
7. R.J. Mstafa, C. Bach, information hiding in images using steganoraphy techniques, in *Conference: Northeast Conference of the American Society for Engineering Education (ASEE)*. At: Norwich University David Crawford School of Engineering. https://doi.org/10.13140/RG.2.1.1350.9360
8. O. Elharrouss, N. Almaadeed, S. Al- Maadeed, *An Image Steganography Approach Based on k-Least Significant Bits (k-LSB)*. Department of Computer Science and Engineering (2020). IEEE, 978-1-7281-4821-2
9. R.K. Thakur, C. Saravanan, Analysis of steganography with various bits of LSB for color images. In: International conference on electrical, electronics, and optimization techniques (ICEEOT) (2016), 978-1-4673-9939-5
10. A. Singh, H. Singh, Animproved LSB based image steganography technique for RGB images, in *2015 IEEE International Conference on Electrical Computer and Communication Technologies (ICECCT)* (2015)
11. K.Thangadurai, G.Sudha Devi, An analysis of LSB based image steganography techniques, in *2014 International Conference on Computer Communication and Informatics (ICCCI-2014)* (IEEE, 2014). 978-1-4799-2352-6
12. L.K. Gupta, A. Singh, A. Kushwaha, A. Vishwakarma, Analysis of image steganography techniques for different image format, in *2021 International Conference on Advances in Electrical, Computing, Communication and Sustainable Technologies (ICAECT)*
13. A.K. Singh, J. Singh, H.V. Singh, Steganography in images using LSB technique. Int. J. Latest Trends Eng. Technology (IJLTET) **5**(1) (2015). ISSN: 2278-621X
14. J. Fridrich, M. Long, Steganalysis of LSB encoding in color images, in *2000 IEEE International Conference on Multimedia and Expo. ICME2000. Proceedings. Latest Advances in the Fast Changing World of Multimedia* (Cat. No. 00TH8532), vol. 3 (IEEE, 2000), pp. 1279–1282
15. L. Rura, B. Issac, M.K. Haldar, Analysis of image steganography techniques in secure online voting, in *2011 International Conference on Computer Science and Network Technology*
16. S. M. Masud Karim, Md. Saifur Rahman, Md. Ismail Hossain, A new approach for LSB based image steganography using secret key, in *Proceedings of 14th International Conference on Computer and Information Technology (ICCIT 2011)*, 22–24 Dec 2011. 987-161284-908
17. N. Jain, S. Meshram, S. Dubey, Image steganography using lsb and edge – detection technique. Int. J. Soft Comput. Eng. (IJSCE) **2**(3) (2012). ISSN: 2231-2307

# Design and Analysis of 2nd-Order Bandpass Filters Using SIW and Microstrip Patch Transitions with Stub Matching

**Rashmita Mishra, Subhrajit Pradhan, and Kailash Chandra Rout**

**Abstract** The analysis of two designs of bandpass filters using substrate integrated waveguide (SIW) and microstrip patch transitions are presented. The SIW based 2nd order filter is having two transmission zeros from 12 to 18 GHz range with a 50-Ω input impedance. By four metallic walls, two-cavity resonators are created to obtain the filtering characteristics. In the second design using patch transition arrangements with a semicircular patch added with stubs, the filtering performances are suitably characterized. The maximum measured fractional impedance BW is 49.66% for Design 1 and 22.74% for Design 2, respectively. The maximum insertion loss (IL) is 0.22 Db using the SIW, technique. Group delay (GD), phase delay (PD), impedance bandwidth (IBW), scattering parameters, and return loss are verified. A comparison analysis is presented showing the designs are suitable candidates for Ku-band X-band applications.

**Keywords** Resonator · SIW · Fractional IBW

## 1 Introduction

With the fast progress of current wireless communication systems over the last several years, antennas and filters have emerged as one of the most essential components of wireless Radio-Frequency (RF) systems. The bandpass filter (BPF) accepts signals within the operational bandwidth and rejects out-of-band signals taking a significant role in the front-end and subsystems. There is always a high demand for making the

R. Mishra (✉) · S. Pradhan
Department of Electronics and Communication Engineering, GIET University, Odisha 765022, India
e-mail: rasmita.mishra@giet.edu

S. Pradhan
e-mail: subhrajitpradhan@giet.edu

K. C. Rout
Department of Electronics and Communication Engineering, Capital Engineering College, Bhubaneswar 752055, India

 
S. Patnaik et al. (eds.), *Smart and Sustainable Technologies: Rural and Tribal Development Using IoT and Cloud Computing*, Advances in Sustainability Science and Technology,
https://doi.org/10.1007/978-981-19-2277-0_16

RF systems compact, simple, and less costly. In light of the requirement for miniaturizing and improved performance characteristics in microwave front-ends, the development of multifunction structures should be a viable option for shrinking the size of these front-end devices. Subsequently, it results in improved microwave front-end performances. SIW has been proved as a prominent technology in designing filters, resonators, and antennas due to its high selectivity, Q-factor, and wide Impedance Bandwidth (IBW) [1].

A compact size SIW based BPF is designed in [2] with low insertion loss. However, the impedance bandwidth is lower and is less than 15%. To improve the bandwidth, common mode and differential mode techniques are used in [3]. A cross-shaped SIW structure is implemented in [4] to overcome the insertion loss and low bandwidth. Several studies on BPFs and filtering antenna design have been reported so far. The two methods which are extensively followed can be described as follows. (a) The co-design method follows the filter synthesis process where the antenna serves as both a radiating element and a final stage resonator [5, 6]. Although this filter synthesis approach exhibits good selectivity, several resonators take up a lot of space and due to high Insertion Loss, gain degrades. (b) Designing the antenna and filter separately and then integrating both of them using impedance matching networks. Tang et al. [7] developed a planar, wideband, and filter structure with 23.05% IBW and two transmission zeros. It was realized using a driven rectangular patch and slots made in the ground plane. The slots helped in widening the operating bandwidth by introducing two radiation zeros in the realized gain response. By loading a T-stub in the ground plane helped in adding one reflection zero and improving the out-of-band selectivity [8].

This work proposes two second-order BPFs analyzing IBW, IL, and GD. The SIW based Design 1 is having two transmission zeros, low insertion loss, and wide IBW with compact size. Design 2 is having a semicircular patch resonator with rectangular slots in the ground plane. Two stubs with semi-circular patches give vital characteristics like wide bandwidth and low insertion loss. Similarly, four open-loop square resonators along with a metallic via are integrated with a rectangular patch radiating element through an open stub which is capacitively coupled. Since the proposed second-order filters are highly selective, thus can be a useful design for microwave applications.

Rest of the paper is prepared as follows: the design parameters of the proposed structures are described in Sects. 2 and 3. Section 4 explains the results and compares the features of Design 1 and Design 2. Section 5 concludes the article followed by an acknowledgment with essential references.

## 2 Design-1

To design a SIW based BPF some important features are considered here [9].

- Required frequency: Ku band with a cutoff frequency at 16.5 GHz.

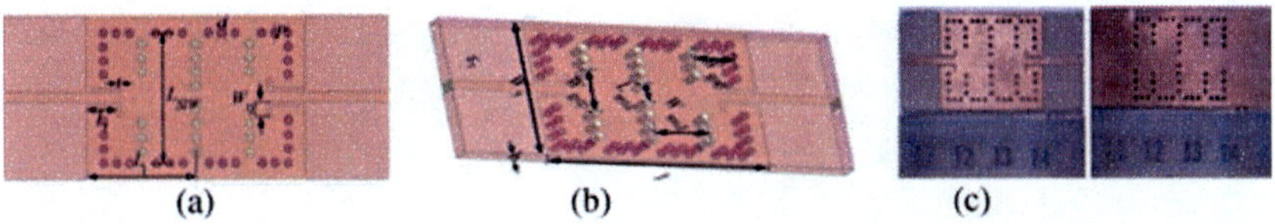


**Fig. 1** Design and fabrication of the design 1. **a** Isometric view of design 1, **b** front view, and **c** fabricated prototype

- Material: Rogers 4232 tm with 3.2 is the dielectric constant having a loss tangent of 0.018 and height = 1.524 mm.
- Vias diameter d = 1.0 mm.

Tuning of parameters like the gap between the vias, length, and width of the cavity in Design 1 is essential to deciding the characteristics of the filter. The design and fabrication are displayed in Fig. 1a, b, c.

Using HFSS 17 the structure is simulated. The parameters are $h$ = 1.524 mm, $y$ = 0.478 mm, €$r$ = 3.2, $W_{\text{gap}}$ = 4.676 mm, $L_{\text{mid}}$ = 2.65, $W_{\text{mid}}$ = 6.524 mm, $W_y$ = 16.65, $L_x$ = 26 mm, $W_{tg}$ = 1.55 mm, $p$ = 1.6 mm, LSIW = 14.8, $d$ = 1.0, $t$ = 0.4 mm, $L_t$ = 13 mm. The $S$ parameter is shown in Fig. 2 shows the effect of $L_{\text{mid}}$ on the $S$-parameter. The surface current, $S$-parameter, the phase plot, and the group delay plot of Design 1 are shown in Figs. 3, 4, 5, and 6, respectively. The impedance of the feed line is kept constant at over 50 Ω. The cutoff frequency is defined according to the length and width of the cavity of the SIW design which is denoted by LSIW and WSIW, respectively [10, 11].

$$f_0 = \frac{c}{2\sqrt{\epsilon_r}}\sqrt{\frac{1}{(L_{\text{eff}})^2} + \frac{1}{(W_{\text{eff}})^2}} \quad (1)$$

where $L_{\text{eff}}$ and $W_{\text{eff}}$ are the length and width of the cavity of the SIW structure.

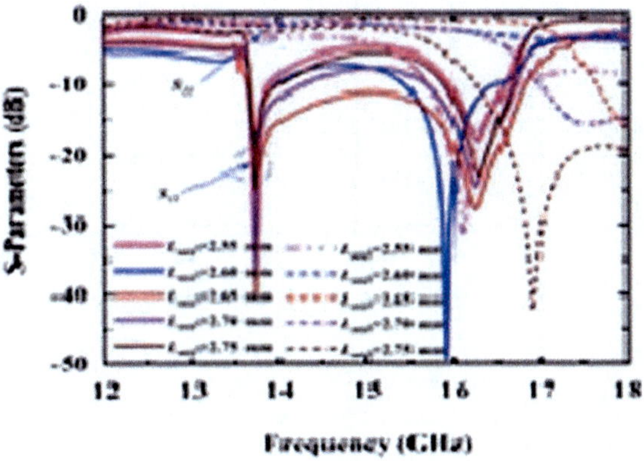


**Fig. 2** Effect of $L_{\text{mid}}$ on $S$-parameter

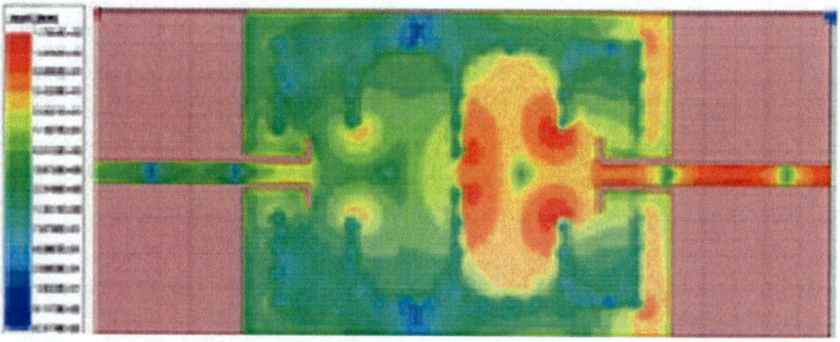

Fig. 3 Surface current distribution of Design 1

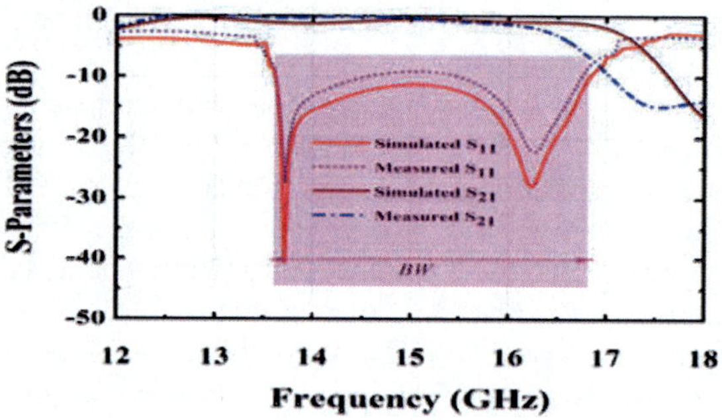


Fig. 4 *S*-parameter of Design 1

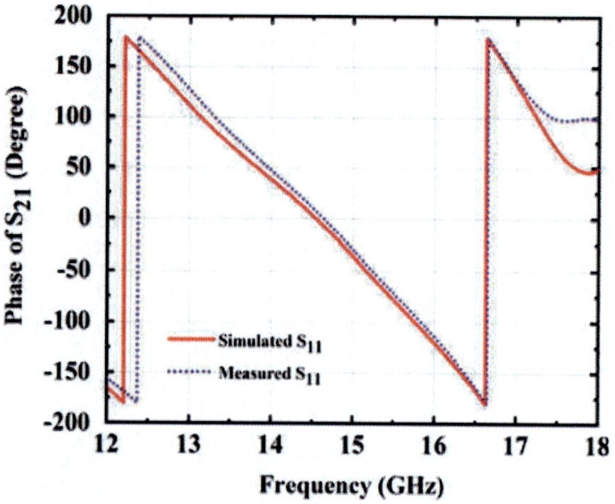


Fig. 5 Phase plot of Design 1

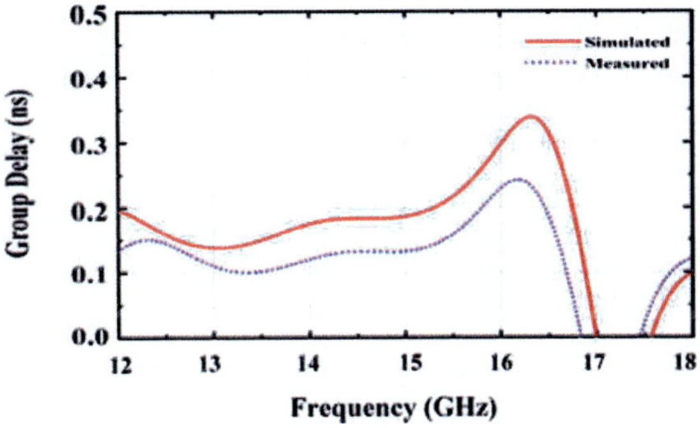


**Fig. 6** Group delay for Design 1

# 3 Design 2

The proposed 2nd order filter is designed by taking FR4 substrate with a semicircular patch and defected ground structure as displayed in Fig. 7a, b. The details of dimensions of the design are shown in Table 1.

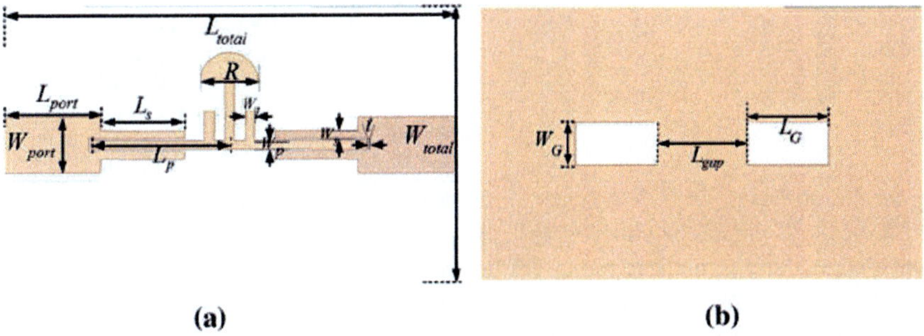


**Fig. 7** Design 2 antenna. **a** Front view, and **b** Rear view

**Table 1** Font sizes dimensional parameters of design 2

| Parameters | Size (mm) | Parameters | Size (mm) |
|---|---|---|---|
| $W_{\text{total}}$ | 13.5 | $R$ | 3.2 |
| $L_{\text{total}}$ | 22 | $R_1$ | 1.6 |
| $L_p$ | 7.14 | $t$ | 0.1 |
| $L_s$ | 4.29 | $W_l$ | 0.5 |
| $L_{\text{gap}}$ | 4.68 | $W_G$ | 2.2 |
| $L_G$ | 4.3 | $W_p$ | 0.4 |
| $H$ | 1.6 | $W_s$ | 0.4 |

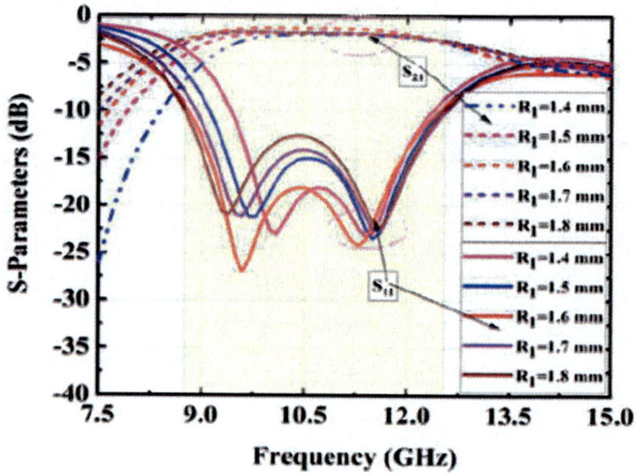


**Fig. 8** Effect of Gap on *S*-parameter

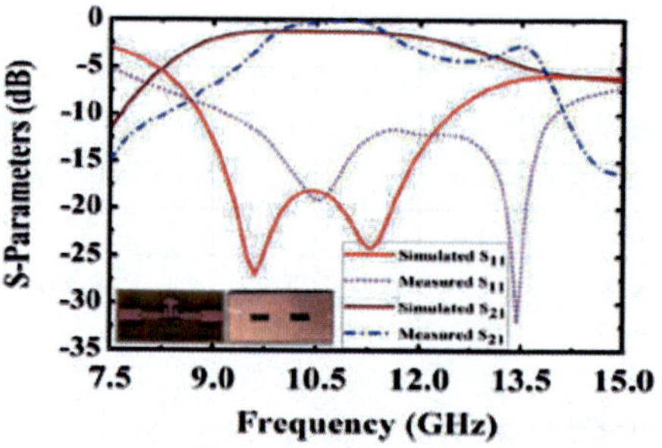


**Fig. 9** *S*-parameter of design 2

In the proposed Design 2, a semicircular patch and four open stubs are considered for the design. The stub length is equivalently taken as per half of the guided wavelength. The semicircle radius *R*1 holds an important parameter that influences the IBW. A parametric study on effect of gap on *S*-parameter of design 2 is shown in Fig. 8. The optimized values are shown in Table 1.

The fabricated design with the simulated result and *S*-parameter values are shown in Fig. 9. The resonance is obtained at 10.53 and 13.43 GHz. The deviation may be due to the fabrication and calibration errors. The surface current distribution plot is illustrated in Fig. 10. The simulated and measured phase plot and group delay plot are displayed in Figs. 11 and 12, respectively.

## 4 Results Analysis

In Design 1, two transmission zeros are obtained at 13.72 GHz, and 16.24 GHz. Two resonator cavities in the structure give TE101 and TE102 modes. The two resonators also play an important role in impedance matching purposes. The SIW based BPF gives a wide impedance bandwidth. The parametric analysis using the gap between

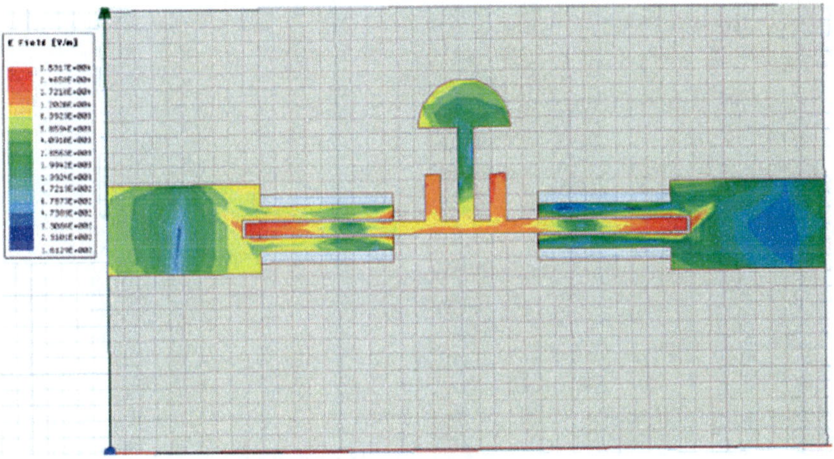

**Fig. 10** The distribution of surface current for the Design 2

**Fig. 11** Phase response of design 2

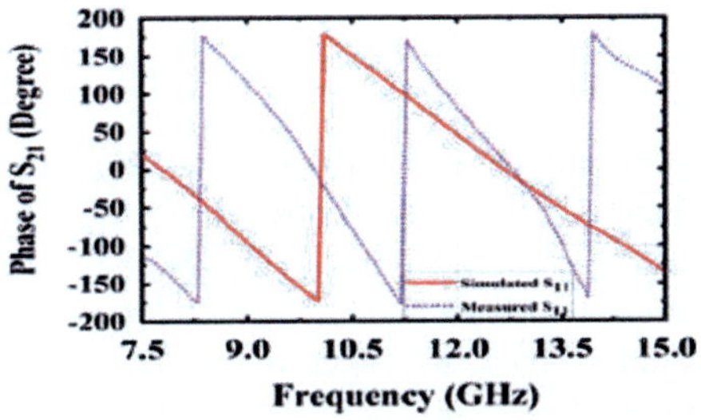


**Fig. 12** Group delay for design 2

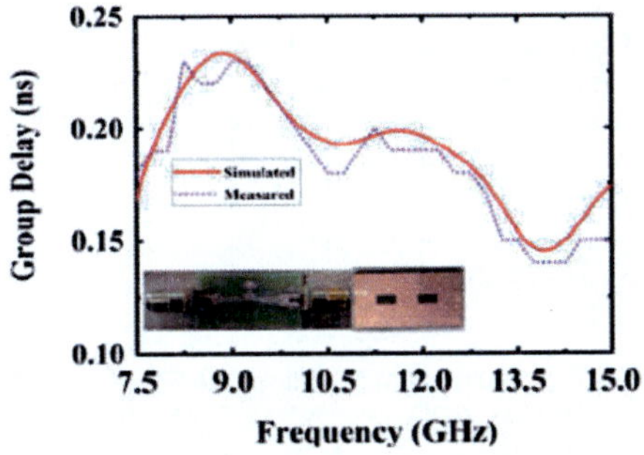


the vias is shown in Fig. 2. A constant value of 2.65 mm is chosen for $L_{\text{mid}}$. Figure 4 shows the simulated diagram and $S$-parameter values for design 1.

In Design 2, the 2nd order BPF is designed using a semicircular shaped patch with a defective ground structure. The phase plot and the group delay are shown in Figs. 11 and 12 where the simulated and the measured responses closely relate to

**Table 2** Analysis between design 1 and design 2

| Features | Design1 | Design2 |
|---|---|---|
| Material | Rogers 4232$^{tm}$ | FR4 |
| $\epsilon_r$ | 3.2 | 4.4 |
| Loss tangent | 0.018 | 0.02 |
| Dimension | $30 \times 16$ mm$^2$ | $13.5 \times 22$ mm$^2$ |
| Cutoff frequencies | 13.72 and 16.24 GHz | 10.53 and 13.43 GHz |
| Insertion loss | <0.22 dB | <2.12 dB |
| Fractional BW (maximum) | 22.74% | 49.66% |

each other. The analysis of features of Design 1 and Design 2 is shown in Table 2. Both the designs are successfully validated with measurement and hence are suitable for microwave applications.

## 5 Conclusions

In this article, two BPF designs are proposed and analyzed. The SIW based BPF is found fractional IBW values of 22.74% and 19.21% at the two resonant frequencies. The IBF values for the second-order BPF in design 2 at the corresponding resonant frequencies are 49.66% and 38.94%. The IL loss is found to be 0.22 dB for Design 1 and less than 2 dB for Design 2. The simulated and the measured results are closely similar to the proposed designs. The wider bandwidth and good frequency selectivity of the BPFs make them suitable for Ku-band and X-band applications.

**Acknowledgements** I want to express my gratefulness to BPPIMT's authorities for permitting me to pursue my Ph.D. at GIET University and for all necessary provisions.

## References

1. R.J. Cameron, C.M. Kudsia, R.R. Mansour, Microwave Filters for Communication Systems: Fundamentals, Design, and Applications (Wiley, NewYork, 2018)
2. D. Deslandes, M. Bozzi, P. Arcioni, W. Ke, Substrate integrated slab waveguide (SISW) for wideband microwave applications, in *IEEE MTT-S International Microwave Symposium Digest* (2003), pp. 1103–1106
3. X. Xu, J. Wang, L. Zhu, A new approach to design differential mode band pass filters on SIW structure. IEEE Microwave Wirel. Compon. Lett. **23**, 635–637 (2013)
4. P. Li, H. Chu, R. Chen, Design of compact band pass filters using quarter-mode and eighth-mode SIW cavities. IEEE Trans. Components Packaging Manuf. Technol. **7**, 956–963 (2017)
5. C. Chuang, S. Chung, Synthesis and design of a new printed filtering antenna. IEEE Trans. Antennas Propag. **59**, 1036–1042 (2011)

6. N. Nie, Z. Tu, Wideband filtering dumbbell-shaped slot antenna with improved frequency selectivity for both band-edges. IEEE Access **8**, 121479–121485 (2020)
7. M.C. Tang, C. Li, Y. Chen, Electrically small, planar, wideband filtering antenna with dual ring patches. Electron. Lett. **56**(2), 61–62 (2020)
8. T.N. Kuo, S.C. Lin, C.H. Chen, Compact ultra-wideband bandpass filters using composite microstrip–coplanar-waveguide structure. IEEE Trans. Microw. Theory Tech. **54**(10), 3772–3778 (2006)
9. D.M. Pozar, *Microwave Engineering* (Wiley, 2004)
10. M. Bozzi, A. Georgiadis, K. Wu, Review of substrate-integrated waveguide circuits and antennas. IET Microwaves Antennas Propag. **5**(8), 909–920 (2011)
11. Nayak, A. K., Yadav, V. S., Patnaik, A.: Wideband transition from tapered microstrip to corrugated SIW, in *IEEE MTT-S International Microwave and RF Conference (IMARC)* (2019), pp 1–4

# Planar Split Ring Resonating Antenna Design

**Raghuvu Budhisagar, G. Anudeep, Pritam Palo, and Ribhu Abhusan Panda**

**Abstract** This work is based on a unique design of planar antenna for 5G applications. The design is a resonating antenna prepared by split rings. There are four concentric rings placed on a substrate of dimension 40 mm × 50 mm with a thickness of 1.6 mm. Each ring consists of cuts arranged at 180° with one another. This novel design resonates at 13 GHz frequency with return loss of −17.5 dB with peak gain of 7.2 dB and radiation efficiency of 0.56. This design is suitable for 5G applications like satellite-based communication devices, RADAR, Military and Wireless devices.

**Keywords** 5G application · Peak gain · Resonant frequency · Return loss · Split rings

## 1 Introduction

In recent times, the communication-based technology has a vast growth because the technology has been improvised day by day. As Antenna is the basic part of the wireless communication devices, the design of antenna has taken center stage of improvisation. So, most of advanced research has been devoted to this section.

The microstrip antenna is a special design as it has advantages like lightweight, miniature, and low-cost. The foremost design of the patch of the antenna is of rectangular shape and circular shape. Later, these shapes have been modified according to

R. Budhisagar (✉) · G. Anudeep · P. Palo · R. A. Panda
Department of Electronics and Communication Engineering, GIET University, Gunupur, Odisha, India
e-mail: 18ece060.rbudhisagar@giet.edu

G. Anudeep
e-mail: 18ece034.ganudeep@giet.edu

P. Palo
e-mail: 18ece026.pritampalo@giet.edu

R. A. Panda
e-mail: ribhuabhusanpanda@giet.edu



S. Patnaik et al. (eds.), *Smart and Sustainable Technologies: Rural and Tribal Development Using IoT and Cloud Computing*, Advances in Sustainability Science and Technology,
https://doi.org/10.1007/978-981-19-2277-0_17

the need and outcome. One of the recent designs of the patch is split ring resonating antenna [1]. This design can have a resonant frequency as desired simply changing the number of rings. Further, a resonant frequency of the other antenna parameter like gain can be enhanced by the application of different methods. Sano et al. [2] have designed a split resonator antenna using small broad-band coupling method. Gandhi Mohan et al. [3] used Z-slot, whereas Khanna et al. [4] used metamaterials for high gain. Hosseini et al. [5] used triangular split ring while Sajith et al. [6] used loaded conductor-based CPW fed antenna. Pradeep et al. [7] had given a comparative study of two shapes of split rings. Again Ramachandran et al. [8] used a MIMO antenna with concentric square ring. Pandeeswari et al. [9] used loaded antenna whereas Sharma et al. [10] used slot antenna with the split ring resonator. Many more improvisation has been established with the split ring resonators in later time [11–13]. In recent past, some modifications have been done in the conventional shapes of antenna [14–20].

In this paper, the design of the split rings is altered with cuts placed at 180° to their adjacent ones. The design has resonant frequency of 13 GHz. As this is the frequency required for 5G communication, this design is hence efficient for wider applications.

## 2 Antenna Design

This design consists of four split rings with cuts at 180° with one another arranged in concentric manner. The radii of the rings are 15 mm, 11 mm, 7 mm, and 3 mm, respectively. The set of the rings is planar. So, placed on substrate of FR-4 epoxy material. The dimensions of the substrate are 40 mm × 50 mm × 1.6 mm. There is a ground plane underneath the substrate whose dimensions are 40 mm × 50 mm × 0.1 mm. The thickness of the both the patch and ground plane are same. The feed of width 2 mm is connected to the outer ring, at the opposite point of the cut (Fig. 1; Table 1).

## 3 Simulation Results

Here In order to prove that our design is having some advantages what we have done is we have found the simulation results at resonating frequency of 13 GHz. Figure 2 indicates the VSWR. Figure 3 indicates the S-parameter plot in which bandwidth is 600 MHz with the resonant frequency 13 GHz. Other simulated parameters like 2D radiation pattern, 3D radiation pattern, Surface current distribution, Peak Gain, Radiation Efficiency and Peak Directivity have been determined from HFSS software (Figs. 4, 5, 6, 7, 8 and 9).

The Simulation result of the proposed antenna has been done by using HFSS software. The simulated parameters are listed in Table 2

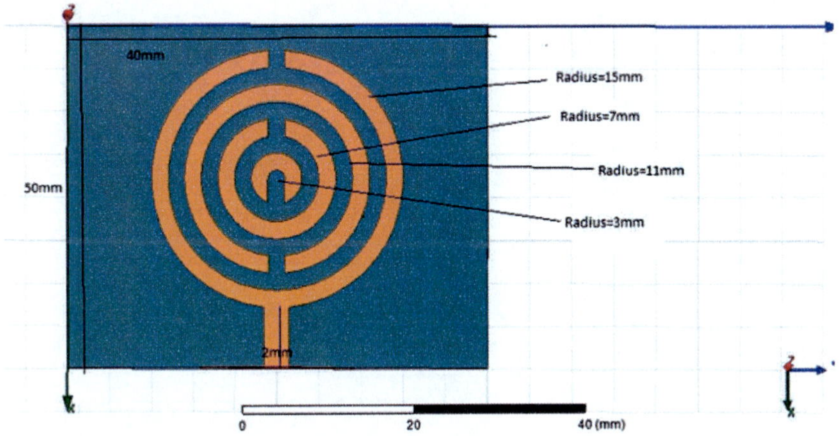


**Fig. 1** Planer split ring resonating antenna

**Table 1** Design parameters of the Antenna design

| Parameters | Values |
|---|---|
| Frequency used | 13 GHz |
| Width of the feed | 2 mm |
| Thickness of the rings | 0.01 mm |

**Fig. 2** VSWR of the proposed antenna

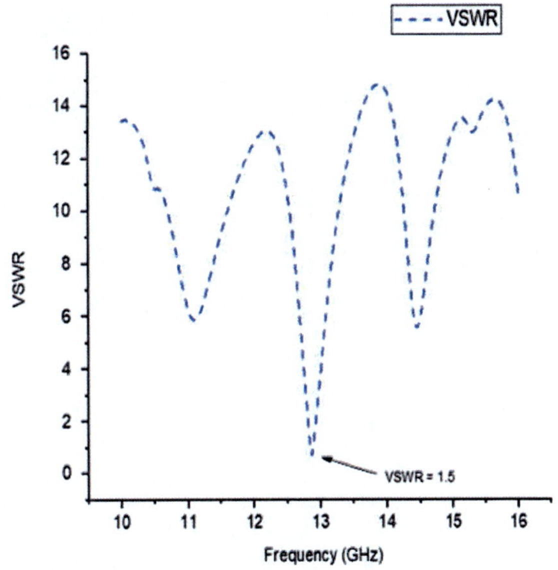

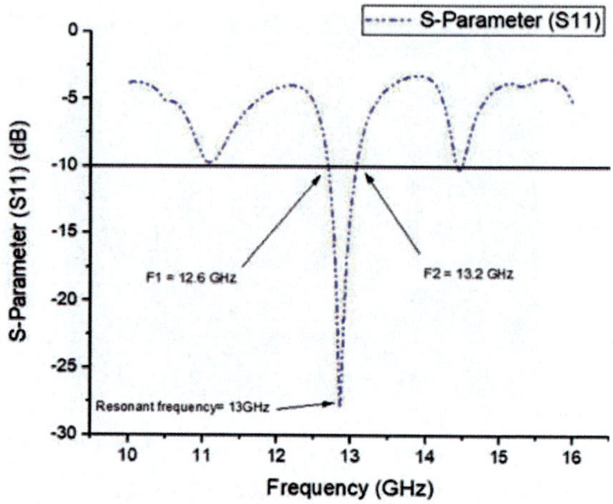


**Fig. 3** S11 parameter of the proposed antenna

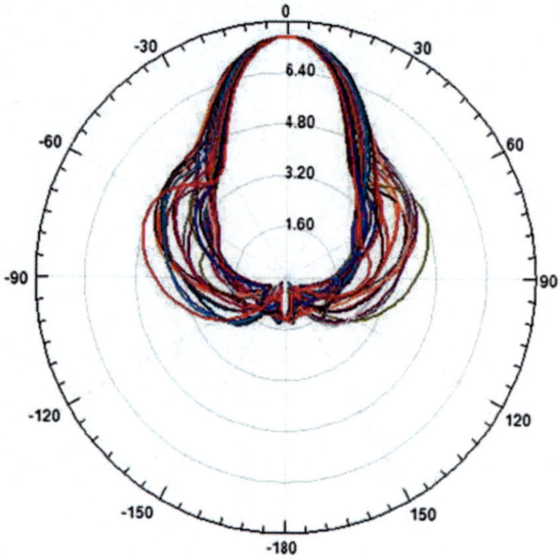

**Fig. 4** 2D radiation pattern of the proposed design

## 4 Conclusion

The proposed antenna has a novel shape and it has a resonating frequency of 13 GHz with high gain ant directivity. With high directivity of 9.6 dB the antenna can be used in many applications like 5G, satellite communication, etc.

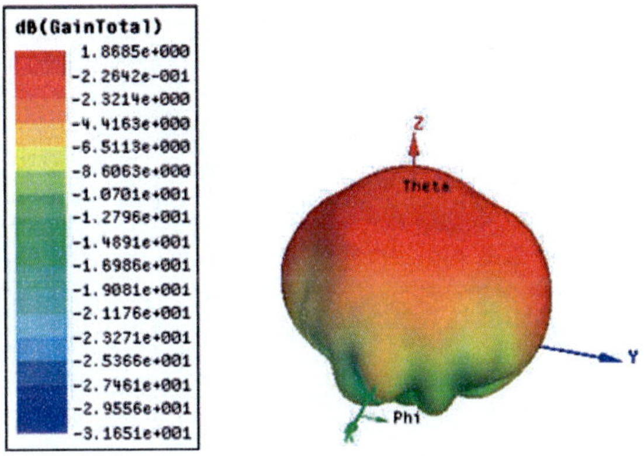


**Fig. 5** 3D radiation pattern of the proposed design

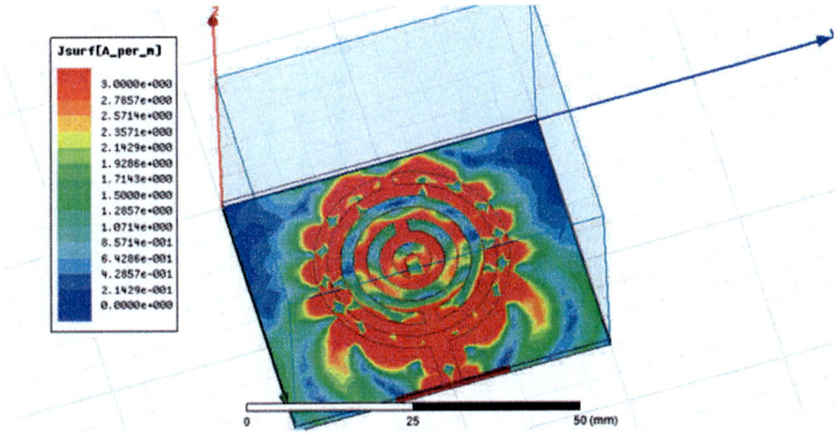


**Fig. 6** Surface charge distribution of the design

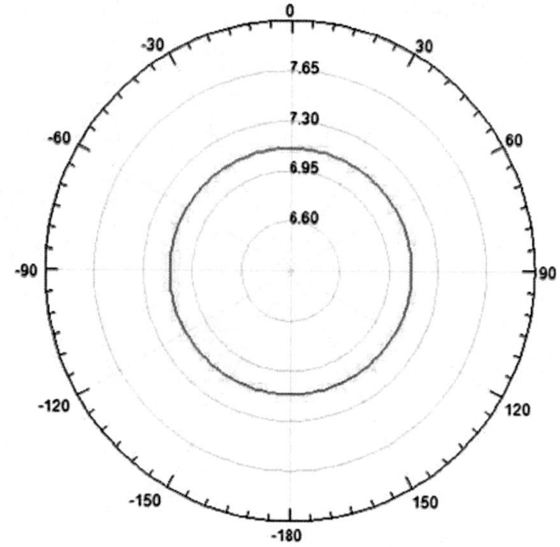


**Fig. 7** Peak gain of the antenna

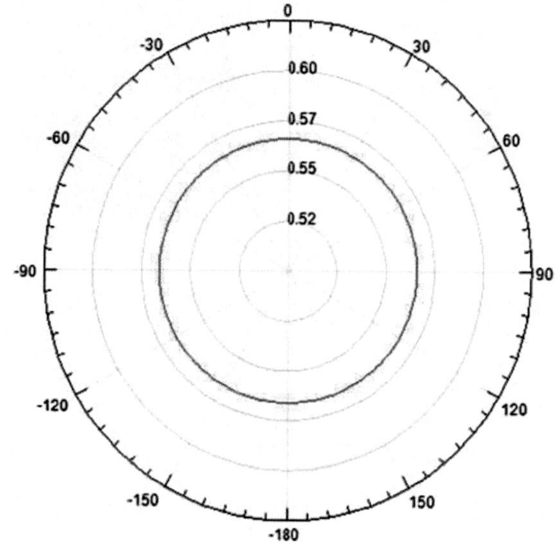


**Fig. 8** Radiation efficiency of the proposed antenna

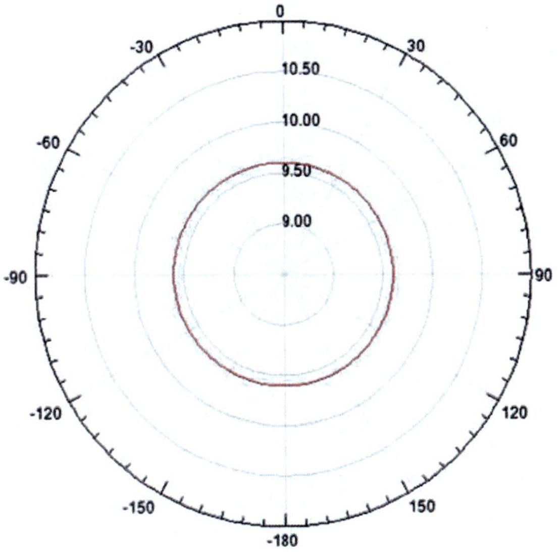


**Fig. 9** Peak Directivity of the proposed antenna

**Table 2** Simulated parameters

| Parameters | Return loss | Resonant frequency | Radiation efficiency | Peak gain |
|---|---|---|---|---|
| Values | −17.50 dB | 13.00 GHz | 56.32% | 7.2 dB |

## References


1. M.A. Rahman, M.R.I. Faruque, M.T. Islam, Circularly split-resonator-based frequency-reconfigurable antenna. Appl. Phys. A **123**(1) (2017). https://doi.org/10.1007/s00339-016-0620-8
2. M. Sano, K. Yamada, M. Higaki, Study of the resonant frequencies of an electrically small broadside-coupled split-ring-resonator antenna, in *2018 IEEE International Symposium on Antennas and Propagation & USNC/URSI National Radio Science Meeting* (2018). https://doi.org/10.1109/apusncurinrsm.2018.8
3. S.J.K. Gandhi Mohan, T. Shanmuganantham, Split ring resonator loaded Z-slots antenna for on-body communication, in *2018 International Conference on Circuits and Systems in Digital Enterprise Technology (ICCSDET)* (2018). https://doi.org/10.1109/iccdset.2018.8821118
4. Y. Khanna, R. Gaur, R. Gupta, Y.K. Awasthi, Design of metamaterial slotted split ring resonator-SSRR for dual frequency band applications, in *2019 6th International Conference on Signal Processing and Integrated Networks (SPIN)* (2019). https://doi.org/10.1109/kbei.2019.8711789
5. S.H. Hosseini, C. Ghobadi, J. Nourinia, P. Faeghi, Novel broadband circularly polarized split triangular ring resonator SIW cavity backed antenna, in *2019 5th Conference on Knowledge Based Engineering and Innovation (KBEI)* (2019). https://doi.org/10.1109/kbei.2019.8735004
6. K. Sajith, J. Gandhimohan, T. Shanmuganantham, The split ring resonator loaded conductor backed CPW Fed antenna for interfacing applications, in *2018 IEEE Indian Conference on Antennas and Propagation (InCAP)* (2018). https://doi.org/10.1109/incap.2018.8770753
7. A.S. Pradeep, G.A. Bidkar, A comparative analysis of circular and hexagonal split ring resonator for reduction of mutual coupling effect, in *2019 International Conference on Recent Advances*

*in Energy-Efficient Computing and Communication (ICRAECC)* (2019). https://doi.org/10.1109/icraecc43874.2019.899
8. Ramachandran, A., Valiyaveettil Pushpakaran, S., Pezholil, M., & Kesavath, V. (2016). A Four-Port MIMO Antenna Using Concentric Square-Ring Patches Loaded with CSRR for High Isolation. IEEE Antennas and Wireless Propagation Letters, 15, 1196-1199. doi:https://doi.org/10.1109/lawp.2015.2499322
9. R. Pandeeswari, S. Raghavan, P.A. Badge, A.K. Chittipothul, A compact multi-split ring resonator loaded antenna, in *2013 International Conference on Communication and Signal Processing* (2013). https://doi.org/10.1109/iccsp.2013.6577168
10. S. Sharma, R. Mehra, Dual-band planner slot antenna loaded with split ring resonators for WLAN/Wi-Max applications, in *2018 International Conference on Smart City and Emerging Technology (ICSCET)* (2018). https://doi.org/10.1109/icscet.2018.8537254
11. P. Paga, H.C. Nagaraj, T.S. Rukmini, Design of a dual band frequency reconfigurable monopole antenna using a circular split ring resonator for Wi-Fi and Wi-Max applications, in *2017 International Conference of Electronics, Communication and Aerospace Technology (ICECA)* (2017). https://doi.org/10.1109/iceca.2017.8212796
12. M.S. Arifianto, Maryadi, A. Munir, Dual-band circular patch antenna incorporated with split ring resonators metamaterials, in *2013 10th International Conference on Electrical Engineering/Electronics, Computer, Telecommunication and Information Technology* (2013). https://doi.org/10.1109/ecticon.2013.6559663
13. R. Guttula, V. Nandanavanam, Microstrip antenna embedded with circular split ring resonator for WLAN applications, in *2017 Innovations in Power and Advanced Computing Technologies (i-PACT)* (2017). https://doi.org/10.1109/ipact.2017.8245202
14. C.A. Balanis, *Antenna Theory: Analysis and Design*, 3rd edn. (Chap.14) (Wiley, New York, 2005)
15. R.A. Panda, D. Mishra, Efficient design of bi-circular patch antenna for 5G communication with mathematical calculations for resonant frequencies. Wireless Pers. Commun. **112**, 717–727 (2020). https://doi.org/10.1007/s11277-020-07069-9
16. R.A. Panda, P. Kumari, J. Naik, P. Negi, D. Mishra, Flower Shaped Patch with Circular Defective Ground Structure for 15 GHz Application, in *Innovations in Bio-Inspired Computing and Applications. IBICA 2019.* Advances in Intelligent Systems and Computing, vol. 1180, ed. by A. Abraham, M. Panda, S. Pradhan, L. Garcia-Hernandez, K. Ma. Springer, Cham (2021). https://doi.org/10.1007/978-3-030-49339-4_24
17. R.A. Panda, B. Hansdah, S. Misra, R.K. Singh, D. Mishra, Gauging trowel shaped patch including an optimized slot for 60 GHz WLAN, in *Innovative Data Communication Technologies and Application. ICIDCA 2019. Lecture Notes on Data Engineering and Communications Technologies*, vol. 46, ed. by J. Raj, A. Bashar, S. Ramson. Springer, Cham. https://doi.org/10.1007/978-3-030-38040-3_9
18. R.A. Panda, D. Mishra, E.P. Panda, N. Patnaik, Reshaped circular patch antenna with optimized circular slot for 5G application, in *Emerging Trends in Electrical, Communications, and Information Technologies*. Lecture Notes in Electrical Engineering, vol. 569, ed. By T. Hitendra Sarma, V. Sankar, R. Shaik (eds). Springer, Singapore (2020). https://doi.org/10.1007/978-981-13-8942-9_63
19. P.R. Meher, B.R. Behera, S.K. Mishra, Design and its state of the art of different shaped DRAs at millimeter wave frequency band. Int. J. RF Microwave Comput. Aided Eng. **30**(7), 1–15 (2020)
20. P.R. Meher, B.R. Behera, S.K. Mishra, A.A. Althuwayb, A chronological review of circularly polarized dielectric resonator antenna: design and developments. Int. J. RF Microwave Comput. Aided Eng. **31**(5), 1–23 (2021)
21. P.R. Meher, B.R. Behera, S.K. Mishra, A compact circularly polarized cubic DRA with unit-step feed for Bluetooth/ISM/Wi-Fi/Wi-MAX applications. AEU-Int. J. Electron. Commun. **128**, 1–8 (2020)

# Pentagonal Microstrip Patch Antenna with Circular Slot for 9 GHz Applications

**Rani Swetashri Naik, Manasi Panda, Saphalya Kumar Sahu, and Ribhu Abhusan Panda**

**Abstract** A pentagonal patch antenna having a rounded slot has been designed. The antenna design is illustrated by taking a pentagonal patch on a FR-4 epoxy substrate. The circular slot is optimized by the optimetric tool in the High-Frequency Structure Simulator (HFSS). The resonating frequency for the proposed design is 9 GHz. The circular slot in pentagonal patch excites circular polarization which enhances the gain and reduces the complexity of the design. The return loss has been found to be −23.2909 dB with a bandwidth of 500 MHz. This design has been evaluated in terms of VSWR, return loss, radiation, gain, etc.

**Keywords** Pentagonal patch · VSWR · Return loss ($S_{11}$) · Radiation · Circular slot · HFSS

## 1 Introduction

Because of the easy fabrication and low profile, the microstrip patch antenna has drawn the awareness of the researchers over past few decades. There are numerous applications of the microstrip patch antenna such as aircraft, spacecraft, and missile applications even in the biomedical field also but the major shortcoming of microstrip patch antenna are its low power, low efficiency, and smaller bandwidth [1, 2]. These can be reduced by using some of the methods like slot, feeding, and optimization.

R. S. Naik · M. Panda · S. K. Sahu · R. A. Panda (✉)
Department of Electronics and Communication Engineering, GIET University, Gunupur, Odisha, India
e-mail: ribhuabhusanpanda@giet.edu

R. S. Naik
e-mail: 18ece062.raniswetashrinaik@giet.edu

M. Panda
e-mail: 18ece068.manasipanda@giet.edu

S. K. Sahu
e-mail: 18ece066.saphalyakumarsahu@giet.edu



S. Patnaik et al. (eds.), *Smart and Sustainable Technologies: Rural and Tribal Development Using IoT and Cloud Computing*, Advances in Sustainability Science and Technology,
https://doi.org/10.1007/978-981-19-2277-0_18

Slots are mainly created to have a new resonance frequency. By the use of the slot, the bandwidth can be increased [3]. There are different types of patch antennas. The usually accessible shapes are rectangular, square, dipolar, circular, triangular, and elliptical. Among these, rectangular and circular shapes are widely used. For some specific applications, pentagonal patches are also being used. This shape reduces the overall size as compared to other shapes. Previously a perturbed pentagon patch for c-band application has been designed [4]. By implementing some additional techniques with the pentagonal patch, novel antennas have been designed for definite applications [5–8]. In past few years alterations have been done in the shape of conventional patch antennas for specific applications [9–17]. This proposed design includes a simple pentagonal patch and a circular slot is implemented on it. The radius of the circular slot has been optimized and the optimum value has been found to be 2.25 mm.

## 2 Design of Pentagonal Patch with Slot

Using microstrip line feed technique, the pentagon-shaped microstrip patch antenna has been designed with HFSS software. The circular slot has a radius of 2.25 mm which has been determined from the optimization. Here, the material for the ground as well as the patch is copper and for substrate FR-4 epoxy has been taken as this is widely available as a PCB base material. It has the low cost and more versatile material. Using the basic equation, the resonant frequency is calculated to be 9 GHz. The proposed patch has been designed on a substrate of dimension 40 mm × 40 mm × 1.6 mm. The ground plane including the copper material has the same length and width as that of the substrate. The dimensions are included in Table 1 (Fig. 1).

**Table 1** Specifications of antenna

| Parameters | Values used |
|---|---|
| Frequency | 9 GHz |
| Width of the feed | 2 mm |
| Thickness of the patch | 0.01 mm |
| Thickness of the ground | 0.01 mm |
| Substrate height | 1.6 mm |
| Circular slot radius | 2.25 mm |
| Feed width | 2 mm |

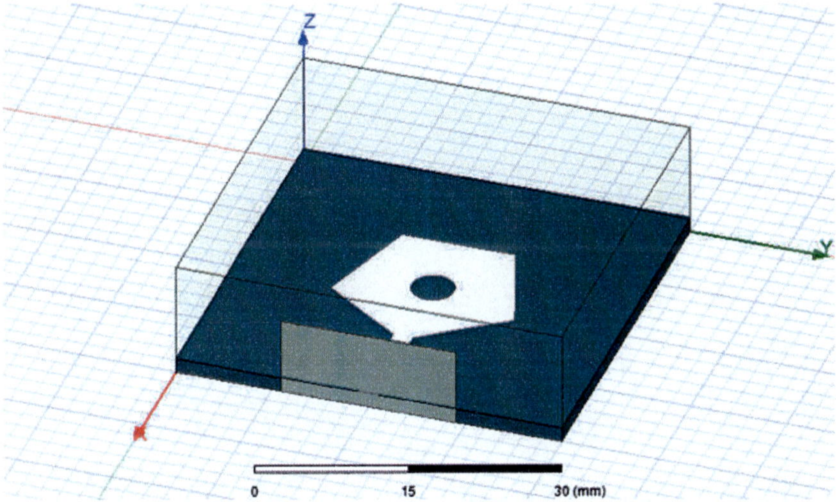


**Fig. 1** Pentagonal patch design using HFSS

# 3 Results of the Proposed Antenna

Implementation of slot increases the bandwidth. Optimization of the radius of the slot has been done with the optimatrics tool in HFSS software. The optimization plot has been illustrated in Fig. 2. The performance parameter like gain, VSWR, bandwidth, return loss, and radiation pattern has been determined by using the HFSS software. Figure 3 shows the S-Parameter plot which indicates that the $S_{11}$ (return loss) is –

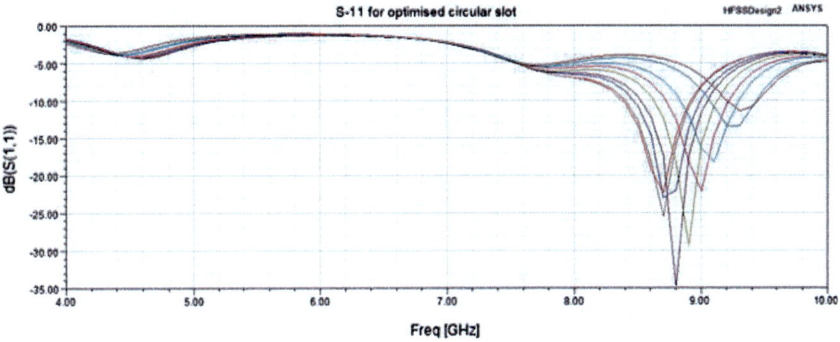


**Fig. 2** Optimization plot for radius of the slot

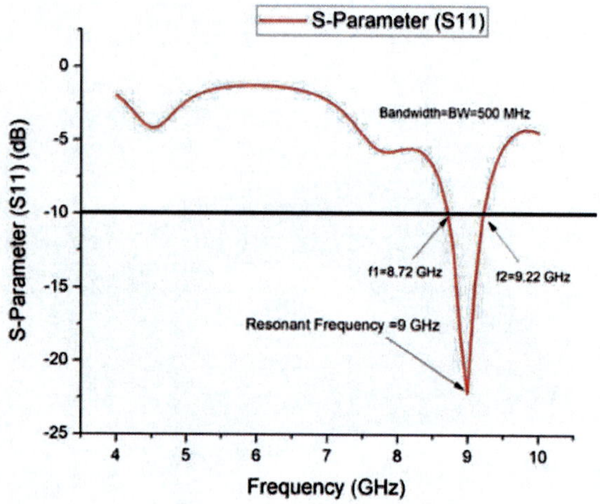


**Fig. 3** $S_{11}$ Plot in accordance with frequency

23.2909 dB at the resonating frequency of 9 GHz. As indicated in Fig. 4, the VSWR is 1.5 which is closely equal to the anticipated value of 1.

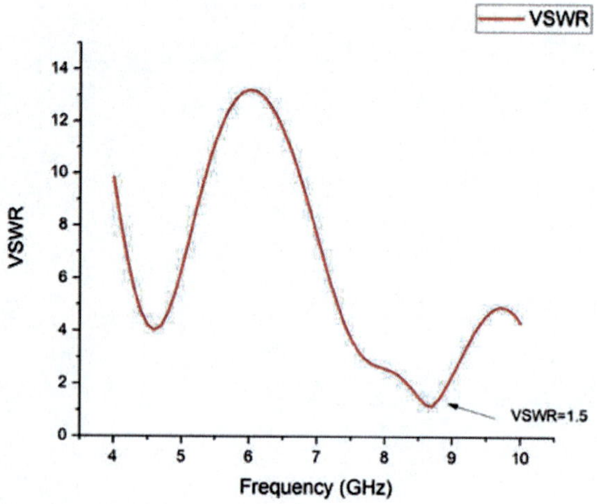


**Fig. 4** VSWR versus frequency plot

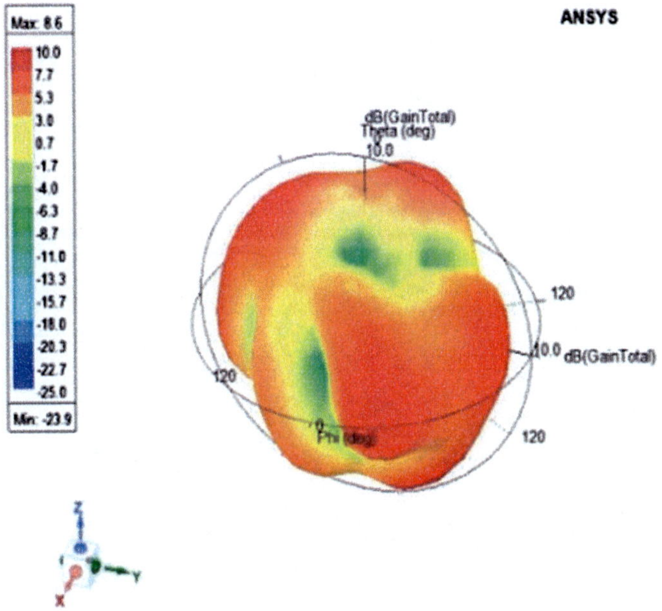


**Fig. 5** 3D radiation pattern

3D radiation pattern, peak gain, peak directivity, and 2D radiation pattern have been illustrated in Figs. 5, 6, 7 and 8, respectively. The simulated parameters are included in Table 2.

## 4 Conclusion

The proposed antenna can be operated at 9 GHz efficiently as it is having a high gain of 8.66497 dB and a directivity of 9.39 dB. There is no complexity associated with this proposed design as a simple optimized circular slot is implemented on the pentagonal patch.

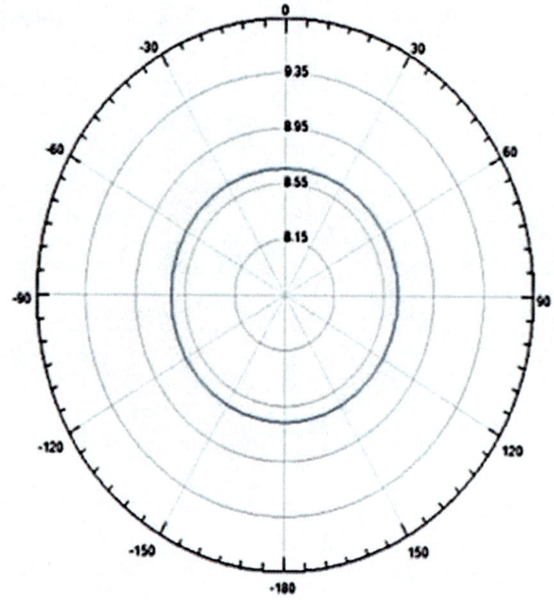


**Fig. 6** Peak gain

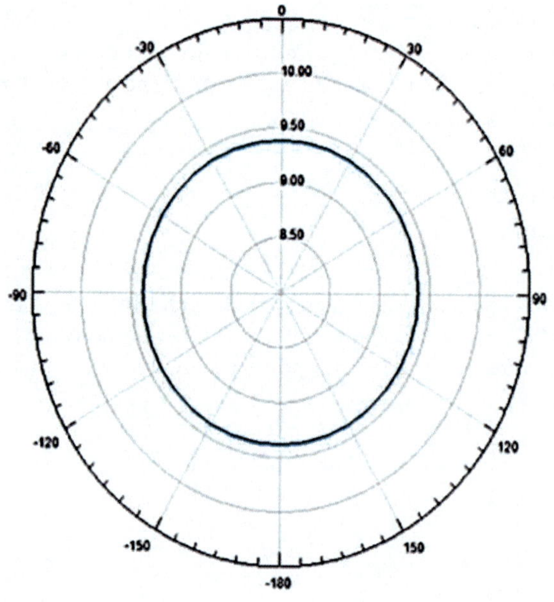


**Fig. 7** Peak directivity

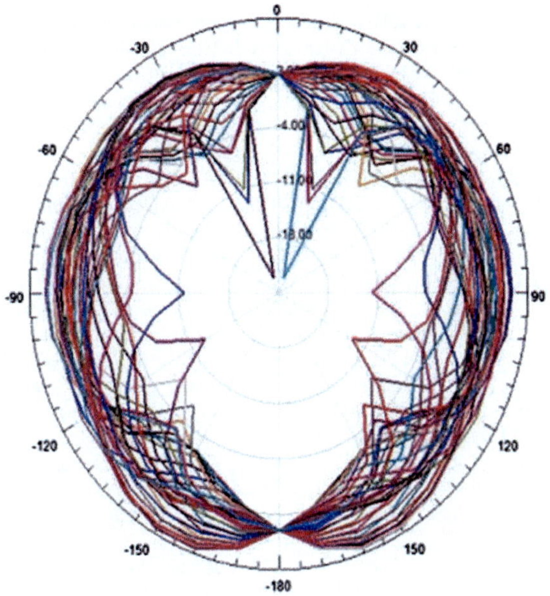


**Fig. 8** 2D radiation pattern

**Table 2** Simulated parameters

| Parameters | Return loss | Peak directivity | Resonant frequency | Bandwidth | Peak gain |
|---|---|---|---|---|---|
| Values | −23.2909 dB | 9.39 dB | 9.00 GHz | 500 MHz | 8.66497 dB |

## References


1. C.A. Balanis, *Antenna Theory Analysis and Design*, 2nd edn. (Wiley India (p.) Ltd., 2007)
2. R. Garg, P. Bhartia, I. Bahl, A. Ittipiboon, *Microstrip antenna design handbook* (Artech House, Boston, 2001)
3. P.A. Ambresh, P.M. Hadalgi, P.V. Hunagund, Planar microstrip slot antenna for S & C band wireless applications. J. Phys: Conf. Ser. **435**(1), 12–22 (2013)
4. M. Rajan Babu, D. Venkatachari, K. Yogitha, K. P. Sai Sri, K. Jagadeesh, M. Lokesh, A low-profile pentagon inscribed in circle shaped wide band antenna design for C-band applications, in *2020 IEEE Bangalore Humanitarian Technology Conference (B-HTC)* (2020), pp. 1–4. https://doi.org/10.1109/B-HTC50970.2020.9298012
5. L. Bhargavi, L. Vardhan Reddy, P. Yashwanth, T. Anil Kumar, Pentagonal shaped Koch fractal monopole slot antenna. ARPN J. Eng. Appl. Sci. (2017)
6. L. Chandiea, K. Anusudha, Pentagon shaped microstrip patch antenna with metamaterial for UWB application, in *2016 International Conference on Control, Instrumentation, Communication and Computational Technologies (ICCICCT)* (2016)
7. T.P. Mithun, K. Sreelakshmi, Design of monopole CPW-fed pentagonal patch antenna, in *2018 International Conference on Electrical, Electronics, Communication, Computer, and Optimization Techniques (ICEECCOT)* (2018), pp. 539–544. https://doi.org/10.1109/ICEECCOT43722.2018.9001594

8. D. Surender, T. Khan, F.A. Talukdar, A pentagon-shaped microstrip patch antenna with slotted ground plane for RF energy harvesting, in *2020 URSI Regional Conference on Radio Science (URSI-RCRS)* (2020), pp. 1–4. https://doi.org/10.23919/URSIRCRS49211.2020.9113536
9. R.A. Panda, D. Mishra, Efficient design of bi-circular patch antenna for 5G communication with mathematical calculations for resonant frequencies. Wireless Pers. Commun. **112**, 717–727 (2020). https://doi.org/10.1007/s11277-020-07069-9
10. R.A. Panda, P. Kumari, J. Naik, P. Negi, D. Mishra (2021) Flower shaped patch with circular defective ground structure for 15 GHz application, in *Innovations in Bio-Inspired Computing and Applications. IBICA 2019*. Advances in Intelligent Systems and Computing, vol. 1180, ed. by A. Abraham, M. Panda, S. Pradhan, L. Garcia-Hernandez, K. Ma (Springer, Cham, 2019). https://doi.org/10.1007/978-3-030-49339-4_24
11. R.A. Panda, B. Hansdah, S. Misra, R.K. Singh, D. Mishra, Gauging trowel shaped patch including an optimized slot for 60 GHz WLAN, in *Innovative Data Communication Technologies and Application. ICIDCA 2019*. Lecture Notes on Data Engineering and Communications Technologies, vol. 46, ed. by J. Raj, A. Bashar, S. Ramson (Springer, Cham, 2020). https://doi.org/10.1007/978-3-030-38040-3_9
12. R.A. Panda, D. Mishra, E.P. Panda, N. Patnaik, Reshaped circular patch antenna with optimized circular slot for 5G application, in *Emerging Trends in Electrical, Communications, and Information Technologies*. Lecture Notes in Electrical Engineering, vol. 569, ed. by T. Hitendra Sarma, V. Sankar, R. Shaik (Springer, Singapore, 2020). https://doi.org/10.1007/978-981-13-8942-9_63
13. P.R. Meher, B.R. Behera, S.K. Mishra, Design and its state of the art of different shaped DRAs at millimeter wave frequency band. Int. J. RF Microwave Comput. Aided Eng. **30**(7), 1–15 (2020)
14. P.R. Meher, B.R. Behera, S.K. Mishra, A.A. Althuwayb, A chronological review of circularly polarized dielectric resonator antenna: design and developments. Int. J. RF Microwave Comput. Aided Eng. **31**(5), 1–23 (2021)
15. P.R. Meher, B.R. Behera, S.K. Mishra, A compact circularly polarized cubic DRA with unit-step feed for Bluetooth/ISM/Wi-Fi/Wi-MAX applications. AEU-Int. J. Electron. Commun. **128**, 1–8 (2020)
16. P.R. Meher, B.R. Behera, S.K. Mishra, Broadband circularly polarized edge feed rectangular dielectric resonator antenna using effective glueless technique. Microw. Opt. Technol. Lett. **62**(10), 1–9 (2020)
17. P.R. Meher, B.R. Behera, S.K. Mishra, A.A. Althuwayb, Design and analysis of a compact circularly polarized DRA for off-body communications. AEU-Int. J. Electron. Commun. **138**, 1–7 (2021)

# Simulation and Analysis of an Optical Communication System Implementing DCF with Various Pulse Generators

**Padmini Mishra, Subhrajit Pradhan, Gopinath Palai, and Partha Sarkar**

**Abstract** Extensive bandwidth with small delay is a major requirement in current-day communication. Optical fiber communication has emerged as one of the major contributors to the existing communication field as it provides a higher bandwidth with minimal latency for wide range of applications. ISI (Inter Symbol Interference) hampers the performance of optical communication drastically, it induces a superabundant increment in pulse duration, rolling out in time and space engrossed by the pulse as it propagates. As the pulse propagates along with the optical fiber, this spread can be so enormous that photodetector can no longer discriminate boundaries between pulses. Due to ISI, path loss increases rapidly and this causes broadening of pulses, and achieving high data rates becomes difficult. To reduce the effect of ISI various pulse shaping technique is used. In this paper, the channeling performance of optical fiber and DCF is estimated by using different pulse generators. Here we collated the performance of various pulse generators in terms of Quality factor and Bit Error Rate in the eye diagram.

**Keywords** Intersymbol interference · DCF · Q-factor · BER · Eye diagram · PIN photodiode · Continuous-wave laser · Optical amplifier

## 1 Introduction

In present days networks and telecommunication play an important role in everyone's life. Telecommunication mainly involves the transmission of voice, data, or radio for a longer distance, and its applications are roll outing nowadays, and arising from

P. Mishra (✉) · S. Pradhan
GIET University, Gunupur, Rayagada, Odisha 765022, India
e-mail: padminimishra@giet.edu

G. Palai
Department of ECE, GITA, Bhubaneswar, Odisha 752054, India

P. Sarkar
JIS College of Engineering, Kalyani, West Bengal 741235, India

 
S. Patnaik et al. (eds.), *Smart and Sustainable Technologies: Rural and Tribal Development Using IoT and Cloud Computing*, Advances in Sustainability Science and Technology,
https://doi.org/10.1007/978-981-19-2277-0_19

global networks to computers. Optical fiber communication system and network being highly geared up and due to its decency nature, it is immensely used in world-wide telecommunication as well as in many regional communication applications [1]. The optical fiber being used as a channel for optical fiber communication has substantially replaced copper wire. The optical fibers which are observed as waveguides can be appeal to light channeling applications.

Along with all these advantages, there lie few tailbacks in fiber optic communication too. Intersymbol interference and dispersion are one of those few major concerns. Intersymbol interferences hamper the performance of optical communication drastically [2, 3]. It includes a superabundant increment in pulse duration, rolling out in time and space engrossed by the pulse as it propagates. As the pulse propagates along with the optical fiber, this spread can be so enormous that photodetector can no longer discriminate boundaries between pulses. This occurs due to band-limited channels and multipath signal propagation [4].

The boarding of light signals while transmitting through the fiber is called dispersion. The dispersion results spreading of output pulse in time domain and changing its shape as a result of which it may merge with the previous or succeeding pulse. This results in the output pulse to be expanded over a longer period. To overcome this problem various dispersion compensation techniques such as fiber Bragg grating (FBG), dispersion compensating fiber (DCF), and auto electronic dispersion compensation are applied. The fibers with negative dispersion coefficient at 1550 nm are called as DCF. DCF is the most accepted technique to combat the effect of dispersion and ISI.

Alberti et al. has compared a numerical versus experimental procedure of various techniques for chromatic dispersion compensation on a terrestrial 40 Gbps system and collected the data [5]. They found that combination of long amplifier spacing and surrounding with G.652 fiber, HDP with DCF gave a better result for terrestrial systems. Peucheret et al. peruse a systematic numerical enhancement of pre and post-compensation normalized sections making use of standard and dispersion compensating fiber for NRZ 10 Gbit/s single-channel system [6]. They found that when they have not used pre-compensation at that time post-compensation gave a better result with compensation ratio of 98% but when they have used pre-compensation at that time, they found that it affects the compensation ratio. Alisha ban et al. analyzed the influence of different electrical pulse shapes on the effectiveness of optical transmission systems and their characteristics are assessed and compared [7]. They found that raised cosine pulse shaping gave better results than another pulse shaping. Meena et al. investigated a new model of four channel DWDM fiber optic transmission system using DCT [8]. They have used DCF in various configurations and compared the performance at various data rates. Kour et al. analyzed the performance of 55 Gbit/s fiber optic communication systems employing externally modulated high-speed CW laser with the help of various modulation techniques [9]. Virendra Kumar et al. has proposed optical transmission system based upon RZ and NRZ with single-channel over single-mode fiber [10]. RZ pulse generator produces better result at input power 5 dBm, within the range of 25 km with attenuation coefficient 0.2 dB/km.

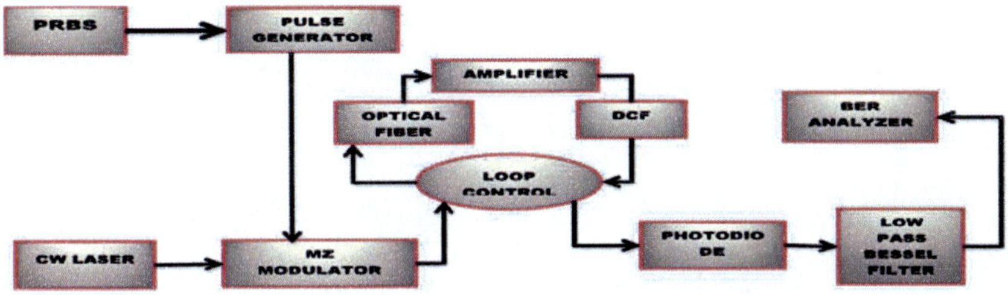


**Fig. 1** Block diagram of simulation model

In this paper, we have evaluated the performance of the different pulse generators (NRZ, RZ, Gaussian, Raised-cosine) while varying the input laser power from −15 to +5 dBm at fiber length 240 km and constant bit rate of 10 Gbps. The performance of various pulse generators for the above set of input parameters is calculated. The performance interpretation variables include eye diagram and min BER for different pulse generators.

## 2 Simulation and Design

The optical transmission link comprises of three parts. Transmitter section comprising of a PRBS, a continuous-wave laser, a pulse generator, and a Mach–Zehnder modulator. Input signal generated by two simulations set up, i.e., PRBS and electrical pulse modulator. When PRBS generates binary bit sequence then that binary sequence is received by electrical pulse generator which converts binary bit sequence into electrical pulses. Both input electrical signal and optical signal generated by CW laser signal go to Mach- Zehnder modulator. The Mach–Zehnder modulator modulates this signal on the high-frequency optical signal. This signal is then transmitted through the fiber optic channel [11]. Optical amplifiers are used in transmission channels in order to reconstruct the signal in long-haul environment [12]. Receiver part comprising of a PIN photodetector, and a low-pass Bessel filter. That optical signal is recognized by the photodetector. Then that electrical signal passes through low-pass Bessel filter in order to suppress the noise accumulated in the channel. This signal is then given to the Bit Error Rate analyzer where the eye diagram and its corresponding parameters can be visualized and analyzed. We have used four different electrical pulse generators for our simulation purpose (Figs. 1 and 2; Tables 1, 2 and 3).

## 3 Result and Discussion

The simulation is done in Opti system 15.0 platform. Simulations were first carried out for individual power of CW laser by keeping the value of data rate constant at

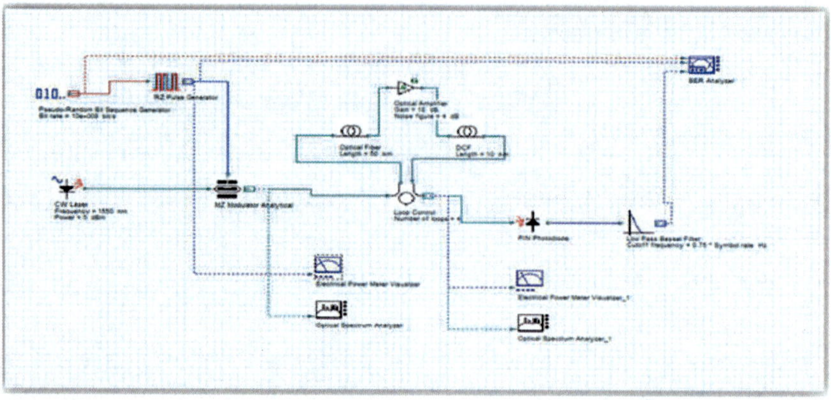

**Fig. 2** Simulation model of RZ pulse generator

**Table 1** Simulation parameters

| Parameters | Value |
|---|---|
| Bit rate | 10 Gbps |
| Sequence length | 1024 Bits |
| Symbol rate | 10e+009 Symbol/s |
| Reference wavelength | 1550 nm |
| Distance | 240 km |

**Table 2** Optical fiber parameters

| Parameters | Value |
|---|---|
| Wavelength | 1550 nm |
| Length | 10 km |
| Attenuation | 0.5 dB/Km |
| Dispersion | 16.75 PS/nm/Km |

**Table 3** DCF parameters

| Parameters | Value |
|---|---|
| Wavelength | 1550 nm |
| Length | 50 km |
| Attenuation | 0.2 dB/Km |
| Dispersion | 16.75 PS/nm/Km |

10Gbps. The input power is then varied from −15 dBm to +5 dBm at a distance of 240 km. The performance of PIN photodiode is verified at reference wavelength of 1550 nm (Figs. 3 and 4; Tables 4, 5, 6 and 7).

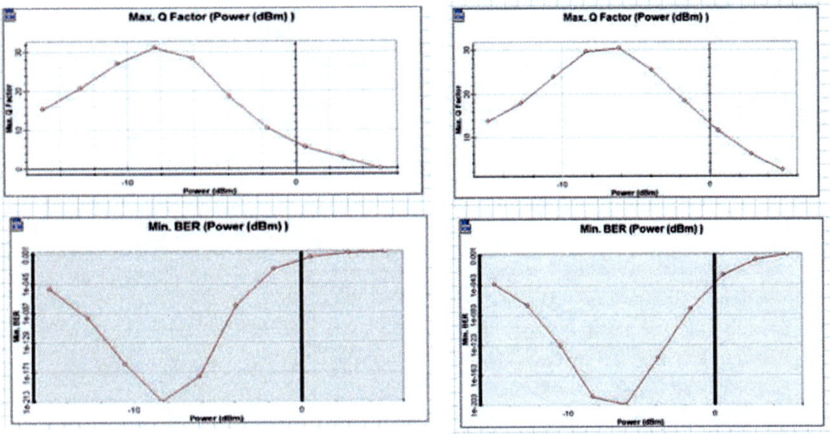


NRZ RZ

**Fig. 3** Graphical representation of max Quality factor and min-Bit Error Rate w.r.t to input power of continuous-wave laser (NRZ, RZ)

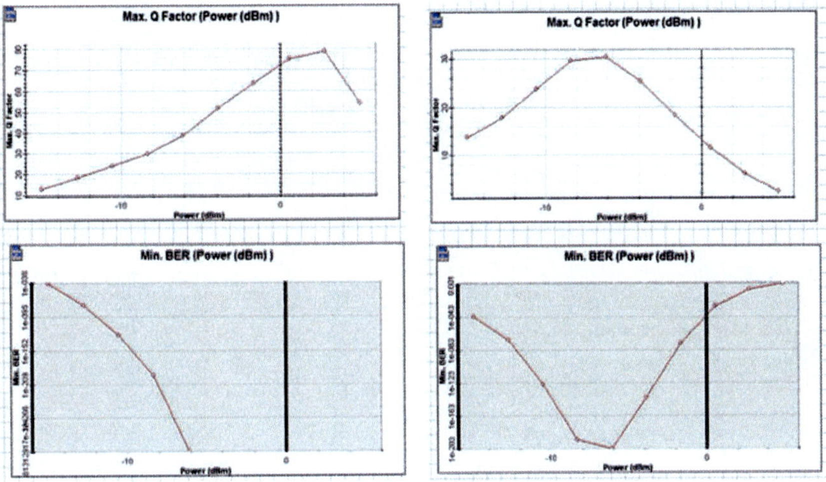


GAUSSIAN RAISED COSINE

**Fig. 4** Graphical representation of max Quality factor and min-Bit Error Rate w.r.t to input power of continuous-wave laser (Gaussian, Raised-cosine)

Here we have performed simulation using different pulse generators and received data using PIN photodiode. We observed that at power −6.11111 dBm, at a distance of 240 km, by keeping constant data rate, i.e., 10 Gbps NRZ gives Q-factor as 28.4504 and min-BER as 1.63E−178, RZ gives $Q$-factor as 38.4201 and min BER

**Table 4** Variation of Q-factor and min BER for NRZ pulse generator at 240 km and data rate of 10 Gbps (PIN)

| Input power | Max Q factor | Min BER |
|---|---|---|
| −15 | 14.871 | 1.87E−50 |
| −12.7778 | 20.4518 | 2.10E−93 |
| −10.5556 | 26.3496 | 1.75E−153 |
| −8.33333 | 32.7068 | 4.40E−235 |
| −6.11111 | 28.9112 | 3.21E−184 |
| −3.88889 | 18.7912 | 3.08E−79 |
| −1.66667 | 10.3971 | 8.78E−26 |
| 0.555556 | 5.49364 | 1.38E−08 |
| 2.77778 | 2.76062 | 0.00231545 |
| 5 | 0 | 1 |

**Table 5** Variation of Q-factor and min BER for RZ pulse generator at 240 km and data rate of 10 Gbps

| Input power | Max Q factor | Min BER |
|---|---|---|
| −15 | 12.3205 | 2.57E−35 |
| −12.7778 | 18.4502 | 1.87E−76 |
| −10.5556 | 24.1694 | 1.60E−129 |
| −8.33333 | 30.047 | 7.85E−199 |
| −6.11111 | 39.2279 | 0.00E+00 |
| −3.88889 | 49.4517 | 0 |
| −1.66667 | 62.1809 | 0 |
| 0.555556 | 75.7434 | 0 |
| 2.77778 | 80.0903 | 0 |
| 5 | 53.1413 | 0 |

**Table 6** Variation of Q-factor and min BER for Gaussian pulse generator at 240 km and data rate of 10 Gbps

| Input power | Max Q FACTOR | Min BER |
|---|---|---|
| −15 | 14.417 | 1.48E−47 |
| −12.7778 | 18.2228 | 1.20E−74 |
| −10.5556 | 23.9046 | 9.12E−127 |
| −8.33333 | 28.3019 | 1.05E−176 |
| −6.11111 | 29.1112 | 7.95E−187 |
| −3.88889 | 24.5687 | 8.39E−134 |
| −1.66667 | 18.0774 | 1.56E−73 |
| 0.555556 | 11.3687 | 2.14E−30 |
| 2.77778 | 6.01653 | 7.71E−10 |
| 5 | 2.28489 | 0.0109707 |

**Table 7** Variation of quality factor and min-bit error rate for raised-cosine pulse generator at 240 km and data rate of 10Gbps

| Input power | Max Q factor | MIN BER |
|---|---|---|
| −15 | 12.4848 | 3.81E−36 |
| −12.7778 | 14.9084 | 1.25E−50 |
| −10.5556 | 17.8333 | 1.76E−71 |
| −8.33333 | 19.1764 | 2.68E−82 |
| −6.11111 | 16.1575 | 4.55E−59 |
| −3.88889 | 12.7941 | 7.95E−38 |
| −1.66667 | 8.68183 | 1.72E−18 |
| 0.555556 | 5.32666 | 4.48E−08 |
| 2.77778 | 2.96413 | 1.43E−03 |
| 5 | 0 | 1 |

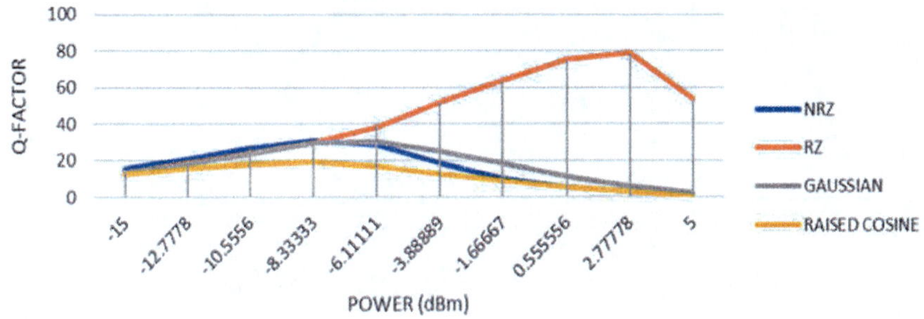


**Fig. 5** Comparison graph for *Q*-factor of different pulse generator w.r.t power in dBm

as 1.97626e−323, Gaussian gives *Q*-factor as 30.0651 and min-BER as 8.72E−204, Raised cosine gives *Q*-factor as 16.395 and min-BER as 9.51E−61 (Fig. 5).

At last, we compare Q-factor of different pulse generators. We observed that RZ pulse generator gives better Q-factor than others (using PIN photodiode) (Figs. 6, 7 and 8).

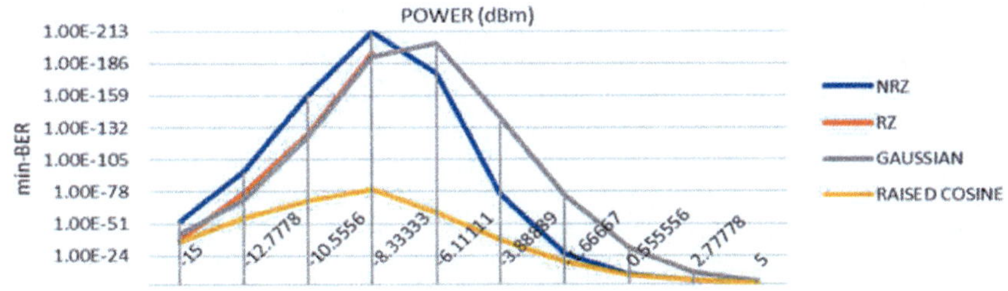


**Fig. 6** Comparison graph for min-BER of different pulse generator w.r.t power in dBm

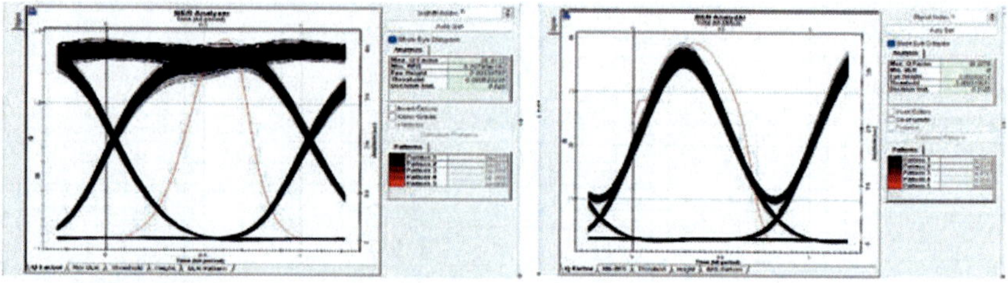

NRZ RZ

**Fig. 7** Eye Diagram of NRZ, RZ pulse generator at same power of −6.11111 dBm

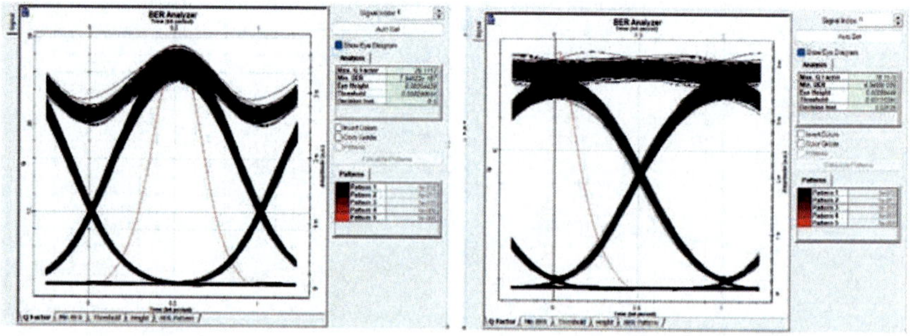

Gaussian Raised cosine

**Fig. 8** Eye diagram of Gaussian pulse and raised cosine generator at power of −6.11111 dBm

# 4 Conclusion

By using different pulse generators (RZ, NRZ, GAUSSIAN, RAISED-COSINE) and by varying the power of continuous-wave laser from −15dBm to +5 dBm having constant bit rate of 10 Gbps at distance of 240 km the performance of PIN photodiodes is observed. We have got highest $Q$-factor of different pulse generators (RZ, NRZ, GAUSSIAN, RAISED-COSINE) at power −6.11111 dBm, i.e., 28.9112, 39.2279, 29.1112, 16.1575 and min BER as 3.21E−184, 0.00E+00, 7.95E−187, 4.55E-59, respectively. RZ pulse generator performs better as compared to other pulse generators.

# References

1. M. Birk, L.M. Lunardi, A.H. Gnauck, H. Schumacher, D. Behammer, 10-Gbit/s RZ pulses using an all-silicon nonlinear transmission line integrated circuit, in *Optical Fiber Communication Conference. Technical Digest Post Conference Edition. Trends in Optics and Photonics*, vol. 37 (IEEE Cat. No. 00CH37079), vol.1 (2000), pp. 248–250

2. G.E. Town, K. Chan, G. Yoffe: Design and performance of high-speed optical pulse-code generators using optical fiber Bragg gratings. IEEE J. Sel. Top. Quant. Electron. **5**(5), 1325–1331 (1999)
3. T. Panda, A.N. Sahu, A. Sinha, Performance analysis of 50km long fiber optic link using fiber Bragg grating for dispersion compensation. Int. Res. J. Eng. Technol. **03**(03) (2016)
4. N. Kikuchi, Intersymbol Interference (ISI) suppression technique for optical binary and multilevel signal generation. J. Lightwave Technol. **25**(8), 2060–2068 (2007)
5. Pizzinat, Numerical and experimental comparison of dispersion compensation techniques on different fibers. IEEE Photon. Technol. Lett. **14**(10), 1415–1417 (2002)
6. C. Peucheret, N. Hanik, R. Freund, L. Molle, P. Jeppesen, Optimization of pre- and post-dispersion compensation schemes for 10-Gbits/s NRZ links using standard and dispersion compensating fibers. IEEE Photon. Technol. Lett. **12**(8), 992–994 (2000)
7. H.J. Abd, A. Ban, Intersymbol interference suppression method based on the selective electric pulses shaping approach. researchgate.net J. Babylon Univ./Eng. Sci. **24**(4) (2016)
8. M.l. Meena, D. Meena, Performance analysis of DWDM optical network with dispersion compensation techniques for 4 × 8 Gbps transmission system. ICTACT J. Microelectron. **04**(02) (2018)
9. M. Kaur, G. Bharti, Performance analysis of optical fiber communication systems for NRZ, RZ and duobinary formats. Int. J. Adv. Eng. Res. Sci. (IJAERS) **1**(5). ISSN: 2349–6495 (2014)
10. V. Kumar, A.K. Jaiswal, M. Kumar, N. Agrawal, R. Saxena, Design and performance analysis of optical transmission system. IOSR J. Eng. (IOSRJEN) **04**(05). ISSN (e): 2250-3021, ISSN (p): 2278-8719, ||V3|| PP 22-26 (2014)
11. P. Mishra, T. Panda, S.S. Rout, G. Palai, Investigation of a 16 channel 40 Gbps varied GVD DWDM system using dispersion compensating fiber, in *2020 International Conference on Computer Science, Engineering and Applications (ICCSEA)* (2020), pp. 1–5
12. T.K. Panda, A.B. Mukesh Kumar Behera, S. Polei, Performance comparison of Fiber Bragg Grating for different Grating length, Apodization function and Chirp function, in *2020 International Conference on Computer Science, Engineering and Applications (ICCSEA), 2020* (2020), pp. 1–5

# Simulation and Analysis of an 8 Channel CWDM Optical Network Suitable for Smart City Applications

**Lalit Kumar Kanoje, A. B. Mukesh Kumar Behera, Tushar Kant Panda, and Sudha Subhalaxmi Muduli**

**Abstract** The ever increasing demand for higher data rates and minimum latency has resulted in a widespread growth of optical fiber communication over the last decade. Dense Wavelength Division Multiplexing (DWDM) technique is one of the most important techniques used to establish a long-distance link with higher data rate transmission. However, when the requirement is to transmit the data inside a metropolitan (shorter range) city, DWDM has certain limitations like wideband change, more power consumption, and high-cost factor for lower channels. CWDM (Coarse Wavelength Division Multiplexing) can easily overcome these disadvantages. In this paper, the performance of 8 channel optical transmission link for Coarse Wavelength Division Multiplexing based optical network has been analyzed and compared. Here we've compared the performances in terms of Q-factor (Quality factor), BER (Bit Error Rate), and eye height in the eye diagram. For experimental setup and simulation, Optisystem-18 software is used.

**Keywords** CWDM · Bitrate · APD photodetector · Low pass filter · Q-factor

## 1 Introduction

In this arise development of internet or intranet, the necessity of bandwidth is becoming more and more serious. For increasing the bandwidth Wavelength Division Multiplexing (WDM) [1] technology is preferred the most. WDM is a type of multiplexing which multiplexes different carrier signals into a single optical fiber thereby increasing the total bandwidth of transmission. The two types of WDM are Coarse wavelength division multiplexing (CWDM) [2, 3] and Dense wavelength division multiplexing (DWDM) [4].

L. K. Kanoje (✉)
Department of ETC, Trident Academy of Technology, Bhubaneswar, Odisha 751024, India
e-mail: lalit.kanoje@tat.ac.in

A. B. Mukesh Kumar Behera · T. K. Panda · S. S. Muduli
Department of ECE, GIET University, Gunupur, Rayagada, Odisha 765022, India



S. Patnaik et al. (eds.), *Smart and Sustainable Technologies: Rural and Tribal Development Using IoT and Cloud Computing*, Advances in Sustainability Science and Technology,
https://doi.org/10.1007/978-981-19-2277-0_20

### 1.1 Multiplexing

Multiplexing and Demultiplexing are techniques used to combine or split multiple signals or the flow of information over a single communications link. When the signal reaches the destination, demultiplexer recovers the separated signals and outputs them in the form of individual lines. Advantages of multiplexing are, that large amount of data can be transmitted in the form of a composite signal over a common medium like radio wave or fiber optic cable [5]. This channel is a physical medium like metallic conductor, coaxial, or a wireless link and has to manage many signals at a time and the cost of the transmission will be reduced. When various signals transmit over the same channel, then it is very important to keep the signals apart for avoiding the interference between them and it will be very easy to get them separated at the receiving end.

### 1.2 DWDM

DWDM or dense wavelength division multiplexing is an advanced form of WDM where the number of input channel multiplexed are quite large as compared to WDM. Deplore of 40 channel systems are common with a few channels being used now, rest left of for future upgrades.

### 1.3 CWDM

Coarse wavelength division multiplexing (CWDM) is an optical transmission technique used for smaller distances of up to 60 km. This allows multiplexing up to 18 channels [2]. It is of two-way communication. The word "coarse" refers to wavelength spacing between channels. CWDM's vast spacing of up to 20 nm, when compare to DWDM's 1.6 nm, can permit much higher temperature variations. CWDM is very useful at 5G as the optics are extremely cheap and can reach an interval of up to 180 km. CWDM is inexpensive to maximize existing fiber by reducing the channel spacing between wavelengths. It also increases bandwidth on existing fiber infrastructure [6].

## 2 Simulation Setup

The picture shown in Fig. 1 represents the basic block diagram of our simulation. Here we've used 8 channels input to the 8 * 1 multiplexer, then the o/p of the mux is passed through the CWDM fiber. The signal then passes through a 1 * 8 Demultiplexer to

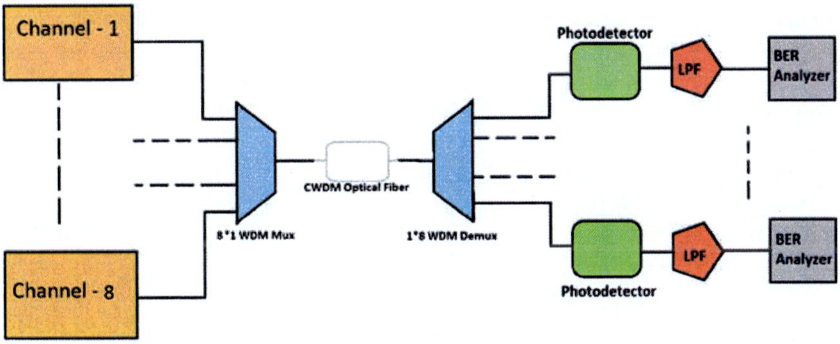


**Fig. 1** Block diagram for the simulation setup

separately be communicated to 8 channels. Here we have compared the performance by using an APD photodetector [7, 8] in terms of its receiver sensitivity. The signal then passes through a Bessel low pass filter [9] to remove the unwanted noise that is added during the propagation through the channel. The performance of the circuit is further analyzed with the help of Q Factor [10] and BER [11].

In the transmitter, we have taken a pseudo-random bit sequence generator to generate 0's and 1's bits randomly. Then the binary bit is given as i/p to a NRZ pulse generator [12] which gives the pulse in the o/p according to the given binary bit and the pulse is then passed through the Mach-Zender (MZ) modulator [13] as the message signal. A CW laser is taken to generate the carrier signal which is again given as i/p to the M-Z modulator and the modulator gives the modulated signal as the o/p. We consider this o/p as a channel o/p. Similarly, we've taken 8 channels and given the o/p of these channels as i/p to a 8 * 1 WDM Mux. The o/p of the multiplexer is passed through the CWDM optical fiber. Here we've taken two optical power meters and two optical spectrum analyzers [14] and connected them to both i/p and o/p of the CWDM optical fiber, respectively. The o/p of the fiber is given i/p to a 1 * 8 WDM DEMUX which receives the multiplexed signal and separates the all 8 signals. The signals are then passed through the photodetectors (APD). Photodetector receives the optical signal and converts it into electrical message signal and the electrical signal is filtered by the low pass filter [15]. After that, the o/p message signal is given as i/p to a BER analyzer.

The simulation parameters and CWDM optical fiber parameters used for the simulation are listed in Tables 1 and 2.

## 3 Result and Discussion

In the simulation, we've taken 8 channel WDM mux and applied i/p modulated signal of bit rate 15 Gbps to each channel. We've varied the distance of the fiber from 10

**Table 1** Simulation parameters used in the setup

| Simulation parameters | Values |
|---|---|
| Bit rate | 15 Gbps |
| Samples per bit | 64 |
| Number of samples | 8192 |
| Sequence length | 128 Bits |
| Interpolation offset | 0.5 nm |
| Power unit | dBm |
| Sensitivity | −100 dBm |
| Resolution | 0.1 nm |
| Interpolation offset | 0.5 nm |

**Table 2** Fiber parameters used in the simulation setup

| Fiber parameters | Values |
|---|---|
| Reference wavelength | 1310 nm |
| Fiber length | 10–60 km |
| Attenuation | 0.2 dB/km |
| Dispersion | 1675 |
| Max. nonlinear phase shift | 3 mrad |
| Differential group delay | 0.2 ps/km |
| PMD coefficient | 0.5 ps/sqrt (km) |
| Nonlinearities effective area | 80 $\mu m^2$ |
| Signal center frequency | 193.1 Thz |

to 60 km. Calculation is carried out till the desired o/p is obtained and the Q-factor and BER readings observed are listed in Table 3.

From the above Q-Factor results of 8 channels, we've considered four channels, i.e., channel 1, channel 3, channel 5, and channel 8 for our comparison and the graph has been plotted in Fig. 2 (Table 3).

**Table 3** Q-factor reading at different CWDM fiber lengths with APD photodetector

| | 10 km | 20 km | 30 km | 40 km | 50 km | 60 km |
|---|---|---|---|---|---|---|
| Channel1 | 26.9595 | 18.7954 | 13.5091 | 11.4812 | 9.1667 | 9.56172 |
| Channel2 | 24.9456 | 18.5632 | 13.0256 | 10.9563 | 9.1986 | 9.25639 |
| Channel3 | 22.015 | 18.0399 | 12.6366 | 10.3394 | 9.26315 | 9.05159 |
| Channel4 | 21.367 | 17.3569 | 12.798 | 10.4369 | 9.34656 | 9.3689 |
| Channel5 | 20.1517 | 16.7149 | 12.885 | 10.5084 | 9.53609 | 9.69215 |
| Channel6 | 23.2569 | 17.3695 | 12.9753 | 10.4983 | 9.41235 | 9.5985 |
| Channel7 | 27.6419 | 18.1289 | 13.5679 | 10.3986 | 9.3423 | 9.5103 |
| Channel8 | 29.2851 | 19.8059 | 14.0438 | 10.3528 | 9.24167 | 9.46335 |

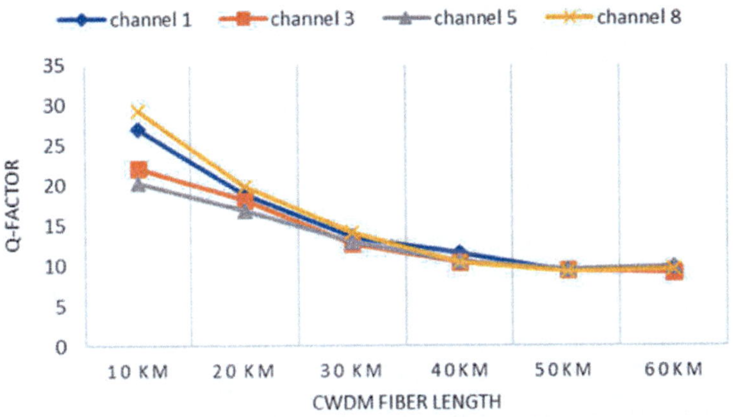


**Fig. 2** The graph shows the variation of Q factor with respect to various lengths of the fiber (10–60 km) for the considered 4 channels when APD photodetector is used at the receiver end

**Table 4** BER reading at different CWDM fiber lengths with APD photodetector

| | 10 km | 20 km | 30 km | 40 km | 50 km | 60 km |
|---|---|---|---|---|---|---|
| Channel 1 | 2.20E−160 | 4.11E−79 | 6.90E−42 | 7.94E−31 | 2.27E−20 | 5.56E−22 |
| Channel 2 | 1.98E−136 | 4.56E−75 | 6.70E−39 | 5.36E−28 | 3.60E−19 | 5.89E−22 |
| Channel 3 | 1.04E−107 | 4.69E−73 | 6.63E−37 | 2.28E−25 | 9.13E−21 | 6.88E−22 |
| Channel 4 | 1.13E−98 | 5.36E−65 | 5.34E−37 | 2.93E−26 | 7.86E−21 | 3.45E−22 |
| Channel 5 | 1.30E−90 | 5.10E−63 | 2.73E−38 | 3.88E−26 | 6.83E−22 | 1.58E−22 |
| Channel 6 | 1.69E−120 | 5.23E−64 | 2.59E−41 | 2.98E−25 | 5.32E−21 | 1.53E−21 |
| Channel 7 | 5.63E−159 | 3.45E−76 | 3.26E−43 | 2.16E−25 | 3.19E−20 | 1.49E−21 |
| Channel 8 | 8.04E−189 | 1.32E−87 | 4.19E−45 | 1.98E−25 | 1.13E−20 | 1.47E−21 |

From the above BER results of 8 channels, we've considered four channels, i.e., channel 1, channel 3, channel 5, and channel 8 for our comparison and the graph has been plotted in Figs. 3, 4, 5, 6 and 7.

## 4 Conclusion

From all our observed data we found that with APD photodetector we have desired o/p at 50 km CWDM optical fiber and we have Q-factor values 9.1667, 9.1986, 9.26315, 9.34656, 9.53609, 9.41235, 9.3423 and 9.24167 and BER values, 2.27E−20, 3.60E−19, 9.13E−21, 7.86E−21, 6.83E−22, 5.32E−21, 3.19E−20 and 1.13E−20 at channel 1, channel 2, channel 3, channel 4, channel 5, channel 6, channel 7 and channel 8, respectively. CWDM along with APD can be a very good candidate

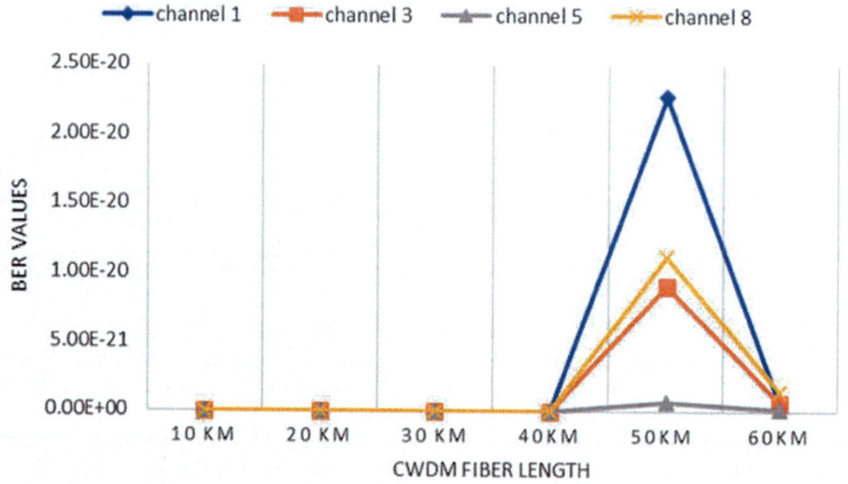


**Fig. 3** The graph shows the variation of BER values with respect to various lengths of fiber (10–60 km) for the considered 4 channels when APD photodetector is used at the receiver end

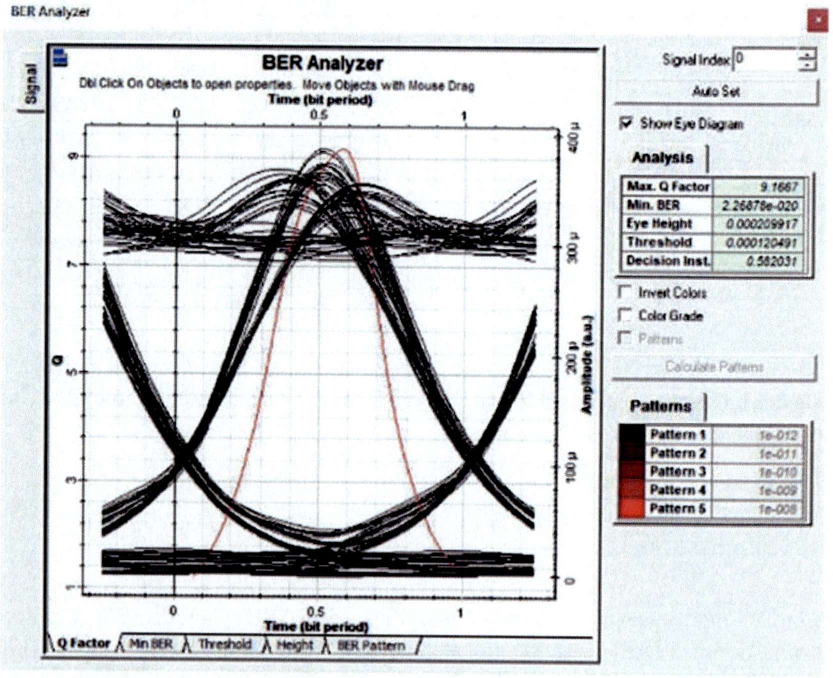


**Fig. 4** Figures show the eye diagrams obtained from the BER analyzer at the channel 1 and channel 3, respectively

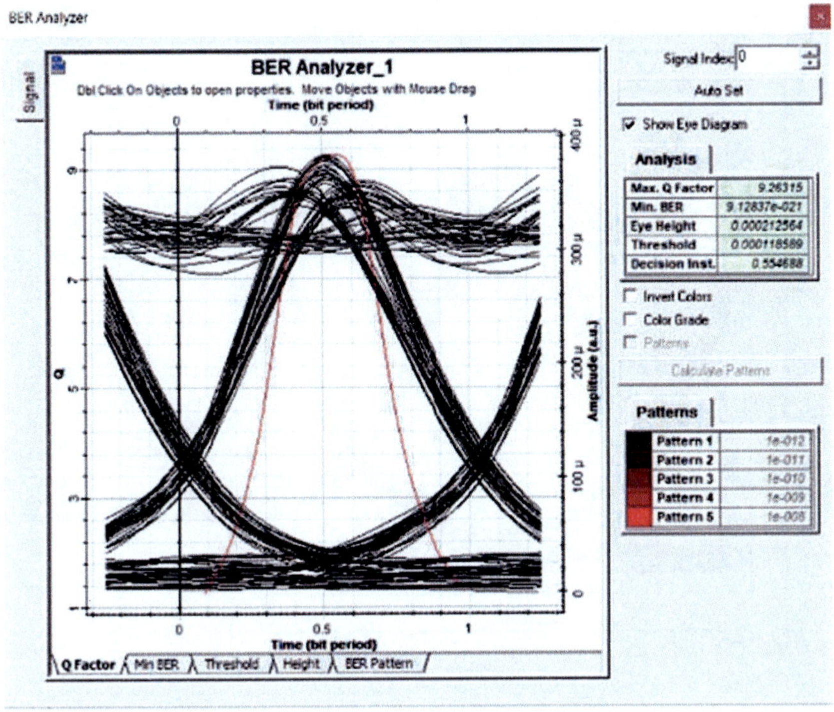


**Fig. 5** Figures show the eye diagrams obtained from the BER analyzer at the channel 1 and channel 3, respectively

for metro cities ranging from 10 to 60 km as the result we found is very much satisfactory for this specified range. However, further works can be carried out to improve this range for bigger cities and there is also a scope for high data rate transmission in an 18 channel CWDM.

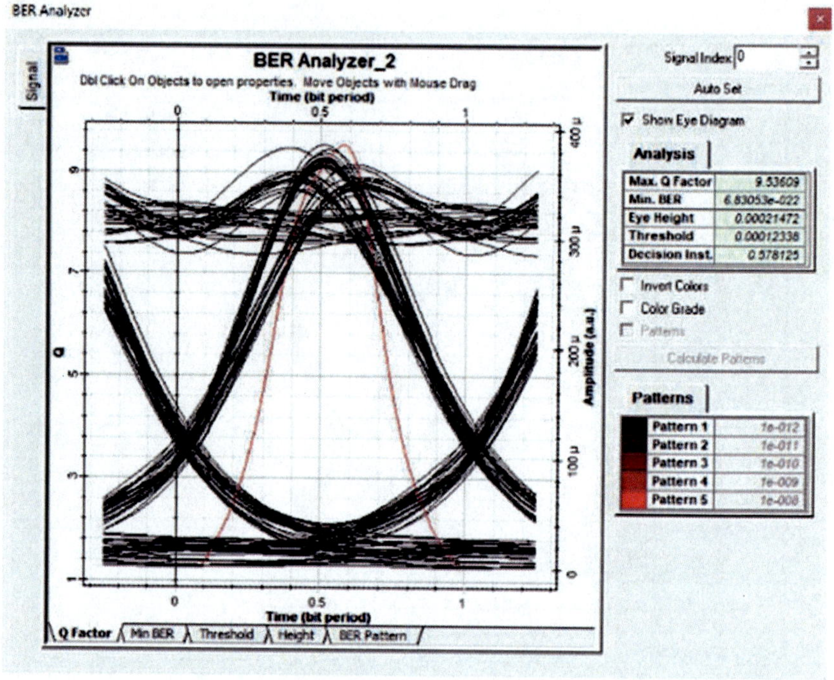


**Fig. 6** Figures show the eye diagrams obtained from the BER analyzer at the channel 5 and channel 8, respectively

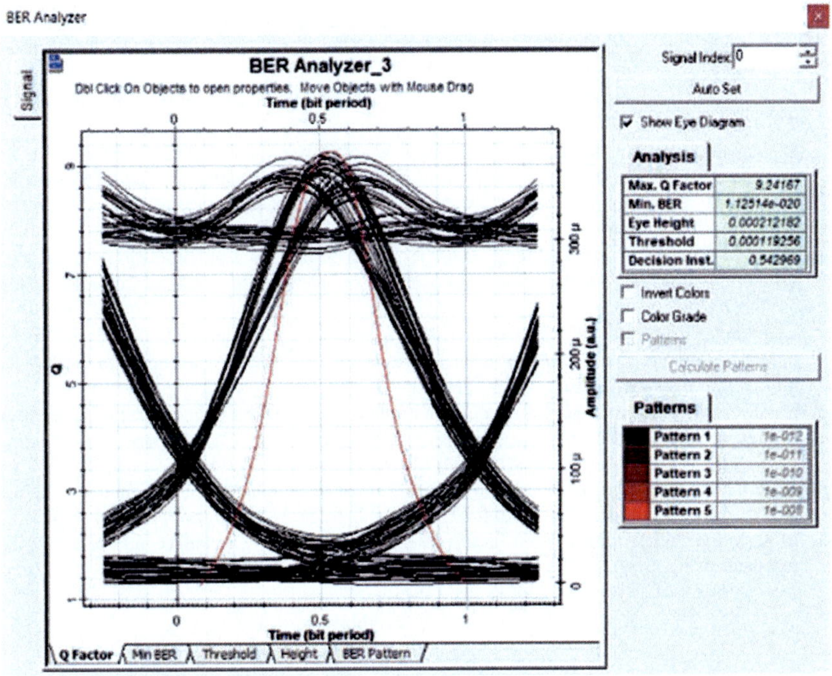


**Fig. 7** Figure show the eye diagrams obtained from the BER analyzer at the channel 5 and channel 8, respectively

## References


1. W. Wu, Y. Yu, S. Hu, B. Zou, X. Zhang, All-optical format conversion for polarization and wavelength division multiplexed system. IEEE Photon. Technol. Lett. **24**(18), 1606–1609 (2012)
2. D. Liu, M. Zhang, Y. Shi, D. Dai, Four-channel CWDM (de)multiplexers using cascaded multimode waveguide gratings. IEEE Photon. Technol. Lett. **32**(4), 192–195 (2020)
3. Y. Zheng, X. Wu, L. Jiang, Y. Wu, J. Duan, Design of 4-channel AWG multiplexer/demultiplexer for CWDM system. In: Optik, vol. 201. ISSN 0030-4026 (2020)
4. F.I. Khatri, Novel architectures for long haul DWDM systems: 2001 Digest of LEOS Summer Topical Meetings (2001), p. 2
5. M.K. Abdullah, G.A. Mahdiraji, A.M. Mohammadi, M. Mokhtar, A.F. Abas, Duty cycle division multiplexing (dcdm); a new electrical multiplexing technique for high speed optical communication systems, in *6th National Conference on Telecommunication Technologies and 2008 2nd Malaysia Conference on Photonics* (2008), pp. 71–74
6. P. Mishra, T. Panda, S.S. Rout, G. Palai, Investigation of a 16 channel 40 Gbps varied GVD DWDM system using dispersion compensating fiber, in *2020 International Conference on Computer Science, Engineering and Applications (ICCSEA)* (2020), pp. 1–5
7. J. Youn, H. Kang, M. Lee, K. Park, W. Choi, High-speed CMOS integrated optical receiver with an avalanche photodetector. IEEE Photon. Technol. Lett. **21**(20), 1553–1555 (2009)

8. X. Liang, Z. Su, S. Bi, Research on the associated feature of APD sensor avalanche gain and photoelectric conversion SNR, in *2012 9th International Conference on Fuzzy Systems and Knowledge Discovery* (2012), pp. 1913–1916
9. A. Kumar, A.K. Verma, Compact low pass Bessel filter using microstrip DGS structure, in *2010 Asia-Pacific Microwave Conference* (2010), pp. 1189–1192
10. J. Vijayamohanan, O. Noakoasteen, A. Gupta, M. Martínez-Ramón, C.G. Christodoulou, On antenna Q-factor characterization with generative adversarial networks, in *2020 IEEE International Symposium on Antennas and Propagation and North American Radio Science Meeting* (2020), pp. 1643–1644
11. X. Li, X. Zhao, P. Zhang, S. Tong, BER performance of FSO communication system with differential signaling over correlated atmospheric turbulence fading. China Commun. **17**(4), 51–65 (2020)
12. Z. Ibrahim, C.B.M. Rashidi, S.A. Aljunid, A.K. Rahman, M.S. Anuar, NRZ and RZ analysis for optical CDMA based on radio over fiber (RoF) technique, in *2016 3rd International Conference on Electronic Design (ICED)* (2016), pp. 151–154
13. M. Khulbe, D. Baisoya, M.R. Tripathy, Photonic circuits for different coding schemes used in high speed Terabit communication, in *2018 4th International Conference on Computing Communication and Automation (ICCCA)* (2018), pp. 1–6
14. Y. Zhong, Impact of out-of-band noise on OSNR measurement using Brillouin optical spectrum analyzer and its mitigation method. IEEE Photon. J. **10**(5), 1–10, Art no. 7204010 (2018)
15. M.K. Mohd Salleh, Out-of-band improvement of the quarter wavelength side-coupled ring resonator using low-pass filter integration, in *2011 IEEE International Conference on System Engineering and Technology* (2011), pp. 128–131

# Design of Low-Power Dynamic Threshold MOSFET (DTMOS) Push–Pull Inverter

**Ganapati Sahu, Baibhab Panigrahi, and Shasanka Sekhar Rout**

**Abstract** This brief represents a low-power push–pull inverter by using dynamic threshold MOSFETs (DTMOSs). The trend in design and manufacturing in semiconductor production units and implementation of million and billion number of transistors in a single chip in very large scale integration is increasing day by day. The demand for Complementary MOSFET (CMOS) integrated circuits (ICs) is continually acceptable due to silent features like low-power consumption, reliable performance, low noise, lesser area, high-speed performance, and outgoing improvement in dealing out technology. However, due to the static circuit standby power and the probability of breakdown in dynamic circuits, a large reduction in the MOS transistor's threshold voltage ($V_t$) occurs, resulting in a reduction in gate overdrive and as a result, the entire circuit's speed of operation. A DTMOS with a low $V_t$ when the MOS turns on and a high $V_t$ when the MOS turns off can be used to overcome this issue. The proposed design is simulated in Cadence tool with 180 nm CMOS technology. It results the average propagation delay of 13.84 ps and average power of 1.39 μW. Therefore, this brief will be an appropriate block for any customized design circuit.

**Keywords** CMOS · Cadence · DTMOS · Threshold voltage · Propagation delay

## 1 Introduction

The demand for high-performance and low-power consumption with low noise circuits in digital systems has grown rapidly day by day. Thus there are some factors that have to consider for increasing the output performance of circuit in this faster growth. For example, during personal computers, laptops, and notebook computers have gained popularity. There have been many developments in ICs with their use in high presentation computing, communication, and consumer electronics requirements. Consequently, the demand for portable applications such as digital watches

G. Sahu · B. Panigrahi · S. S. Rout (✉)
GIET University, Gunupur, India
e-mail: ssrout@giet.edu



S. Patnaik et al. (eds.), *Smart and Sustainable Technologies: Rural and Tribal Development Using IoT and Cloud Computing*, Advances in Sustainability Science and Technology,
https://doi.org/10.1007/978-981-19-2277-0_21

and calculators also increased due to their high output performance. The desire for portability of the innovative systems restricts the size of their batteries, the placement of all equipment in a smaller space, their weight, and their power consumption. Second, non-portable CMOS-based systems' speed, density, and size have all increased dramatically in recent years [1]. As a result, power consumption, which was previously unimportant in these systems, has now become a vital metric. Power dissipation is a problem of increasing concern to designers of VLSI circuits. The switching power dissipation of digital VLSI systems can be derived, which is shown in Eq. (1).

$$P_{\text{avg}} = C_l V_{\text{dd}}^2 f \tag{1}$$

where $C_l$ is the total effective or switching capacitance, $V_{dd}$ is the supply voltage and $f$ is the working frequency.

Technologies scaling has resulted in an exponential increase in discharge current as we go closer to nanoscale technology. As a result, several discharge minimization techniques [2] have been developed, including Sleepy CMOS-Sleepy Stack Technique (SCSS) [3], Sleepy Stack Technique (SST) [4], Sleepy Keeper Technique (SKT) [5], Multi Threshold CMOS Technique (MTCMOS) [6], Super Cut-off CMOS Technique (SCCMOS) [7], Dual Threshold CMOS Technique (DTCMOS) [8], Conventional CMOS Technique (CCT), Transistor Stacking Technique [9], Variable Threshold CMOS Technique (VTMOS) [10–12] have been implemented on CMOS inverter.

In this paper, a low-power push–pull inverter circuit is proposed by using DTMOS concept. The main focus of this article is to implement the design with low-power consumption, with lesser area which will consequently give the high performance.

The paper is designed as following stages. The circuit design with DTMOS technique is described in the concise discussion in Sect. 2. The equivalent simulation outcomes are reported in Sect. 3. Finally, it is concluded in Sect. 4.

## 2 Proposed Design and Analysis

Figure 1 shows the schematic diagram of a DTMOS-based push–pull inverter. The inverter circuit is made up of an enhancement-type nMOS transistor and an enhancement-type pMOS transistor. For high input, the nMOS transistor drives (pulls down) the output node while the pMOS transistor acts as the load, and for low input, the pMOS transistor drives (pulls up) the output node while the nMOS transistor acts as the load [3]. As a result, both devices have an equal influence on the circuit's performance.

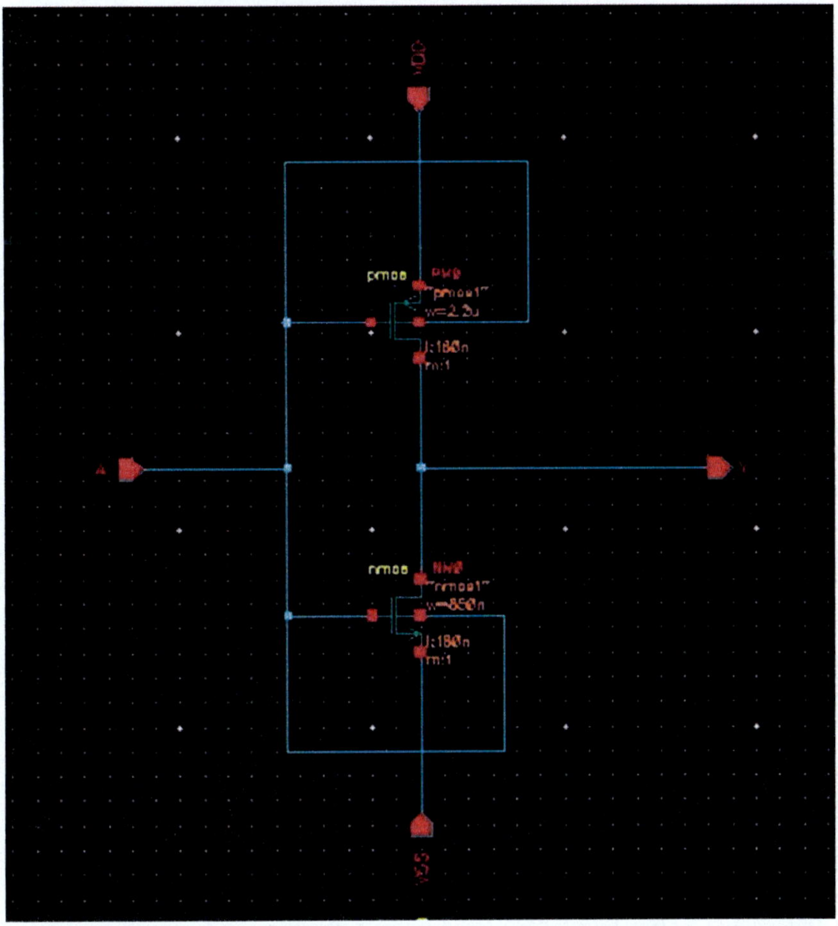


**Fig. 1** Proposed DTMOS inverter schematic

The DTMOS [13] utilizes high $V_t$ potential difference transistors on paths that are noncritical to lessen discharge power. Low $V_t$ transistor is used which helps keep circuit performance on critical path. Algorithm is required by this approach to search for gates at which high $V_t$ potential difference devices can be taken into use. This is popularly called Dual $V_t$ CMOS. When the transistor is OFF, to have low discharge, every transistor's body and gate are connected together in DTMOS [13]. If transistor is ON, current will be large. The concept of DTMOS says that the body terminal of both PMOS and NMOS will be connected together to the gate terminal. Since both gate and body terminals are connected to each other, by increasing the input voltage at the gate terminal the $V_T$ of the derived DTMOS is reduced hence the output current will be increased.

In this paper, by applying the concept of DTMOS connecting together the body terminal to the gate terminal of both PMOS transistor and NMOS transistor we are developing a new model of push–pull inverter referred to as Dynamic Threshold MOSFET (DTMOS) [14] push–pull inverter. Since both gate and body terminals are connected to each other, by increasing the input voltage at the gate terminal the $V_{\mathrm{T}}$ of the derived DTMOS is reduced according to threshold voltage Eq. (2):

$$V_T = V_{T0} + \gamma\sqrt{2\emptyset_F - V_{BS}} - \gamma\sqrt{2\emptyset_F} \quad (2)$$

By decreasing in threshold voltage significantly increases the current at the output terminal. However, because of the need for standby power in static circuits and the risk of failure in dynamic circuits, a large reduction in the MOS transistor's threshold voltage ($V_{\mathrm{t}}$) occurs, resulting in a reduction in gate overdrive and, as a result, the entire circuit's speed of operation. A DTMOS with a low $V_{\mathrm{t}}$ when the MOS turns on and a high $V_{\mathrm{t}}$ when the MOS turns off can be used to overcome this issue. A DTMOS transistor exhibits this precise behavior by linking the gate to the body of transistor. DTMOS has a lower $V_{\mathrm{t}}$ due to a drop in the depletion charge, which produces an enhancement in the inversion charge and so, higher current drive results less power consumption.

## 3 Simulation Results

The proposed work has been implemented in Cadence tool by using 180 nm CMOS technology. The transient analysis and voltage transfer characteristics (VTC) of DTMOS push–pull inverter are done for the analysis. Figure 2 represents the Transient analysis and Fig. 3 shows the VTC of proposed design. From the simulation results, the high-low propagation delay ($t_{\mathrm{PHL}}$) is found 14.795 ps, low–high propagation delay ($t_{\mathrm{PLH}}$) is found 12.881 ps. It results the average propagation delay of 13.84 ps. The average power of the design is calculated as 1.39 $\mu$w.

After performing the analysis, output of DTMOS push–pull inverter acts like a normal CMOS inverter, i.e., when we are applying HIGH input to the circuit it provides LOW as output and vice versa. Apart from that, the output of this circuit results higher current drive with less power consumption. Figure 4 represents the layout of DTMOS push–pull inverter circuit.

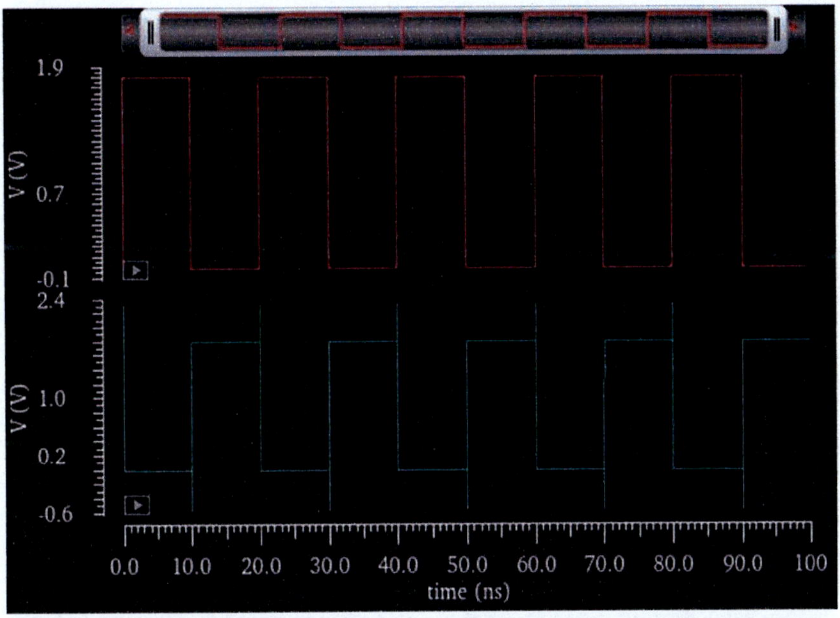


**Fig. 2** Transient of DTMOS inverter

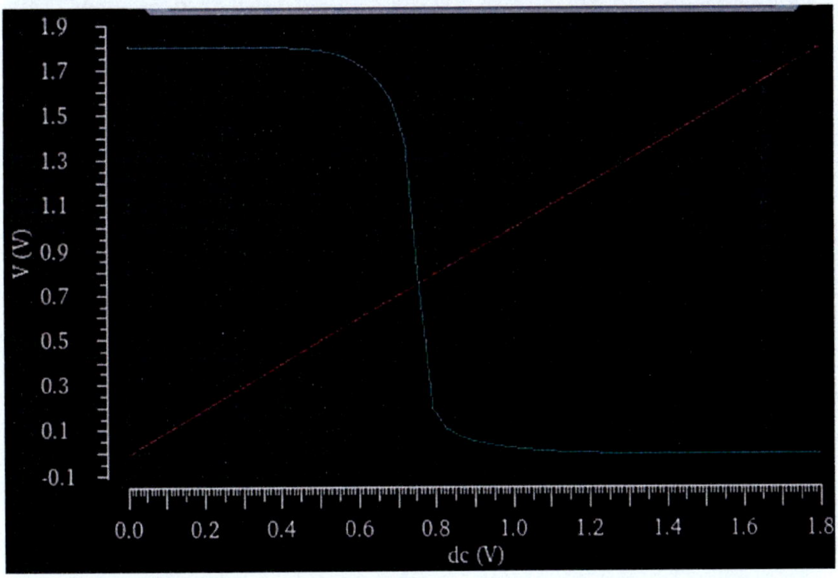


**Fig. 3** VTC of DTMOS push–pull inverter circuit

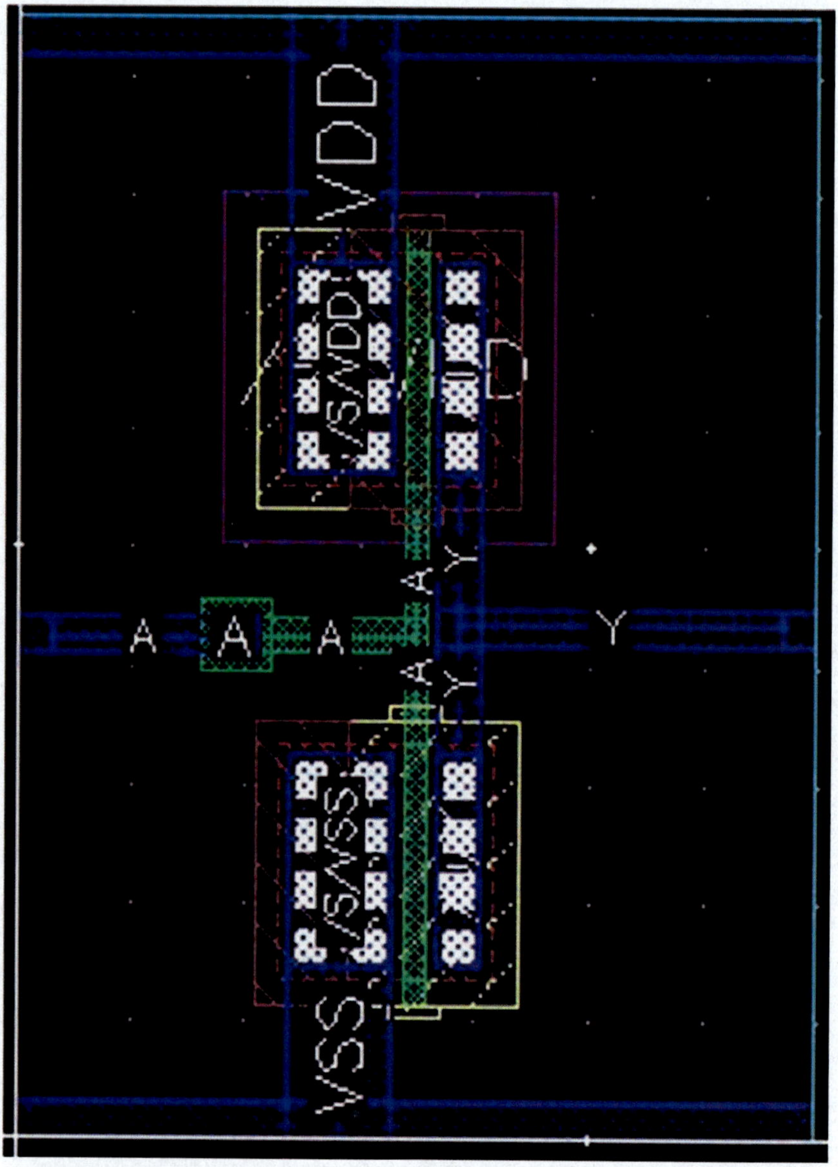


**Fig. 4** Layout of DTMOS push–pull inverter circuit

## 4 Conclusion

The proposed circuit is designed with the help of DTMOS and simulated in 180 nm CMOS technology. The DTMOS is a good technique for the analysis of low voltage and low-power operations. DTMOS also clears the floating body issues of SOI MOSFET's such as transient analysis and $V_t$ stability. From the analysis, the high-low propagation delay ($t_{PHL}$) is found 14.795 ps, low–high propagation delay ($t_{PLH}$) is found 12.881 ps. The average propagation delay and the average power of the design are calculated as 13.84 ps and 1.39 μw respectively. Therefore this proposed design can be used in low poser VLSI full custom design.

## References

1. L. Garg, V. Sahula, Efficient CMOS subthreshold discharge analysis with improved stack based models in presence of parameter variation. Electron. Lett. **49**(19), 644–646 (2013)
2. M.G. Priya, K. Baskaran, D. Krishnaveni, Discharge power reduction techniques in deep submicron technologies for VLSI applications. Procedia Eng. **30**, 1163–1170 (2012). https://doi.org/10.1016/j.proeng.2012.01.976
3. A. Sharma, A.P. Goyal, H. Sohal, H.J. Kaur, Sleepy CMOS-sleepy stack (SC-SS): a novel high speed, area and power efficient technique for VLSI circuit design. J. Circuits Syst. Comput. **28**(12), 1950197 (2019)
4. J.C. Park, V.J. Mooney, Sleepy stack discharge reduction. IEEE Trans. Very Large Scale Integr. (VLSI) Sys. **14**(11), 1250–1263 (2006)
5. K. Gnana, K. Deepika, K., P.K. Mariya, D.S. Raj, Sleepy Keeper approach for power performance tuning in VLSI design. Int. J. Electron. Commun. Eng. **6**(1), 17–28 (2013)
6. S. Mutoh, T. Douseki, Y. Matsuya, T. Aoki, S. Shigematsu, J. Yamada, 1-V power supply high-speed digital circuit technology with multithreshold-potential difference CMOS. IEEE J. Solid-State Circuits **30**(8), 847–854 (1995)
7. H. Kawaguchi, K. Nose, T. Sakurai, A super cut-off CMOS (SCCMOS) scheme for 0.5-V supply potential difference with Pico Ampere stand-by current. IEEE J. Solid-States Circuits **35**(10), 1498–1502 (2000)
8. F. Assaderaghi, D. Sinitsky, S.A. Parke, J. Borkar, P.K. Ko, C. Hu, Dynamic threshold potential difference MOSFET (DTMOS) for ultra-low potential difference VLSI. IEEE Trans. Electron Dev. **44**(3), 414–422 (1997)
9. S. Narendran, S. Borkar, V. De, D. Antoniadisn, A. Chandrakasann, Scaling of stack effect and its application for discharge reduction, in *Proceedings of ISLPED* (2001), pp. 195–200
10. A. Abdollahi, F. Fallah, M. Pedram, Discharge current reduction in CMOS VLSI circuits by input vector control. IEEE Transa. Very Large Scale Integr. (VLSI) Syst. **12**(2), 140–154 (2004)
11. S.S. Rout, K. Sethi, DTMOS based Gilbert mixer design for MICS receiver using current source helpers and switched biasing technique. Sadhana **45**(1), 1–7 (2020)
12. S. Nehra, P.K. Ghosh, Design of a low power XNOR gate using MTCMOS technique, in *Advance in Electronic and Electric Engineering* (2013), pp. 2231–1297
13. S.S. Rout, S. Acharya, K. Sethi, Design of a low power and low phase noise VCO using active resistor and DTMOS, in *Proceedings on Advances in Electrical Control and Signal Systems (AECSS-2019)*, vol. 665 (2020), pp. 671–679
14. S.S. Rout, S. Acharya, K. Sethi, A low phase noise $g_m$-boosted DTMOS VCO design in 180 nm CMOS technology. Karbala Int. J. Mod. Sci. **4**(2), 228–236 (2018)

# Epileptic Seizure Detection Using Deep Learning Architecture

**Nagavarapu Sowmya, Subhrajit Pradhan, Pradyut Kumar Biswal, Sudeep Kumar Panda, and Vishnu Priya Misra**

**Abstract** Epilepsy is one of the dangerous and a serious issues in the field of biomedical application. The exact prediction of seizures enables the patients to get over the weakness and ambiguity. Electroencephalogram (EEG) is a method that is generally used to diagnose conditions like epilepsy, sleep disorder, and brain tumor along with that it studies the changes in frequency and electrical activity of the brain during seizures. According to traditional approach, the features were extracted from EEG signals manually and conventional Machine learning (ML) based techniques have been used to identify this neurological disorder. Though the scientists succeeded in terms of effectiveness but multiple class classification using automatic feature extraction could not be achieved. In this paper, we have analyzed the five different EEG datasets and implemented the dataset which is recording the seizure activity using Deep learning model with Keras in Python to make the detection process more memory efficient and time-efficient. An accuracy of 97.46% is achieved with the help of proposed model and classification of other EEG datasets using deep learning would be analyzed in near future.

**Keywords** Epilepsy · EEG · Machine learning · Deep learning

## 1 Introduction

Epilepsy is an abnormal brain activity that most of the people fight with all over the world. The patients with epilepsy undergo impulsive situations irrespective of time. The International League Against Epilepsy (ILAE) has come up with latest footings in order to explain and categorize kinds of seizures in an efficient manner. There are three criteria that need to be considered while deciding on the class of seizure

N. Sowmya (✉) · S. Pradhan · S. K. Panda · V. P. Misra
GIET University, Gunupur, India
e-mail: nagavarapu.sowmya@giet.edu

P. K. Biswal
IIIT, Bhubaneswar, India



S. Patnaik et al. (eds.), *Smart and Sustainable Technologies: Rural and Tribal Development Using IoT and Cloud Computing*, Advances in Sustainability Science and Technology,
https://doi.org/10.1007/978-981-19-2277-0_22

disorder. At first it describes seizure onset. Then second is level of awareness that the patient has, which can indicate what type of seizure it is. And third is the presence of movement during the seizure [1].

Brain signals can be detected by using various methods like computerized X-ray imaging, Magnetic Resonance Imaging (MRI). Similarly, EEG is another method that helps to scan brain and its actions based on the corresponding frequency of a signal [2].

In current time, EEG-based activity analysis and cataloging have been extensively utilized in several applications starting from clinical to brain-machine interface and robotics. Brain wave test is a suitable device for sensing neurological disorders. During this process, the probes were kept on brain scalp via glue. The recordings and samples of victim taken during seizures differ from an ordinary man. Doctors study and go through the EEG recordings and takes decisions based on the pattern [3].

The diagnosis of epilepsy with EEG signals is time-consuming as the neurologist needs to examine the patterns very carefully. And, there is a chance of human error, and thus, emerging into computer-based detection might overcome these problems. Many machine learning algorithms have been evolved by means of arithmetical, frequency domain to encounter this neurological disorder. As per traditional approach, the identification of features and classifiers is done by hit and trial method. And, one needs to have complete information on signal processing and data mining to build a robust model. Such frameworks work for limited data. Nowadays, the availability of data is very huge for which conventional machine learning methods may not perform very well. Hence, the deep learning techniques have been introduced and the deep learning methods are usually developed using Python language with open-source toolboxes. Deep Neural Networks are known for instinctive feature abstraction and because of their capability to categorize non-linear models between the input and output, they have been considered as the powerful tool for EEG classification in recent times. Psychic seizure disorder occurs which are same as epileptic ones but lacks medical evidence. Patients with psychogenic confiscations might tumble or quiver and experience a momentary loss of attention. The symptoms start from complete paralysis and not speaking to intense spanking also shuddering actions of arms, legs, or head. The patients' quality of life can be improved by automatic detection of seizures. During seizure attacks, many life-threatening emergencies arise which causes the patients to injure themselves and hence creating many other health-related issues. Head injuries, breathing difficulty, fractures in bones, and fits are some of serious situations caused by seizures. Generally as compared to general population, patients dying with seizures are 1.6 to 3 times higher, which is sudden and most probably unexpected. Even though it is not a frequent issue, people need to be aware of this risk, and measures must be taken accordingly. The important traits of EEG comprise non-corrosive, painless, and interpretate accurately diseases like trauma, amnesia, and mental illness. In addition to that, these waves are categorized based on their respective frequencies used for several circumstances such as movement of eyelid, shut eyes, tightening of fingers, etc. These signals are related with particular frequency ranges from 0 to 100 Hz. Dementia is one of the most frequent types of disease which involves brain deterioration that is irreparable and slowly ends up with complete brain failure. Aging and other circumstances upsurge the probabilities of neurons disintegration which increases the load on mind [4–11].

Machine Learning is one of the basic concepts in the field of artificial intelligence. It is used in various fields like speech recognition, image and pattern recognition, web search, covid data forecast, and so on. The tasks are widely divided into two types: Supervised and Unsupervised learning. Former is about labeled data and latter provides an idea of unlabeled data. Semi-supervised is between categorized and non-categorized data. Reinforcement denotes that agent is rewarded for good actions or gets punishment [12].

Human brain controls all body movements, emotions, and complicated tasks. When it comes to face recognition and learning language, we can solve such issues easily with much-enhanced efficiency and better performance than the supercomputers. As a result, brain-inspired computation has attracted most of the researchers to provide an interesting architectural solution compared to von Neumann machines. Seizure disorder is one of the most common types which is experienced by most of the people in current time. As this type of abnormal brain activity is irregular thus the analysis, diagnosis, and forecast results in a tedious and time-consuming process. It requires an expert to analyze the EEG signal and the reason behind the occurrence of confiscations. Seizure is the second most critical nervous system disease after stroke. EEG based on the computer-aided methodologies was found very effectual in detecting the epilepsy as compared with other techniques [13–15].

The paper is planned as follows: The literature review regarding Deep learning model is summarized in Sect. 2. Method initiation with improved techniques is described in Sect. 3. The proposed model is depicted in Sect. 4. Result and discussion are shown in Sect. 5. Finally, Sect. 6 concludes it.

## 2 Literature Review

Soroor [1] demonstrated estimate of confiscations using Heart rate variability along with the brain-based procedures. Here, the latest sensing and forecast methods for diagnosing the inactivity of human brain through cyclic length variability is defined.

Sriram et al. [7] described that neurological disorder detection and forecast give innovative and intended prospects for keeping a track of epilepsy. Moreover, these structures could be utilized in closed loops to halt the confiscation. Besides, they evaluate the risk factor while monitoring any abnormality in brain.

Kenneth et al. [8] summarized that known menaces linked with refractory epilepsy give medical endorsements and specify areas for forthcoming investigation.

Nandy et al. [3] presented a complete computerized system for seizure detection through physical conditions. Initially, this makes the signal eliminate the commotion and objects via a complex waveform then takes out the traits. Traits were taken from several domains. Moreover, integration, as well as energy-related characteristics, were considered for feature selection. Then the important elements are chosen via broad set by a multifunction metamorphic procedure. Support Vector Machine (SVM) is used for sorting resolution. A Bayesian optimization method has been used to enhance the parameters of SVM. The suggested methodology is tested online using MIT database, comparison is made and accuracy is determined as 97.05%.

Satheesh Kumar et al. [2] elaborated that Electroencephalography (EEG) is an efficient tool that helps to get the brain signals with respect to different positions by the cranium shallow region. Alpha, beta, gamma, etc. are categorized as different waveforms based on signal frequencies ranging from 0.1 to 100 Hz. This paper concentrated on EEG signals and their classification via several states of human body.

Sandeep et al. [10] proposed a new altered particle swarm optimization approach that aims at radial basis function semantic net to categorize the epileptic seizures. And this method is carried out on a diverse data point that is eye state prediction. In this paper, discrete wavelet transform was utilized for investigation and arithmetical feature abstraction from EEG datasets for seizure disorder.

Hamid et al. [5] depicted that artificial neural networks particularly convolution neural networks (CNN) give effective and precise outcomes in different pattern-related classification issues. This paper highlights the alcoholism detection using EEG signals. It is observed that CNN on raw EEG data has been used to show how 98% average precision is attained on UCI-ML dataset by enhancing the proposed model in terms of recall, AUC, etc. It shows the line-by-line improvement of baseline model by means of dropout, batch standardization, and kernel regulation methods and also provides comparison with respect to classification problems. This research also describes that there is no limitation of applying these methods to EEG categorization only but also can be applied to wide range of CNN applications.

Yang [12] highlighted several machine learning approaches for analysis of EEG data during seizures and also focused on diverse applications with respect to automatic detection of abnormality in brain, future prediction along with the pros and cons of the above-mentioned technique.

Shoeibi et al. [4] studied complete synopsis of works concentrating on computerized epileptic seizure detection using deep learning methodologies and neuroimaging sensory systems. Several methods were used to detect the confiscations by means of EEG, MRI, and along with that some recuperation systems have also been established to sense this disorder for suitable investigation and provided a summary on the same. Rehabilitation tools comprise cloud computing systems and hardware required for operation of deep learning procedures. The pros and cons of the proposed techniques for diagnosis of this neurological disorder have also been presented in this paper.

Lorna et al. [6] aim was to study and investigate the analytical and clinical variances among three groups of patients identified with psychic non-epileptic seizures: those with shell shock; those who have antiquity of shock but not battle fatigue; and those without history of trauma.

Aditya et al. [9] have come up with machine learning techniques for diagnosing Alzheimer's disease. In this paper, supervised learning model has been used to enumerate the sternness of mental illness on a normal assessable gauge. Additionally, this scale of measurement can be used for medical dementia evaluation.

Jiang et al. [11] proposed diagnosis of seizure disorder by combining two diverse methods called complementary ensemble empirical mode decomposition and extreme gradient boosting namely CEEMD-XGBoost. Former is used to extract the raw EEG data and later helped in categorizing the model to identify both seizure and non-seizure EEG signals. Moreover, using these techniques the cataloging performance has been enhanced.

Kai et al. [14] described a new way of detecting the seizures using Hilbert-Huang transform and support vector machine. HHT based time–frequency representation has been taken into consideration as time–frequency image and segmentation has been employed on the basis of frequency bands. Statistical traits like mean, variance, etc. have been extracted. The classification accuracy and receiver operating characteristics have been utilized for estimation of performance of EEG classifier.

Qian et al. [13] presented FPGA based digital neuromorphic processor and various parallel designs. The planned assembly talked about several issues relating to effective parallelization in estimation of membrane potential, loading of synaptic weights, and integration of arithmetic units. The comparison between throughput, cost, and power for several configurations has been investigated. Training and recognition speed of 13.5$\times$ and 25.8$\times$ is achieved, respectively whose degree of parallelism is 32.

Hong et al. [15] portrayed a classification procedure for epilepsy EEG signals with a dictionary learning based on the sparse illustration, rather than using a simple training sample as a lexicon. Then the comparison is made between sparse and original test samples along with that minimization of error is considered for enhancing the recognition accuracy to meet the requirements of automatic seizures detection.

In our paper, deep learning model has been proposed to detect the epileptic seizure. The data of five EEG datasets has been taken from MIT database for seizure classification. Analysis of the data is made and confirmed that subjects falling under class 1 records seizure activity. Implementation of Deep Neural Networks with Keras has been employed in our model to attain the best accuracy.

## 3 Method Initiation with Improved Techniques

Basically, data is the key factor while building the model. Collection of information is an important job because on the basis of that, we can develop our model efficiently. Once the required information is grouped, we would pre-process the information as the actual statistics are deafening, inadequate, and unreliable. After processing we would categorize our data into training and testing datasets where former is large and latter is smaller as compared to antecedent. The next step is selecting the prototype that matches with our datum and conditions. The last step is to train and validate the data. Detailed description of different EEG datasets, feature selection, data preparation, classification, and performance evaluation for seizure detection is illustrated in Fig. 1.

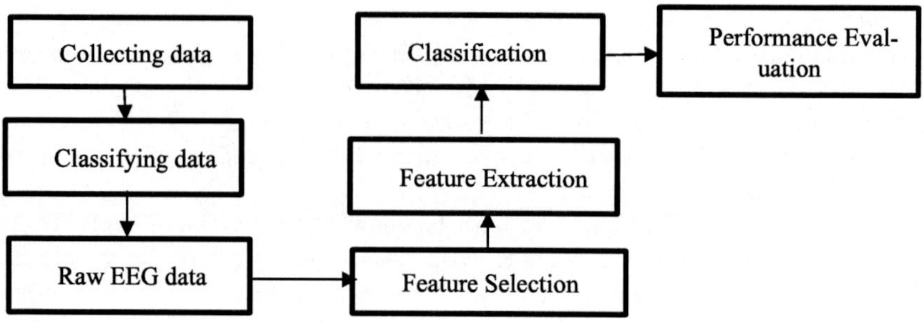


**Fig. 1** Detailed analysis of epileptic seizure using various EEG datasets

The implementation by means of Deep learning model with Keras on TensorFlow platform has been proposed to overcome conventional machine learning problems also to enhance the efficiency and accuracy of our model. Using Deep learning with Keras would help in better sub-classing and effective classification of EEG signals for epileptic seizure detection.

## *3.1 Data Extraction*

The data has been taken into consideration from CHB-MIT database and Kaggle for implementing our model. Hence, the original data file is of five diverse wrappers with hundred records per individual. Among that, single file keeps a track of brain activity for 23.6 flashes. And respective time-sequence data tuned into 4097 facts. Individual datapoint signifies tracking of brain signals with respect to various intervals. Thus, we had 500 entities where each is carrying 4097 points till 23.5 moments.

## *3.2 Data Classification*

In this stage, we have separated and scuffled each 4097 data points in terms of 23 chunks, single portion covers 178 facts till one moment and apiece datum gives reading of brain activity for various time intervals. Thus, $23 \times 500 = 11{,}500$ bits represents (row), 178 datapoints is (column), and label Y signifies $\{1, 2, 3, 4, 5\}$.

$X$—Descriptive variables ($X1, X2 \ldots X178$).

$Y$—Response variable $\{1, 2, 3, 4, 5\}$.

which means $Y$ comprises the category of 178-dimensional input vector.

5—Eyes open, it means the patients had their eyes open during the recording of EEG signal of the brain.

4—Eyes closed, is observed during the diagnosis of EEG signals.

3—Locate the region of tumor and record the signals from the healthier fragment.

2—Record the EEG signal where the tumor is located.

1—Records the seizure action.

The individuals under classes {2, 3, 4, 5} do not have epileptic seizures. Only the subject falling under class 1 have epileptic seizures.

The main reason behind this type of creation of data is streamlining the accessibility toward information through producing a.csv file for proper maintenance of datum.

## 4 Proposed Model

The deep learning architecture model in the Python language using Keras API with TensorFlow as the backend has been implemented. The model visualization of Deep Learning is shown in Fig. 2. Following are the steps considered while developing the model:

1. Imported all the necessary library files for Python.
2. Data spitted into features ($X$) and labels ($Y$).
3. Dataset is divided into 70:30 ratio. Former is for training and latter is for testing and authentication.
4. 178 features were used as an input from training set.
5. Created the deep learning model.

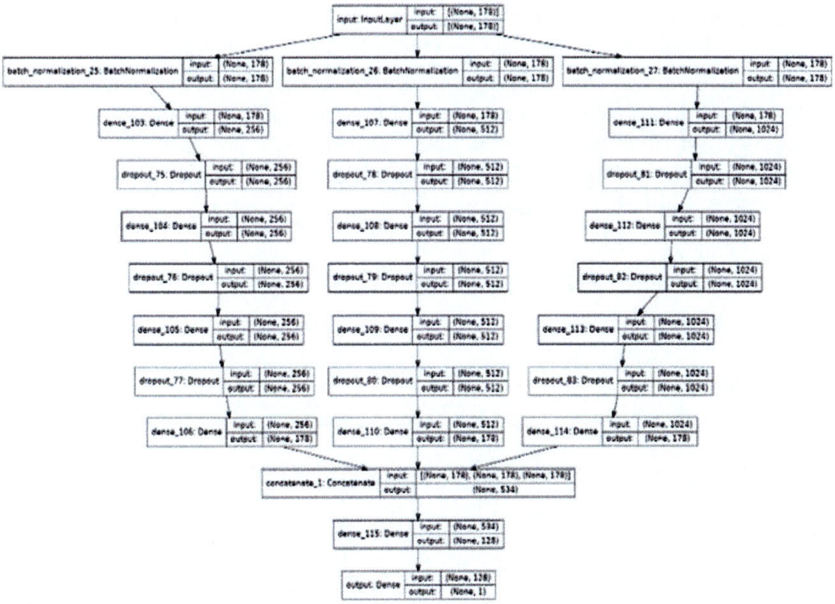


**Fig. 2** Architecture of deep learning model

- Input Layer: Defined the input layer with 256 neurons.
- Hidden Layer: It has 512 nodes (neurons).
- Output Layer: It consists of 1024 nodes.

6. Once the input is given to the input layer and the nodes are fully connected to other layers, we added an activation function for getting the probability of the output. Hence, we used Sigmoid to estimate the probability of output.

$$F = 1/(1 + e^{-x})$$

7. Two regularizations called Dropout and Batch normalization is used. Dropout is used to minimize the overfitting whereas Batch normalization makes sure that data is managed by layers and altered into a wide range of values that need to be stabilized.
8. Dense Layer is deeply connected to its preceding layer for smooth data transfer from one layer to next layer.
9. The model is compiled and verified the loss and accuracy parameters.
10. Finally, Adam optimizer is used to train the network in order to fasten the gradient descent method by taking the exponentially weighted average of the gradients.

## 5 Result and Discussion

The proposed model shows a test accuracy of 97.46% which means only 2% of cases would not be classified properly. The precision of the model after testing and validation is shown in Fig. 3. Moreover, this result shows that our model is quite precise by considering all the classes and which sets a new benchmark for existing machine learning models. In addition to that model, loss is also negligible. The loss against the epochs having multiclass dataset is illustrated in Fig. 4.

Accuracy on different Data Splits:

- Training Accuracy: 99.48%
- Validation Accuracy: 96.12%
- Testing Accuracy: 97.46%

**Fig. 3** Accuracy of the Deep Learning model

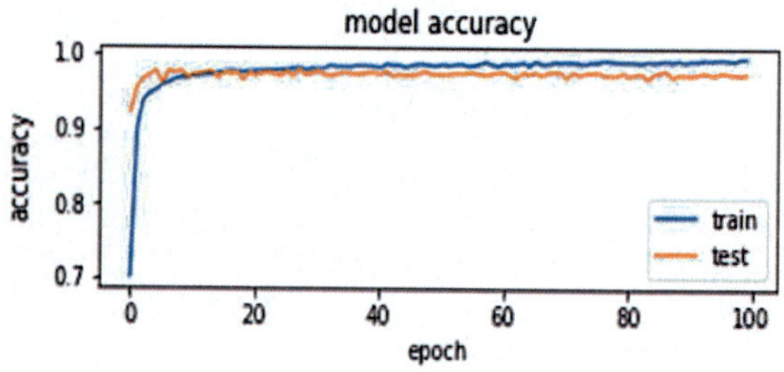

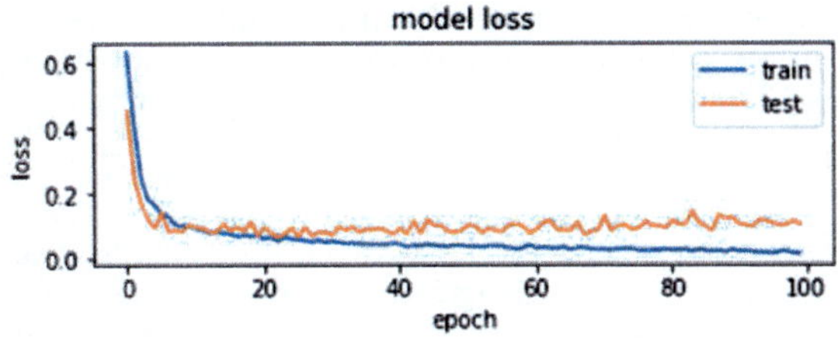


**Fig. 4** Loss against the epochs of multiclass dataset

## 6 Conclusion

It is very important to ensure that whether a brain wave test shows uneven or consistent cerebral function during identification of any nervous system disorder. As traditional method of setting the probes on brain scalp is costlier and time taking procedure, automated cataloging techniques would reduce the delay in therapy and lessen the burden of supervision. Hence, a new Deep learning system design has been demonstrated which is apt for analyzing time series data of brain signals besides perceiving an individual has epilepsy or not in a very well-mannered way. Therefore, the proposed structure produces best accuracy compared to conventional methods of machine learning algorithms. Therefore, the estimation of suggested technique by means of other EEG datasets would be taken into consideration for future analysis.

## References

1. B. Soroor, A review of significant research on epileptic seizure detection and prediction using heart rate variability. Turk Kardiyol Dern. Ars. 46(5):414–421 (2018)
2. J. SatheeshKumar, P. Bhuvaneswari, Analysis of electroencephalography (EEG) signals and its categorization—a study. Procedia Eng. **38**, 2525–2536 (2012)
3. A. Nandy, S.M. Nasim Uddin, et al., Feature extraction and classification of EEG signals for seizure detection, in *International Conference on Robotics, Electrical and Signal Processing Techniques (ICREST)* (2019)
4. A. Shoeibi, F. Amir, et al., Epileptic seizures detection using deep learning techniques: a review (2021).
5. M. Hamid, M.Q. Saeed, et al., Deep convolutional neural network regularization for alcoholism detection using EEG signals. Deep Learn. Big Data Healthc. Ind. **21**(16)
6. M. Lorna, Z. Ruifan et al., Cognitive differences between patients who have psychogenic nonepileptic seizures (PNESs) and posttraumatic stress disorder (PTSD) and patients who have PNESs without PTSD. Epilepsy Behav. **37**, 82–86 (2014)
7. R. Sriram, J. Michele et al., Seizure detection, seizure prediction, and closed-loop warning systems in epilepsy. Epilepsy Behav. **37**, 291–307 (2014)
8. D.L. Kenneth, T. Eugen et al., The consequences of refractory epilepsy and its treatment. Epilepsy Behav. **37**, 59–70 (2014)
9. C.R. Aditya, M.B. Sanjay Pande, Devising an interpretable calibrated scale to quantitatively assess the dementia stage of subjects with Alzheimer's disease: a machine learning approach. Inform. Med. Unlocked **6**, 28–35 (2017)
10. S. Sandeep et al., EEG signal classification using PSO trained RBF neural network for epilepsy identification. Inf. Med. Unlocked **6**, 1–11 (2017)

11. W. Jiang, Z. Tengfei, et al., Detecting epileptic seizures in EEG signals with complementary ensemble empirical mode decomposition and extreme gradient boosting. Entropy 22 (2020)
12. S. Yang, Machine learning applications for electroencephalograph signals in epilepsy: a quick review (2020), pp. 2–5
13. W. Qian, L. Youjie et al., Energy efficient parallel neuromorphic architectures with approximate arithmetic on FPGA. Neurocomputing **221**, 146–158 (2017)
14. F. Kai, Q. Jianfeng et al., Classification of seizure based on the time-frequency image of EEG signals using HHT and SVM. Biomed. Signal Process. Control **13**, 15–22 (2014)
15. Hong, P., Cancheng, L., et al.: A novel automatic classification detection for epileptic seizure based on dictionary learning and sparse representation. Neurocomputing (2019)

# Ultra High Rate Inter-Satellite Optical Wireless Transmission Using DP-QPSK

**Bibhu Prasad, Ami Kumar Parida, Jagana Bihari Padhy, Bandana Mallick, and Ajit Kumar Patro**

**Abstract** This paper proposes a design for inter-satellite system with optical wireless communication using dual polarized quadrature phase shift keying (DP-QPSK) modulation format. The issues related to the IsOWC system such as bandwidth management, pointing error are targeted to minimize using advanced modulation technique. A comparative study of few advanced modulation formats like DP-QPSK, quadrature phase shift keying (QPSK) and differential phase shift keying (DPSK) has been carried out for achieving the maximum link distance. The performance matrices like bit error rates (BER), received signal power, constellation diagram are used for result analysis of the proposed model. It is observed that, DP-QPSK is a best suitable modulation format for IsOWC-based system with data rate of 100 Gbps and beyond. Due to dual polarization technique with higher OSNR tolerance, a coverage distance of 13,000 km is obtained with the proposed inter-satellite optical wireless (IsOWC) at 100 Gbps for a targeted BER of $3.8 \times 10^{-3}$.

**Keywords** DP-QP SK · IsOWC · ISL · BER

## 1 Introduction

A major upswing is observed in the area of data and multimedia services technologies in the recent years. An unprecedented demand is noticed to accommodate more number of data and multimedia services. As a result, the network suffers from spectrum congestion. To overcome this, optical wireless communication is the best possible way to alter radio frequency (RF) spectrum [1]. The concept of inter-satellite optical wireless communication (IsOWC) is becoming attractive because of their ability to perform efficient exchange of huge information within the satellite network, as a reliable transmission medium [2]. Space is used as a transmitting medium for data communication between satellites. Due to light weight structure, considerable low mass, low launch cost, less transmitted power, higher data rate and unlicensed

B. Prasad · A. K. Parida · J. B. Padhy (✉) · B. Mallick · A. K. Patro
Department of Electronics and Communication Engineering, GIET University, Gunupur, India
e-mail: jaganpadhy@giet.edu



S. Patnaik et al. (eds.), *Smart and Sustainable Technologies: Rural and Tribal Development Using IoT and Cloud Computing*, Advances in Sustainability Science and Technology,
https://doi.org/10.1007/978-981-19-2277-0_23

spectrum of optical system make it more feasible inter-satellite communication [3]. IsOWC operates through line of sight (LOS) propagation by narrow coherent laser beam [1, 4]. IsOWC may be associated with either inter- or intra-orbital communication. Inspite of various advantages as mention above, IsOWC-based systems have limited performance. The limiting factors such as inter-satellite pointing error, satellite vibration background noise in the solar system [5]. To mitigate the above-mentioned challenges, various technological strategies have been published in the related state of the art. Out if this performance improvement using advanced modulation technique has put a major importance. Patnaik et al. [6] have demonstrated an IsOWC-based system with advanced modulation technique of QPSK modulation scheme. The proposed set up is executed with different data rates of ITU-T standards. Transmission path length of 4767 km, 7542 km and 9532 km has been achieved at various datarate such as 400, 160 and 100 Gbps respectively. Sodnik et al. [7] have presented a binary phase shift keying (BPSK)-based IsOWC system for a bit rate of 5.6 Gbps and able to transmit data up to a link distance of 6000 km. Kumari et al. [8] have proposed an IsOWC-based system with data carrying capacity of 10 Gbps. DPSK modulation format is used in the proposed system. The achieved propagation length of laser signal is 6000 km for data transmission.

After undergone the detailed study in the related field of IsOWC, an ultra-high rate IsOWC-based system is proposed. To improve the spectral efficiency, dual polarization-based QPSK modulation (DP-QPSK) technique is used. DP-QPSK is famous for its high tolerance in optical signal to noise ratio (OSNR) at the receiver. As a result, the error performance of the system is improved. A comparative study is also carried out with related IsOWC-based system with QPSK and DPSK modulation scheme. Finally, the maximum allowable bit rate for the IsOWC system using DP-QPSK modulation has been obtained for intra-satellite links as well as for inter-satellite links.

## 2 Modulation Format and Inter-Satellite Link

The proper utilization of channel bandwidth is a prime issue for the current scenario of communication. Addressing to this, there may be a tradeoff between bandwidth efficient and spectral efficient modulation scheme [9]. Performance of the system is directly influenced by the modulation technique used. Depending upon the OSNR tolerance and spectral efficiency, modulation techniques have such as BPSK, DPSK, QPSK, on–off been proposed the past literature for optical wireless communication system [2]. Out of these modulation formats QPSK is more bandwidth and energy efficient. As in IsOWC system the propagation length is very high compared to terrestrial communication, hence high receiver sensitivity is required which is accomplished by QPSK modulation scheme with coherent detection. The performance can further be enhanced by incorporating the dual polarization technique with QPSK modulation. Here laser carrier signal is split into two orthogonal polarized signals. Then both the polarized signal is modulated externally with QPSK sequence [9]. Due to orthogonal polarization, the inter-channel interference is zero hence improves the performance. Mathematical interpretation of DP-QPSK signal is given by [9]

$$E_{\text{DP-QPSK},X} = E_X(t)\exp\big(j2\pi f_0 t + j\theta_{r,X}(t)\big) \tag{1}$$

$$E_{\text{DP-QPSK},Y} = E_Y(t)\exp\big(j2\pi f_0 t + j\theta_{r,Y}(t)\big) \tag{2}$$

where, $E_X(t)$ is $X$-polarized amplitude, $E_Y(t)$ is the $Y$-polarized amplitude, $f_0$ is the carrier frequency, $\theta_{r,X}$ and $\theta_{r,Y}$ phase of $X$ and $Y$ polarized electric signal. The received output signal at the detector is given by [8],

$$\begin{aligned} E'_{X/Y,I} = {} & R\sqrt{2P_{\text{LO}}}E(t)\cos[2\pi(f_0 - f_{\text{LO}})t \\ & + \theta_{r,X/Y}(t)n\theta_{\text{LO}}(t)] \end{aligned} \tag{3}$$

$$\begin{aligned} E'_{X/Y,Q} = {} & R\sqrt{2P_{\text{LO}}}E(t)\sin[2\pi(f_0 - f_{\text{LO}})t \\ & + \theta_{r,X/Y}(t) - \theta_{\text{LO}}(t)] \end{aligned} \tag{4}$$

where, $E'_{X/Y,I}$ and $E'_{X/Y,Q}$ is the in-phase and quadrature component of received electrical signal. $P_{\text{LO}}$ is the local oscillator power, $f_{\text{LO}}$ is local oscillator frequency, $R$ is the responsivity of photo diode, $\theta_{LO}$ is phase of local oscillator. Communication through satellite has gained a lot of attention over the years due to its tracking system around the Globe. Use of Optical Inter-Satellite Link (ISL) adds a benefit to this communication system. The inter-satellite link may be of line of sight (LOS) or diffused link type. Out of which, LOS type link is very much suitable for higher transmission capacity. Although IsOWC is less affected by turbulence and weather limitation but suffers from various constraints such as point angle ahead, acquisition and tracking, background noise radiation and satellite vibration which reduce the performance. The power budget can be evaluated from the received power of satellite link and is given by [6],

$$P_{\text{Rec}} = P_{\text{Tra}}\eta_{\text{Tra}}\eta_{\text{Rec}}G_{\text{Tra}}G_{\text{Rec}}E_{\text{EL}}\left(\frac{\lambda_p}{4\pi Z_p}\right)^2 \tag{5}$$

where, $P_{\text{Rec}}, P_{\text{Tra}}$ are the received and transmitted power. $\eta_{\text{Tra}}, \eta_{\text{Rec}}$ are the optical efficiency of transmitting and receiving antenna, $G_{\text{Tra}}, G_{\text{Rec}}$ are electrical gain of transmitting and receiving antenna, $E_{\text{EL}}$ represents the extra losses associated with the system such as pointing loss and background radiation noise. $\lambda_p$ is the wavelength of transmission and $Z_p$ is the propagation length between transmitting and receiving antenna. The gain of transmitting antenna is given by [6],

$$G_{\text{tra}} = \left(\pi D_{\text{tra}}/\lambda_p\right)^2 \tag{6}$$

where, $D_{\text{tra}}$ is transmitting telescope diameter.

$$G_{\text{rec}} = \left(\pi D_{\text{rec}}/\lambda_p\right)^2 \tag{7}$$

where, $D_{\text{rec}}$ is receiving telescope diameter. The extra loss is associated with the pointing loss factor $L_{\text{Tra}}$ and loss factor of reciver $L_{\text{Rec}}$ are given by, [6]

$$L_{\text{Tra}} = \exp\left(-G_{\text{Tra}}\theta_{\text{Tra}}^2\right) \tag{8}$$

$$L_{\text{Rec}} = \exp\left(-G_{\text{Rec}}\theta_{\text{Rec}}^2\right) \tag{9}$$

where, $\theta_{\text{Tra}}$ and $\theta_{\text{Rec}}$ are the pointing error angle of transmitter and receiver respectively.

## 3 Proposed Set up

Figure 1 shows the proposed IsOWC set up. It is designed and simulated in Optisystem environment. The model has been organized in three major sections such as transmitter, inter-satellite space channel and coherent detector. The message signal is generated at the pseudo-random bit sequence generator (PRBS) at a rate of 100 Gbps. The laser source in the DP-QPSK transmitter as shown in Fig. 2, splits into two orthogonal light beams termed as X-polarized light signal and Y- polarized light signal. Then the individual light beams are modulated optically by QPSK sequence. The detailed working principle of the dual polarized modulation format has been

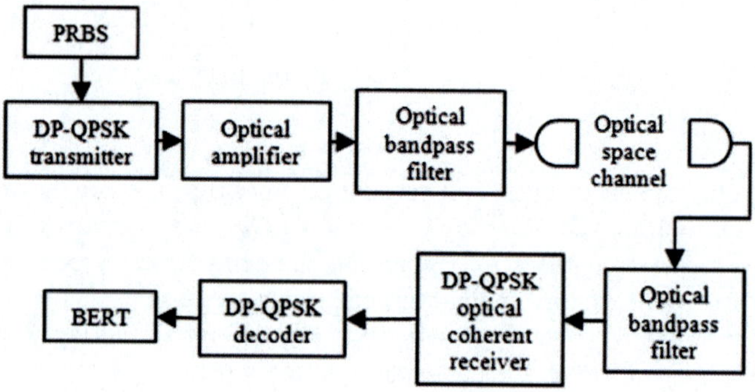


**Fig. 1** Proposed model using IsOWC

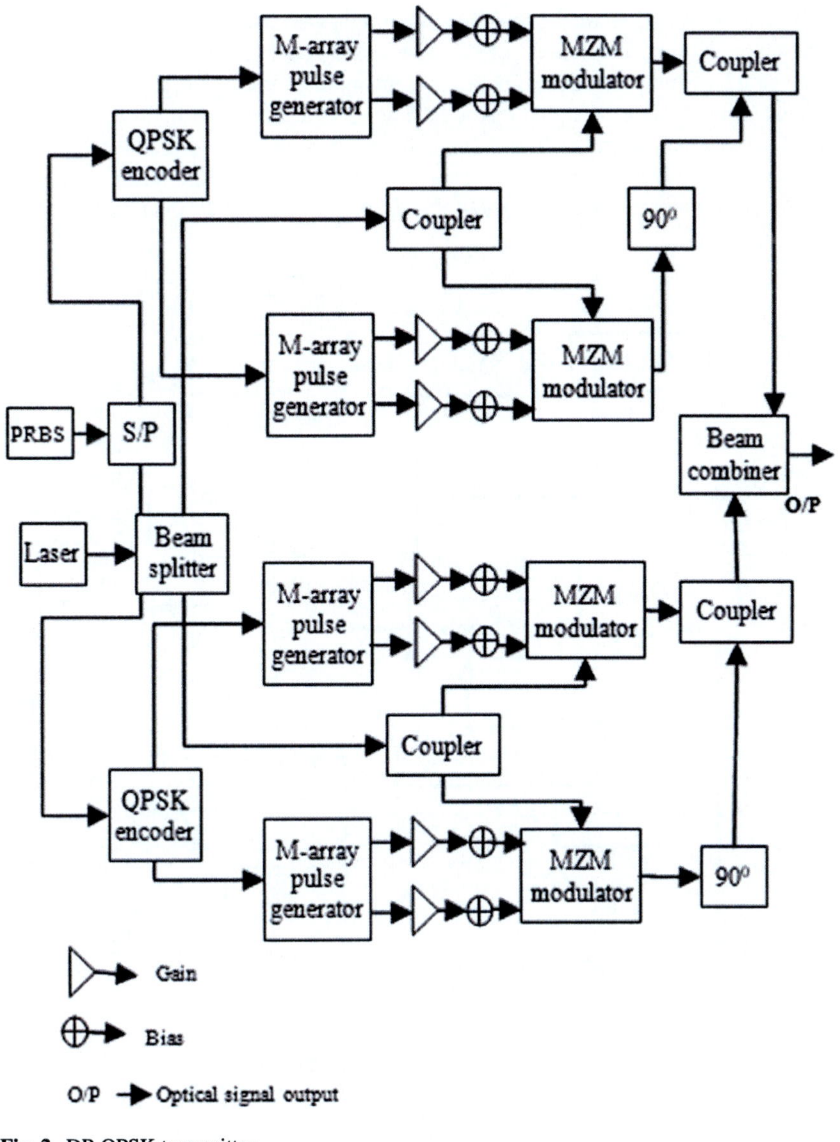


**Fig. 2** DP-QPSK transmitter

suggested from [9]. The gain optical modulated signal is stabilized by erbium doped fiber amplifier (EDFA) of 10 dB gains. To minimize the co-channel interference, the optical signal is passed through an optical Gaussian band pass filter with frequency bandwidth of 0.75* symbol rate. The extra losses in the system are considered as 5 dB. Extra losses include the insertion loss, optical loss, background radiation loss and pointing loss. Then the optical signal is propagated in either inter-orbit or intra-orbit space channel. The losses such as optical loss, background radiation loss, free space path loss and pointing loss have been encountered during propagation in space channel which makes the signal distorted. The optical receiver collects the distorted signal for recovery of information. The coherent type detector is used here because of the good receiver sensitivity as shown in Fig. 3. The receiver section is organized in two major parts such as dual polarized coherent detection and recovery of

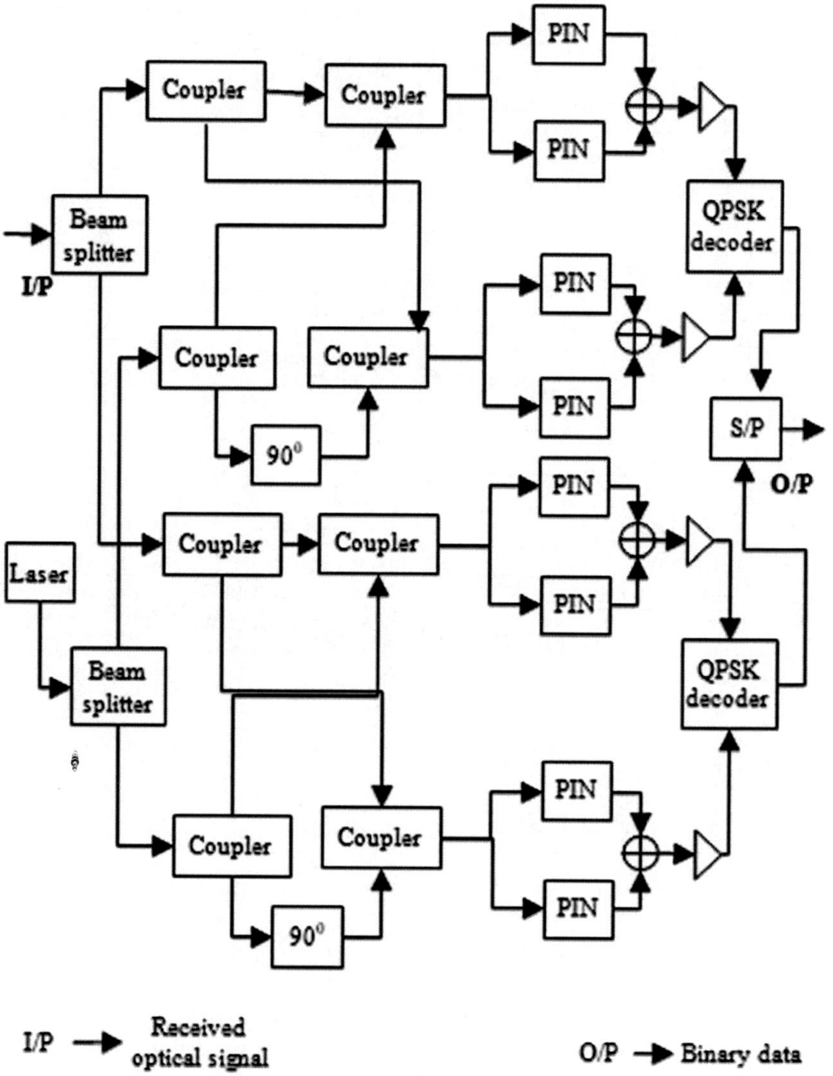


Fig. 3 DP-QPSK receiver

information from the received signal for error measurement. The detailed working principle of the dual polarized modulation format has been suggested from [9]. Due to the hybrid integration of the DP-QPSK receiver makes it possible to reduce the skew and receiver sensitivity issues of conventional system. The error estimation can further be improved by incorporating of digital processing algorithms at the receiving end. After information decode at receiver end, the error bits can be measured using BER test set (BERT).

## 4 Results and Discussion

The simulation of the proposed DP-QPSK-based IsOWC model is carried out with parameters considered as per practical scenario as shown in Table 1 [6, 7].

The performance matrices such as BER, received signal power and constellation diagrams are considered for comparative analysis of the proposed model. Figure 4 shows the BER response versus OSNR for the proposed set up. Due to the high OSNR tolerance of DP-QSK modulation compared to other modulation technique, the targeted BER of is achieved at 14 dB for DP-QPSK modulation. Figure 5 shows the BER response with respect to link distance (km). It is observed that, at 30 dBm laser power the link distance achieved for DP-QPSK modulation is 13,000 km. The link distance achieved for QPSK modulation is 10,200 km and for DPSK modulation is 7700 km. From this we can say that, DP-QPSK modulation-based IsOWC system provides best error performance compared to its counterpart. The electrical power received versus the link length (km) has been plotted at a data rate of 100 Gbps as shown in Fig. 6. Generally, the required OSNR is inversely proportional to inter-symbol distance in the constellation diagram.

**Table 1** Simulation parameters

| Serial number | Parameters | Values |
|---|---|---|
| 1 | IsOWC channel type | LOS |
| 2 | Data rate (Gbps) | 100 |
| 3 | Symbol rate (GS/s) | 25 |
| 4 | Operating frequency (THz) | 193.1 |
| 5 | Sequence length | 65,536 |
| 6 | Transmitted power (dBm) | 30 |
| 7 | Aperture diameter (cm) of transmitter | 5 |
| 8 | Aperture diameter (cm) of receiver | 15 |
| 9 | Extra losses (dB) | 5 |
| 10 | Pointing error (μrad) at transmitter | 1.1 |
| 11 | Pointing error (μrad) at receiver | 1.1 |
| 12 | Extinction ratio (dB) | 60 |

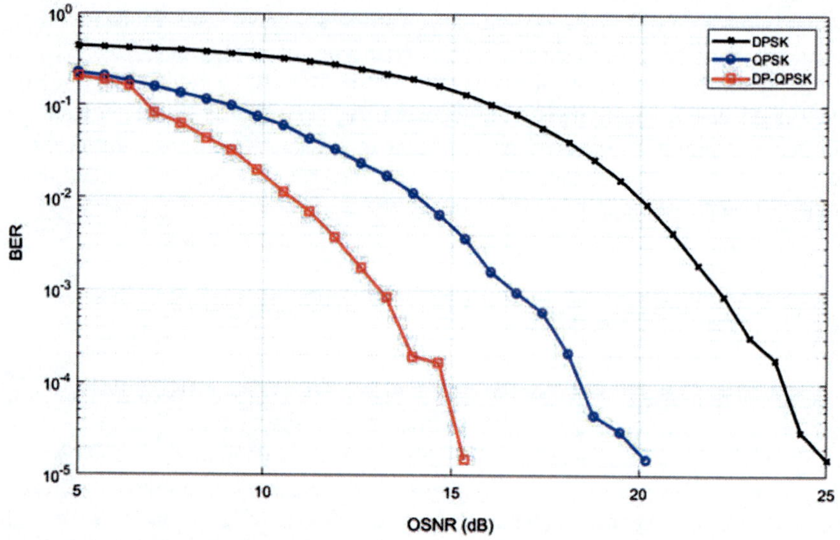


**Fig. 4** BER versus OSNR (dB)

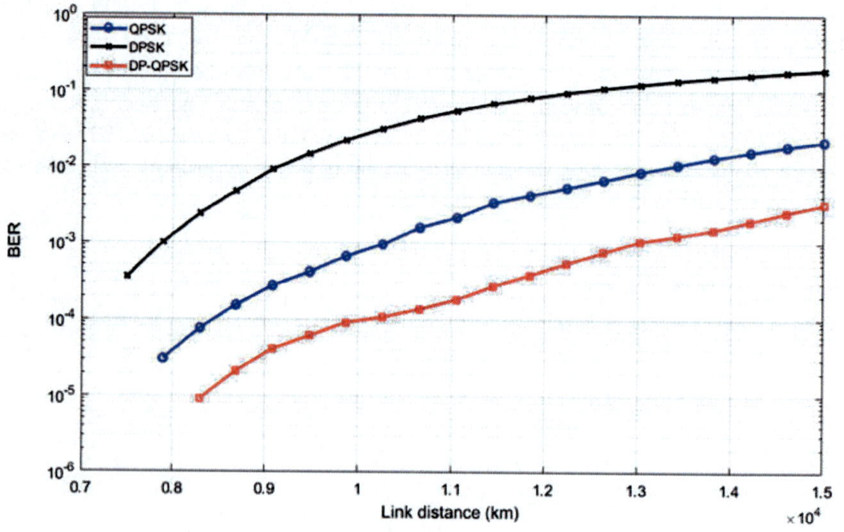


**Fig. 5** BER versus link distances (km)

Due to the dual polarization of transmitted bit in the optical transmitter, the power penalty is less at the receiver for DP-QPSK modulation-based system. Figure 7 shows the scattered diagram of DP-QPSK and QPSK at link distance of 13,000 km. It is

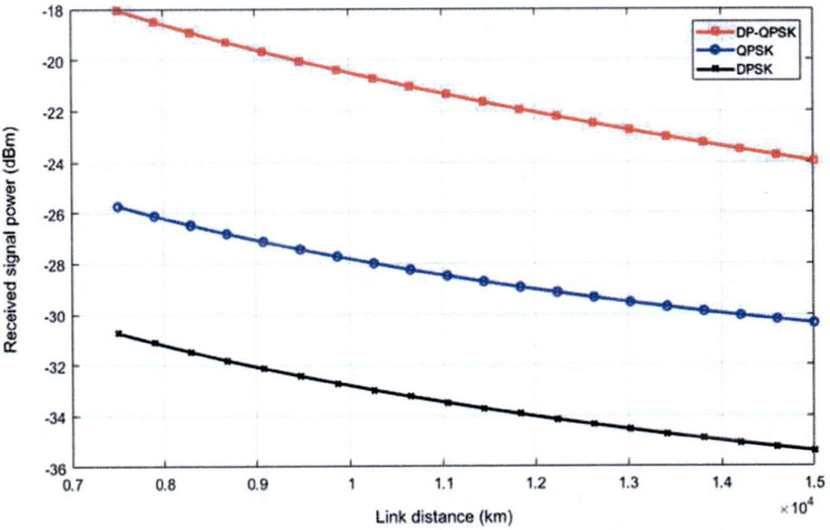


Fig. 6 Electrical power received (dBm) versus propagation length (km)

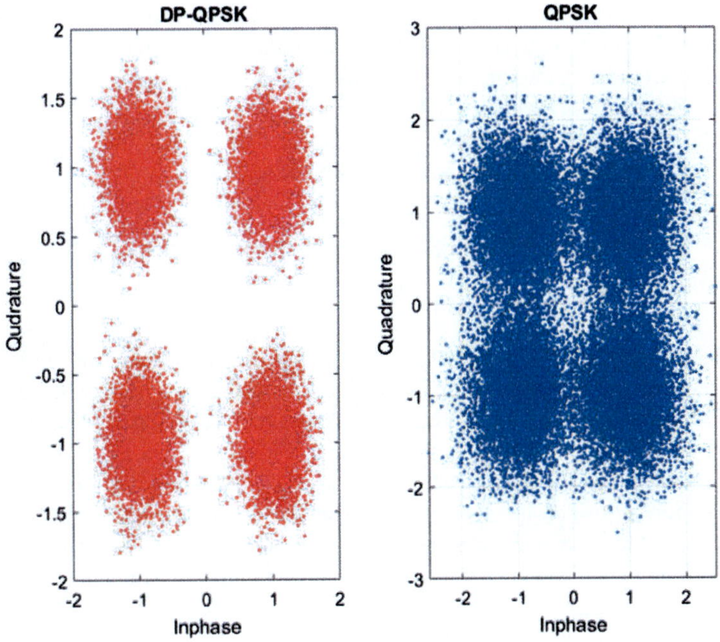


Fig. 7 Scatter diagram at the receiver end

**Table 2** Comparative study with related research works

| Satellite link | Input power (dBm) | Link distance (km) | Patnaik et al. [6] data rate (Gbps) | Proposed model data rate (Gbps) |
|---|---|---|---|---|
| LEO-LEO | 30 | 5,860 | 265 | 640 |
| MEO-MEO | 30 | 18,460 | 27 | 80 |
| LEO-GEO | 30 | 45,000 | 4.19 | 11 |

observed that, the constellation points in DP-QPSK modulation are very prominently separated compared to QPSK modulation.

From Fig. 7, we can say that, the transmitted symbols can be recovered at the receiver with less error in the proposed system than QPSK-based IsOWC system. The system is further simulated for predefined link distances for various intra-orbital links as well as for inter-orbital links. The acceptable bit rates are obtained for the possible inter-satellite communication as shown in Table 2. A comparative analysis is carried out with related IsOWC works as shown in Table 2. It is found that, due to the use of dual polarization of the optical carrier at the transmitter an improvement of OSNR tolerance as well as spectrum resolution is obtained. As a result, the acceptable bit rate is improved in the proposed model as shown in Table 2. For the LEO-LEO inter-satellite link the targeted BER of is achieved at a data rate of 640 Gbps for a link distance of 5860 km. A data rate of 80 Gbps is obtained for MEO-MEO inter-satellite link. Similarly, a data rate of 11 Gbps is achieved for the accepted communication between GEO-LEO and vice versa. From the above results, we can say that due to the DP-QPSK modulation used in the proposed model, system performance is increased in term of BER, received signal power and OSNR tolerance.

## 5 Conclusion

DP-QPSK-based IsOWC system has been proposed and analyzed in presence of pointing error and background noise of space channel. The system is executed at an ultra-high data rate of 100 Gbps. A link distance of 13,000 km is obtained for a transmitted power of 30 dBm at targeted BER of $3.8 \times 10^{-3}$. A comparative analysis has been conducted between DPSK, QPSK and DP-QPSK modulation techniques. The maximum allowable bit rate for the proposed IsOWC system using DP-QPSK has been obtained for communication between intra-satellite links. The further analysis may be done on the system by incorporating with multiple numbers of channels. In addition, the error performance may also be improved by adding various error correction or line coding techniques.

## References


1. H. Kaushal, G. Kaddoum, Optical communication in space: challenges and mitigation techniques. IEEE Commun. Sur. Tutorials **19**(1), 57–96 (2017)
2. Z. Ghassemlooy, W. Popoola, S. Rajbhandari, *Optical Wireless Communications* (CRC Press LLC, Milton, 2019)
3. S. Kumar, S. Gill, K. Singh, Performance investigation of inter-satellite optical wireless communication (IsOWC) system employing multiplexing techniques. Wireless Pers. Commun. **98**(1), 1461–1472 (2017)
4. Z. Zhu, S. Zhao, Y. Li, X. Li, Performance comparison of analogue inter-satellite microwave photonics link using intensity modulation with direct detection and phase modulation with interferometric detection. IET Optoelectron. **9**(2), 88–95 (2015)
5. A. Polishuk, S. Arnon, Optimization of a laser satellite communication system with an optical preamplifier. J. Opt. Soc. Am. A **21**(7), 1307 (2004)
6. B. Patnaik, P. Sahu, Inter-satellite optical wireless communication system design and simulation. IET Commun. **6**(16), 2561–2567 (2012)
7. Z. Sodnik, B. Furch, H. Lutz, Optical Inter-satellite communication. IEEE J. Sel. Top. Quantum Electron. **16**(5), 1051–1057 (2010)
8. L. Wang, Y. Zhao, C. Shi, Z. Chen, Y. Xu, W. Liu, An effective integration solution for 100 Gb/s DP-QPSK coherent receiver with a small Skew. IEEE Photonics Technol. Lett. **26**(3), 227–230 (2014)
9. J. Padhy, B. Patnaik, 100 Gbps multiplexed inter-satellite optical wireless communication system. Opt. Quantum Electron. **51**(7) (2019)

# Investigation on Reactive Power Compensation Using STATCOM

**Debani Prasad Mishra, Prangya Panda, Sukanya Sahoo, and Swosti Pradhan**

**Abstract** The power consumption when increased causes transmission lines to get nearer or sometimes beyond their transfer capacities which results in overload lines and blockage. The shunt connected FACTS devices are the solution for compensation of reactive power and also reduction in problems related to transmission lines. STATCOM, i.e. Static Synchronous Compensator is a variety of shunt FACTS devices and has been used for compensation of reactive power. The difficulties of compensation of reactive power can be overcome with the use of STATCOM. The paper basically explains how a STATCOM can compensate the reactive power which is not very much useful for the machine.

**Keywords** Static synchronous compensator (STATCOM) · Pulse width · Modulation (PWM) · Voltage source converter (VSC) · Reactive power (RP) · Point of common coupling (PCC) · Flexible AC transmission system (FACTS) · Power factor (pf)

## 1 Introduction

The demand for electricity can be satisfied by either developing new transmissions lines or by enhancing the capability of power transfer of transmission line [1–4]. Load compensation and voltage support are the two features to the difficult task of compensation of reactive power [5–7]. Power factor improvement, real power balancing which has been taken from the supply, better regulation of voltage, etc. are included in load compensation. The parameters are modified using above methods to produce enhanced VAR compensation [8–10]. The inductors absorb or store the reactive power while the capacitor develops reactive power. Power compensation in a system is mainly done because of regulation of voltage, proper usage of the connected

D. P. Mishra (✉) · P. Panda · S. Sahoo · S. Pradhan
Department of Electrical and Electronics Engineering, International Institute of Information and Technology, Bhubaneswar, India
e-mail: debani@iiit-bh.ac.in



S. Patnaik et al. (eds.), *Smart and Sustainable Technologies: Rural and Tribal Development Using IoT and Cloud Computing*, Advances in Sustainability Science and Technology,
https://doi.org/10.1007/978-981-19-2277-0_24

machines in a system, the enhancing of the stability of the system, reduction in losses of the system, to avoid voltage sag and collapsing of voltage.

STATCOM manages interchange of reactive power and during dynamic disturbances, it gives voltage support to buses by emphasizing voltage of buses. It comprises of 3 phase inverter i.e. a PWM generator which uses either SCRs or MOSFETs or IGBTs, dc voltage for inverter is given by dc capacitor. The dc link capacitor while charging results in increase in dc voltage which further by STATCOM, increases the absorption of reactive power.

The STATCOM current, three-phase voltage and current are first transformed into equivalent system of two phase which is called as stationary reference and thereafter it again gets converted into rotating reference frame with the help of Phase Locked Loop circuit (PLL). The component of q-axis indicates the active power and the d -axis shows the component of RP. The converter currents and line currents are also converted into d-q transformation [11–13].

## 2 Research Method

The shunt connected device STATCOM is utilized for compensation of RP. By using STATCOM, the improvement of pf and voltage of the transmission system is done. AC filter, DC link capacitor, 3-phase Inverter (using MOSFET/IGBT/SCR) are the parts of the STATCOM. At PCC, generation or absorption of power by STATCOM is done for suitable reactive power with the help of above-mentioned technique.

When grid is connected to a load, it takes two varieties of power, i.e. the reactive power and the active power. The reactive power is changed to zero, so that the power factor will be unity. The STATCOM is then linked at the load point of the system [14, 15]. The RP needed by the load is given by STATCOM and the grid only gives active power [16, 17]. If the inverter voltage value is equal to supply voltage, there will not any reactive power present in the system and STATCOM will not induce or absorb RP. If voltage of inverter is greater than supply voltage, the current passage from ac STATCOM to the ac system through the coupling inductor and the device produce reactive power which is capacitive in nature and if it is less than supply voltage, passage of the current starts flowing from an ac system to STATCOM which results in inductive reactive power.

### *2.1 Block Description and Operational Strategy*

The grid voltage, inverter current and load current are converted into *d-q* form by using alpha–beta to *d-q* transformation blocks. Grid voltage is converted into alpha–beta for finding $\omega t$ value by using PLL (Phase Locked Loop) and then it is converted to *d-q*. PLL is a control system that is used to synchronize an output oscillator signal with reference signal which has been presented in Fig. 1.

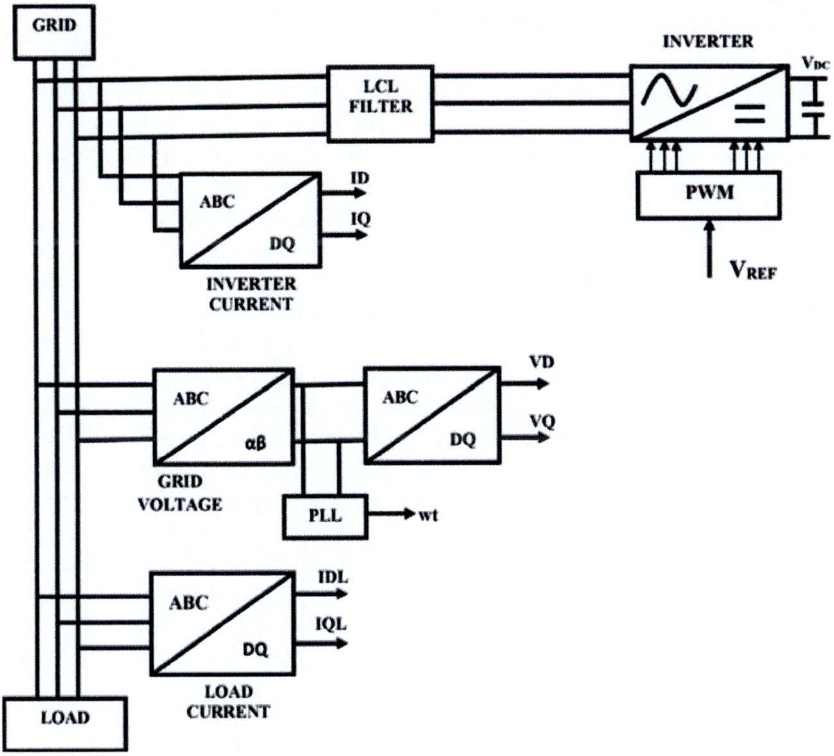


**Fig. 1** Inverter bridge and generation of transformed values block diagram

Using the transformed values, the *dc* bus controller, current controller of *d*-axis and the *q*-axis current controller are implemented as given in Fig. 2. The final result of the controller ($V_{REF}$) is supplied to PWM generation block.

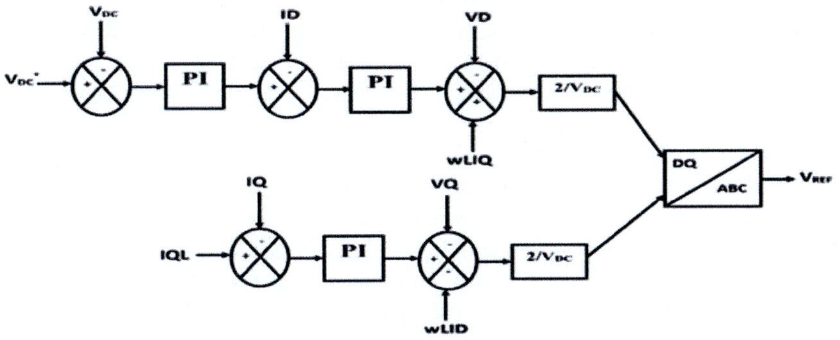


**Fig. 2** Block diagram of alpha–beta and *d*-*q* conversion

## 2.2 Network Description and Simulation

The specification of components are as follows, i.e. the grid voltage is taken as 415 V, the grid frequency has been assigned as 50 Hz, the voltage of DC bus is taken as 800 V, switching frequency has been given 10 kHz, six numbers of IGBT/Diodes are taken, DC link capacitor has been taken as 5600 μF, the active power is assigned the value of 100 Kw and the inductive reactive power is taken as 200 Kw.

The components used for designing the inverter circuit are IGBT/Diode, Series RLC branch, Connection Port and the PWM. The components utilized for its designing are three phase source, three phase parallel RLC load, three phase VI measurement, series RLC branch, from and goto blocks, voltage measurement block. The inverter circuit is connected to the power circuit at the output.

PLL is connected to the designed circuit which consists of components like, alpha–beta to *dq* transformation block, PI Controller, constant block, sum block, demux and integrator. Voltage and current controller blocks are then connected to the aforementioned designed circuit and the components required here are constant block, sum, PI Controller, from and goto, *d-q* to abc converter, product and gain blocks. The PWM blocks are connected thereafter which include components like relational operator, out, logic operator, repeating sequence and demux [18–20].

# 3 Results and Discussion

Simulations are done on MATLAB 2017a. We have set a time delay of 0.5 s by using a step input block. STATCOM comes into action after 0.5 s. When STATCOM is linked to the system at interval of an interval of 0.5 s, the voltage of DC link is made constant so that DC capacitor will induce power (capacitive) and will drive reactive power into the system to improve the pf to unity.

Fig. 3 depicts that voltage of DC Link is maintained at 800 V after connection of STATCOM at 0.5 s to supply constant DC voltage to the VSC.

As demonstrated in Fig. 4, from 0 to 0.5 s source current lags source voltage as load is inductive in nature, but after 0.5 s i.e. after connection of STATCOM, voltage and current of the source become in phase with one another which signifies that power factor is improved to unity. This shows that STATCOM can be used for improving pf.

As shown in Fig. 5, from 0 to 0.5 s STATCOM is off and for rest of the period it is on. Here switching on of STATCOM does not affect the active power supply to load by source. Most active power consumed by load is supplied by source.

As shown in Fig. 6, from 0 to 0.5 s STATCOM is off and for rest of the period it is on. Here from 0 to 0.5 s most of the RP consumed by load is provided by source, but after 0.5 s RP consumed by the load is supplied by inverter and RP given by source becomes zero.

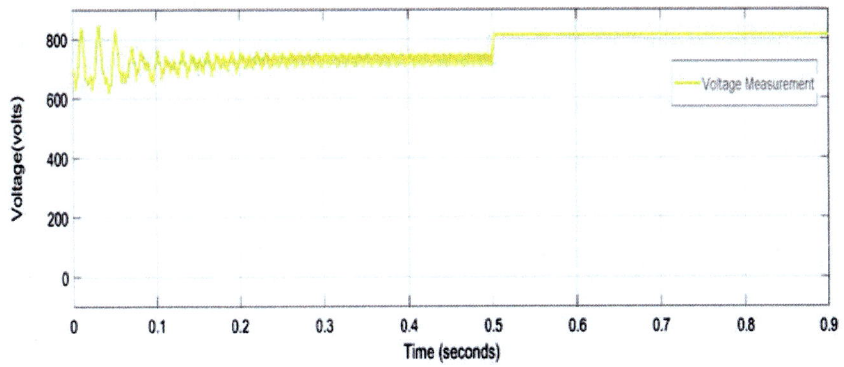


**Fig. 3** DC link voltage

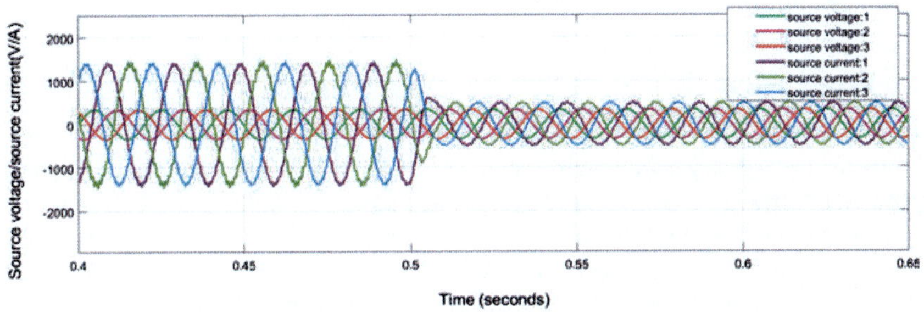


**Fig. 4** Load current and load voltage waveform

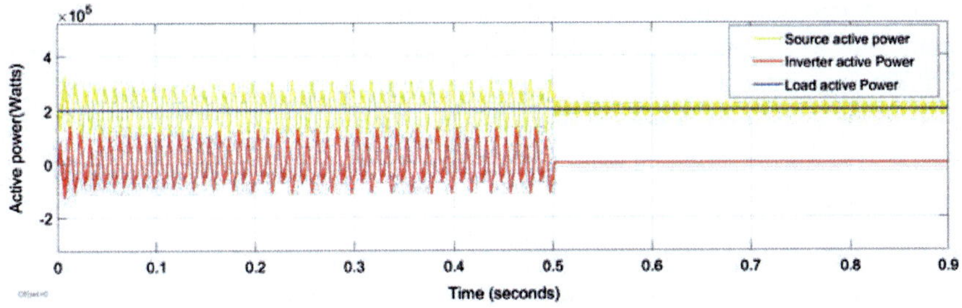


**Fig. 5** Active power of inverter, source and load

From the above results it is perceived that the total RP which is required by the load is provided by STATCOM by maintaining voltage of dc link as constant. Hence, the RP given by the grid is closely equal to zero which decreases load on grid and helps to maintain the stability.

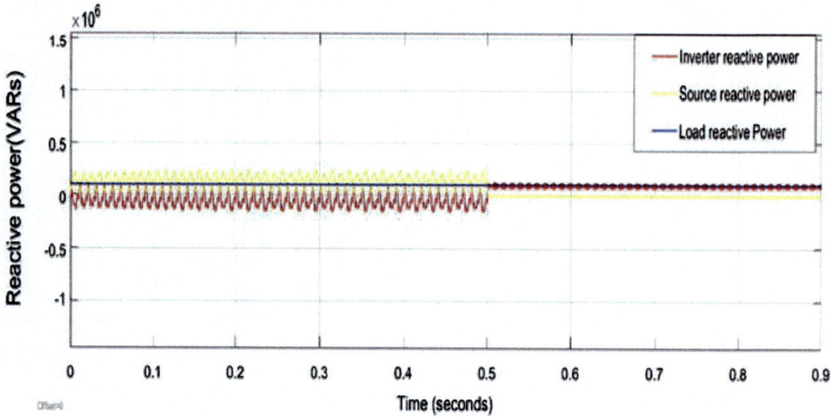


**Fig. 6** Reactive power of inverter, source and load

# 4 Conclusion

From this analysis we conclude that STATCOM is a controlled reactive power generator which provides required generation and absorption of reactive power.

Conventional methods include shunt and series capacitors and reactors which are fixed or mechanically switched and synchronous generators. So to overcome these problems STATCOM i.e. a technology utilizing solid state devices like thyristors are developed which have fast response time characteristics, no mechanical parts, compact design, higher operational adaptability, fantastic dynamic characteristics under different operating conditions.

Analysis of this paper shows that by using STATCOM in the system its voltage profile can be improved, pf of the system can be enhanced and overall voltage disturbances can be diminished.

# References


1. N. Agashe, R.D. Kulkarni, A.R. Thorat, Simulation study for reactive power compensation using STATCOM. Int J. Adv. Res. Electr. Electron. Instrum. Eng. **04**(06), (2015). online-ISSN: 2340-3765, online-ISSN:2278-8875, https://doi.org/10.15662/ijareeie.2015.0406065
2. M. Rastogi, A.H. Bhat, A. A. Zargar, Performance evaluation of D-STATCOM for mitigation of power quality problems at different source and load side fault conditions, in *2016 IEEE 7th Power India International Conference (PIICON)*, pp. 1–6. https://doi.org/10.1109/POWERI.2016.8077379
3. R. Dehini, C. Ben, A. Bassou, Simulation of distribution static compensator (D-STATCOM) to improve power quality. Arab. J. Sci. Eng. (2012). https://doi.org/10.1007/s13369-012-0461-5

4. D.P. Mishra, S. Chakraborty, T.K. Tripathy, Voltage regulation of PV cell using PID controller, in *2018 International Conference on Recent Innovations in Electrical, Electronics & Communication Engineering (ICRIEECE)* (2018), pp. 206–211. https://doi.org/10.1109/ICRIEECE44171.2018.9008536
5. Y. Zhao, B. Sun, Research on STATCOM principle and control technology, in *2011 International Conference on Consumer Electronics, Communications and Networks (CECNet)* (2011), pp. 1593–1596. https://doi.org/10.1109/CECNET.2011.5769064
6. M.G.G. Titus, Reactive power compensation using STATCOM for single phase distribution system, in *2015 International Conference on Circuits, Power and Computing Technologies [ICCPCT-2015], 2015*, pp. 1–5. https://doi.org/10.1109/ICCPCT.2015.7159530
7. H.S. Saini, T. Srinivas, D.M. Vinod Kumar, K.S. Chandragupta Mauryan, Innovations in Electrical and Electronics Engineering, in *Proceedings of the 4th ICIEEE 2019* (Springer Nature, 2020), p. 865
8. S.K. Sah et al., Reactive power control of modified IEEE 14 bus system using STATCOM. Second Int. Conf. Intell. Comput. Control Syst. (ICICCS) **2018**, 335–339 (2018). https://doi.org/10.1109/ICCONS.2018.8663177
9. M.T.L. Gayatri, A.M. Parimi, A.V. Pavan Kumar, A review of reactive power compensation techniques in microgrids. Renew. Sustain. Energy Rev. **81**(Part 1), 1030–1036. ISSN 1364-0321
10. A.P. Hota, S. Mishra, D.P. Mishra, S.R. Salkuti, Allocating active power loss with network reconfiguration in electrical power distribution systems. Int. J. Power Electron. Drive Syst. (IJPEDS) **12**(1), 130–138, Mar 2021, ISSN: 2088–8694. https://doi.org/10.11591/ijpeds.v12.i1
11. V.S. Pal, Joy, Reactive power compensation and in a lossless transmission line by the application of STATCOM, in *Proceedings of Second IRF International Conference*, Cochin, India, pp 68–70, Aug 2014
12. A.M. Eltamaly, Y. Sayed, A.N.A. Elghaffar, Power flow control for distribution generator in egypt using facts devices. Acta Technica Corviniensis Bull. Eng. ISSN: 2067-3809
13. S.R. Das, D.P. Mishra, P.K. Ray, S.R. Salkuti, A.K. Sahoo, Power quality improvement using fuzzy logic based compensation in a hybrid power system. Int. J. Power Electron. Drive Syst. (IJPEDS) **12**(1), 576–584 (2021)
14. N. Zhi, X. Xiao, P.G. Tian et al., Research and prospect of multi-microgrid control strategies. Electr. Power Autom. Equipment **36**(4), 107–115 (2016)
15. G.Z. Tian, S.T. Wang, Y. Pang et al., Mechanism analysis and research on control strategy of dynamic torque improvement for wind farm with squirrel cage induction generator using STATCOM. Acta Energiae Solaris Sin. **37**(9), 2201–2208 (2016)
16. P.K. Panda, A. Sahoo, A. Samal, D.P. Mishra, S.R. Salkuti, Voltage control of AC hybrid microgrid. Int. J. Power Electron. Drive Syst. (IJPEDS) **12**(2), 793–802 (2021). https://doi.org/10.11591/ijpeds.v12.i2.pp793-802
17. M. Patel, D.S. Patel, M.J. Chapadiya, K.K. Naik, B.K. Patel, Reactive power compensation using STATCOM. Int. Res. J. Eng. Technol. (IRJET) **04**(04), Apr 2017, e-ISSN: 2395-0056, p-ISSN: 2395–0072
18. U.K. Singh, A. Basak, Performance study of different PLL schemes under unbalanced grid voltage, in *2019 IEEE Region 10 Symposium (TENSYMP)* (2019), pp. 66–71. https://doi.org/10.1109/TENSYMP46218.2019.8971048
19. B. Das, P.K. Panigrahi, S.R. Das, D.P. Mishra, S.R. Salkuti, Power quality improvement in a photovoltaic based microgrid integrated network using multilevel inverter. Int. J. Emerg. Electr. Power Syst. 2021, 000010151520210040. https://doi.org/10.1515/ijeeps-2021-0040
20. A.P. Hota, S. Mishra, D. Prasad, S.R. Salkuti, A novel methodology for power loss allocation of both passive and active power distribution systems. Int. J. Emerg. Electr. Power Syst. 20210087 (2021). https://doi.org/10.1515/ijeeps-2021-0087

# Smart Energy Systems

# Optimization of Set-Point Deviation and Current Ripple of PMSM by CCS-MPC Method

**Shaswat Chirantan, Deba Prasad Dash, and Bibhuti Bhusan Pati**

**Abstract** In modern technology it is essential to use inverter fed drive with high power density. As the conventional voltage source inverter has the issues of smooth torque and current control, a model predictive control (MPC) technique has been proposed for smooth control of permanent magnet synchronous motors (PMSM) with field-oriented control. In this research work a continuous control set (CCS) predictive current controller (PCC) based on cost minimization principle is used to provide the switching states to the inverter. The characteristics of electromagnetic torque developed and load currents in *d-q* reference frame of predictive current control method are analyzed for system performance study. All these simulations are carried out using MATLAB/SIMULINK platform.

**Keywords** Continuous control set · Model predictive control · Predictive current controller · Permanent magnet synchronous motors · Pulse width modulation

## 1 Introduction

Permanent magnet synchronous motors (PMSMs) are largely used as machine tool servos and in speed control technique. PMSMs are popular because of very good dynamic response, high density of power, high reliability and high efficiency [1]. Due to the above advantages PMSMs are widely used in automobiles, biomedical instruments, aero-craft, military and in large industries where smooth speed control is one of the major aspects.

Field-oriented control (FOC) is mainly used as control method in PMSMs. The advantages of this method are smooth torque, wide speed range and instant acceleration-declaration using current controller methods [2]. There are some

S. Chirantan (✉) · D. P. Dash · B. B. Pati
Department of Electrical Engineering, Veer Surendra Sai University of Technology, Burla, Sambalpur 768018, India
e-mail: shaswat.chirantan443@gmail.com

B. B. Pati
e-mail: bbpati_ee@vssut.ac.in



S. Patnaik et al. (eds.), *Smart and Sustainable Technologies: Rural and Tribal Development Using IoT and Cloud Computing*, Advances in Sustainability Science and Technology,
https://doi.org/10.1007/978-981-19-2277-0_25

conventional methods used for current control of inverter fed three phase loads such as proportional plus integral (PI), controller-based sinusoidal pulse width modulation (SPWM) and hysteresis current controller (HCC) method [3, 4]. The MPC methods are categorized in two ways. The first one is continuous-control-set MPC (CCS-MPC) and the second one is finite-control-set MPC (FCS-MPC) [5]. The algorithm for continuous control set MPC is based on the Receding horizon control principle with forward Euler approximation for discretized load model [6]. FCS method applies voltage vectors directly to the PMSM whereas CCS method uses voltage modulation technique like SVPWM. Both the methods have similar performance but the CCS method is more versatile [7]. A MPC-based model of five phase PMSM is demonstrated in [8] in order to limit the load current harmonics. Nowadays torque control of PMSM can also be achieved by a predictive controller and further implemented on electric vehicles [9] via switching optimization. In [10], various control schemes such as PID and predictive controllers for electric drives and power converters are implemented with MATLAB platform. The above literature further encourages to implement predictive control technique for optimum control of desired system parameters. In this proposed method, CCS-MPC-based current control scheme is applied for inverter fed discretized PMSM model. The torque and current obtained from the PMSM output are further compared with the set reference parameter for optimum deviation analysis.

## 2 PMSM Model Description

The rotor of PMSM is made of permanent magnet. So, there is no need of DC supply as magnetizing current to rotor. In surface inset type PMSM with smooth rotor surface, the value of inductances in both d-axis and $q$-axis is same. It is assumed that the induced back electromagnetic field (EMF) is sinusoidal, field current is absent and saturations are neglected for the model developed on rotor reference frame. Based on the above assumption, the most commonly adopted PMSM model is described below

$$\frac{\mathrm{d}i_d}{\mathrm{d}t} = \frac{1}{L_d}\left(\vartheta_d - (R.i_d) + \left(\omega_e.L_q.i_q\right)\right) \tag{1}$$

$$\frac{\mathrm{d}i_q}{\mathrm{d}t} = \frac{1}{L_q}\left(\vartheta_q - \left(R.i_q\right) - (\omega_e.L_d.i_d) - \left(\omega_e.\varphi_{mg}\right)\right) \tag{2}$$

$$\frac{\mathrm{d}\omega_e}{\mathrm{d}t} = \frac{P}{J}\left(T_e - \left(\frac{B}{P}\omega_e\right) - (T_L)\right) \tag{3}$$

$$T_e = \frac{3}{2}P\left(i_q.\varphi_{mg} + \left(L_d - L_q\right)i_d i_q\right) \tag{4}$$

where

| | |
|---|---|
| $i_d$ | Stator current in $d$-axis in ampere |
| $i_q$ | Stator current in $q$-axis in ampere |
| $\omega_e$ | Electrical speed in radians per second $= P.\omega_m$ |
| $\omega_m$ | Mechanical speed in radians per second |
| $\varphi_{mg}$ | Amplitude of flux linkage established by the rotor in weber |
| $P$ | No. of pair of poles |
| $\vartheta_d$ | Stator voltage in $d$-axis in volts |
| $\vartheta_q$ | Stator voltage in $q$-axis in volts |
| $T_L$ | Load torque in $N$-$m$ |
| $T_e$ | Electromagnetic torque in $N$-$m$ |
| $L_d$, $L_q$ | Stator inductances in $d$-axis and $q$-axis respectively in henrys |
| $R$ | Stator resistance in ohms |
| $J$ | Moment of inertia in kg-m$^2$ |
| $B$ | Coefficient of friction in newton meter per second. |

In case of $i_d = 0$ or surface inset PMSM with $L_d = L_q$, then the electromagnetic torque will be modified as follows.

$$T_e = \frac{3}{2}P\left(i_q.\varphi_{mg}\right) \tag{5}$$

The load torque applied to the motor is a step input of initial value of 3 N-meter for half of the simulation time i.e., 100 ms and final value of -3 N-meter for the last half of the simulation time i.e., (100–200) milliseconds. The simulation model is discrete type and sample time is 10 microseconds. PMSM model parameters are given in Table 1.

**Table 1** PMSM model parameters [10]

| Parameter | Values |
|---|---|
| Stator resistance ($R$) | 2.98 Ohms |
| Inductance in $d$-axis ($L_d$) | 0.007 Henrys |
| Inductances in $q$-axis ($L_q$) | 0.007 Henrys |
| Rotor flux constant | 0.125 voltage per radian per second |
| Moment of inertia ($J$) | 1e-5 kg-meter square |
| Friction viscous gain ($B$) | 11e-5 N meter per radian per second |
| Number of poles ($P$) | 4 |

## 3 Model Description of CCS-MPC Method

MPC is an optimization approach that executes current control action by minimizing the error between desired value and predicted value. The state space model-based algorithm is mostly favored [6]. A PWM inverter is used as modulator to get the optimal output voltage. The structure of CCS-MPC method used for the PMSM model is shown as schematic diagram in the Fig. 1.

The three-phase inverter shown in the model converts 300 V DC to three phase AC, the load model used here is a discretized PMSM model. Here the inverter is considered as a non-linear discrete system with 180° conduction mode. It has 7 different possible output states with 8 configurations. For simplicity modeling dead time, IGBT saturation voltage and diode forward voltage drops are not considered. The switching states of the converter are determined by the gating signals $S_a$, $S_b$ and $S_c$ as follows [6]:

$$S_a = \begin{cases} 1, & \text{if Switch}_1 \text{ on and Switch}_4 \text{off} \\ 0, & \text{if Switch}_1 \text{ off and Switch}_4 \text{ on} \end{cases}$$

$$S_b = \begin{cases} 1, & \text{if Switch}_2 \text{ on and Switch}_5 \text{off} \\ 0, & \text{if Switch}_2 \text{offand Switch}_5 \text{on} \end{cases}$$

$$S_c = \begin{cases} 1, & \text{if Switch}_3 \text{on and Switch}_6 \text{off} \\ 0, & \text{if Switch}_3 \text{off and Switch}_6 \text{on} \end{cases}$$

The output current of the load is compared to a reference current of $i_d = 0$ and $i_q = \pm 2$. The sampling time of the predictive algorithm is 10 microseconds. The frequency of reference current is 50 Hz. Conversion of rotating frame to alpha–beta frame is done to avoid complexity in machine modeling.

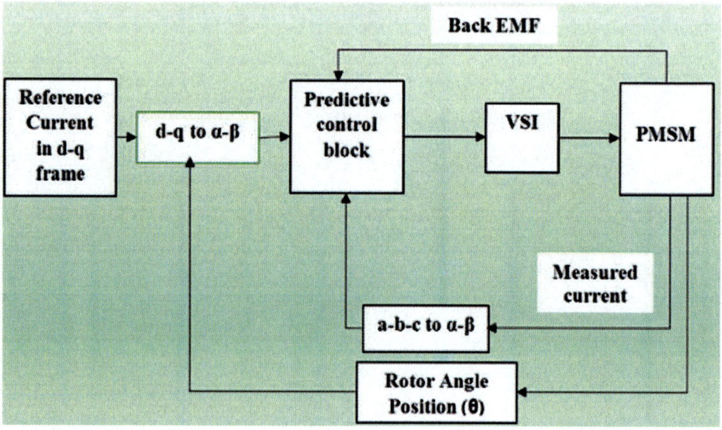


**Fig. 1** The structure of the CCS-MPC method

Considering the unitary vector, $a = e^{-j(2\pi/3)} = -\frac{1}{2} + j\frac{\sqrt{3}}{2}$, which represents the 120° phase displacement between phases, the current vector can be defined as

$$i = \frac{2}{3}\left(i_a + ai_b + a^2 i_c\right) \tag{6}$$

Converting the above into real and imaginary part

$$\text{Real part of } i = i_\alpha = \frac{2}{3}\left(i_a - \frac{1}{2}i_b - \frac{1}{2}i_c\right) \tag{7}$$

$$\text{Imaginary part of } i = i_\beta = \frac{2}{3}\left(\frac{\sqrt{3}}{2}i_b - \frac{\sqrt{3}}{2}i_c\right) \tag{8}$$

From forward Euler approximation,

$$\frac{di}{dt} \approx \frac{i(k+1) - i(k)}{T_s} \tag{9}$$

From the above equation predicted current can be defined as.

$$i^p(k+1) = \left(1 - \frac{RT_s}{L}\right)i(k) + \frac{T_s}{L}(v(k)) \tag{10}$$

Hence the objective function can be expressed as

$$J = \left|i_\alpha^*(t_{i+1}) - i_\alpha(t_{i+1})\right| + \left|i_\beta^*(t_{i+1}) - i_\beta(t_{i+1})\right| \tag{11}$$

In this algorithm using old values of load current next state of the current is predicted for 8 different configurations of inverter switching. The predictive current for 8 different switching states are compared with the reference current assuming the change in reference current is very small. For the above switching states, the absolute difference between these currents is estimated. The state of minimum absolute difference is stored and switching states are sent to the inverter circuit accordingly (Table 2).

## 4 Results and Discussion

In this section, the experimental results are obtained with given simulation parameters. The simulation outputs such as electromagnetic torque and stator currents in d-axis and q-axis are shown for the CCS-MPC method validated in Simulink (Table 3; Figs. 2 and 3).

**Table 2** Switching states and voltage vectors [6]

| $S_a$ | $S_b$ | $S_c$ | Voltage vector $\vec{v}$ |
|---|---|---|---|
| 0 | 0 | 0 | $\vec{v_0} = 0$ |
| 1 | 0 | 0 | $\vec{v_1} = \frac{2}{3} v_{dc}$ |
| 1 | 1 | 0 | $\vec{v_2} = \frac{1}{3} v_{dc} + j\frac{\sqrt{3}}{3} v_{dc}$ |
| 0 | 1 | 0 | $\vec{v_3} = -\frac{1}{3} v_{dc} + j\frac{\sqrt{3}}{3} v_{dc}$ |
| 0 | 1 | 1 | $\vec{v_4} = -\frac{2}{3} v_{dc}$ |
| 0 | 0 | 1 | $\vec{v_5} = -\frac{1}{3} v_{dc} - j\frac{\sqrt{3}}{3} v_{dc}$ |
| 1 | 0 | 1 | $\vec{v_6} = \frac{1}{3} v_{dc} - j\frac{\sqrt{3}}{3} v_{dc}$ |
| 1 | 1 | 1 | $\vec{v_7} = 0$ |

**Table 3** Simulation parameters [10]

| Parameters | Start time | Stop time | Simulation type | Sample time |
|---|---|---|---|---|
| Values | 0.0 [s] | 0.2 [s] | Discrete | 1e–05 [s] |

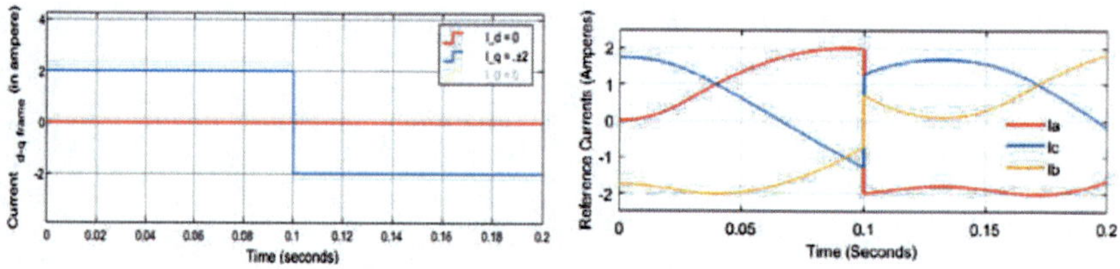


**Fig. 2** Three phase reference currents

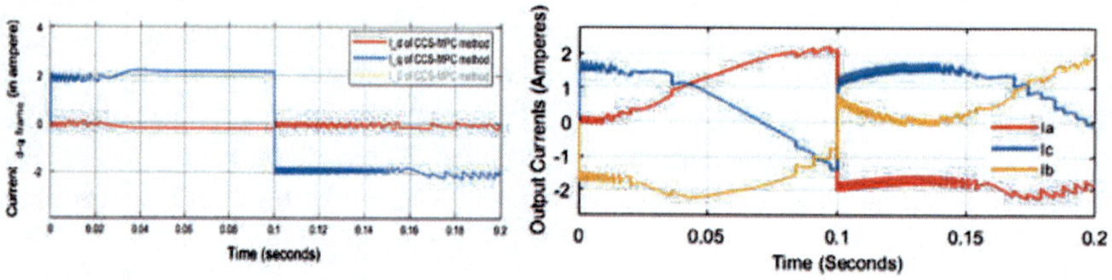


**Fig. 3** Output currents of CCS-MPC method

The step reference and measured output torque of PMSM model for CCS-MPC method are shown below (Figs. 4 and 5).

The performance of a method can be described by analyzing the current ripple, peak overshoot and how electromagnetic torque follows load torque. From the above depicted current and torque characteristics the performance of the proposed MPC scheme can be visualized. The quadrature axis current and load torque plots clearly

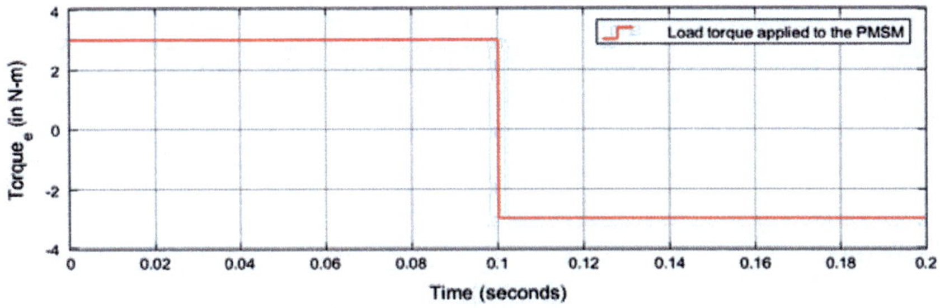


**Fig. 4** The load toque applied to the PMSM model

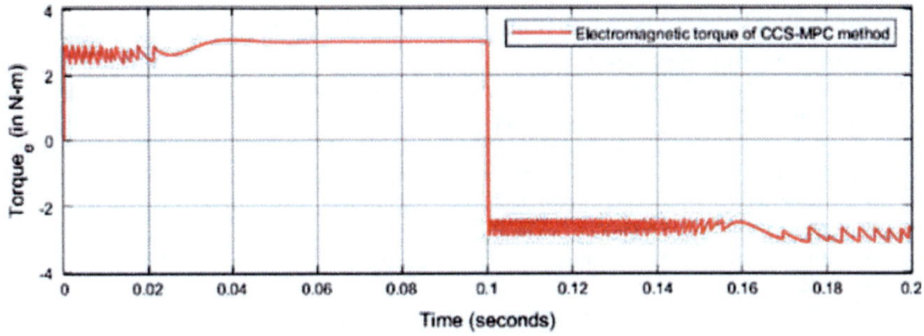


**Fig. 5** Electromagnetic torque developed by CCS-MPC method

show how the obtained load parameters follow the set reference point value as mentioned earlier. Thus we can optimize the deviation of selected parameters from the set point and hence the overall system performance. Further CCS-MPC technique can be compared with other predictive schemes, PWM and hysteresis control to evaluate the adaptability, reliability and flexibility of the controlling approach.

## 5 Conclusion

Continuous control set MPC has been applied to PMSM with current control method. The major advantage of this simulation lies in the ability to reduce torque and current ripple. Both the reference current in q-axis and the reference load torque are applied as step function to observe the dynamic behavior of PMSM as the torque and the current in $q$-axis of PMSM are proportional to each other with the value of current in $d$-axis is zero. Because of unit time delay there exists some steady state error. Although the performance of PMSM is improved by CCS-MPC method, the CCS-MPC method may be applied with some compensation to overcome steady state error. There are possibilities to lowering harmonics, switching power loss for multi-phase PMSM with considering non-linear characteristics of power converter switches. Further improvement in the load current harmonics can be achieved by using filters. Apart from this some other predictive schemes such as Finite Control Set (FCS-MPC), Integral Finite Control Set (IFCS-MPC) can also be implemented to overcome the shortcomings of CCS-MPC.

## References

1. M.K. Waghmare, S.V. Patil, Speed control strategy of permanent magnet synchronous motor drive using SPWM technique, in *2019 4th International Conference on Recent Trends on Electronics, Information, Communication & Technology (RTEICT)*, pp. 1328–1332
2. P. Maji, G.K. Panda, P.K. Saha, Field oriented control of permanent magnet synchronous motor using PID controller. Int. J. Adv. Res. Electr. Electron. Instrum. Energy **4**, 632–639 (2015)
3. S. Chirantan, B.B. Pati, Model predictive control of three phase inverter fed RL load, in *2020 International Conference on Computational Intelligence for Smart Power System and Sustainable Energy (CISPSSE)* (2020), pp. 1–6
4. S. Chirantan, B.B. Pati, MPC And HB-PWM based current control techniques for three phase inverter fed RL load, in *2020 IEEE International Symposium on Sustainable Energy, Signal Processing and Cyber Security (SSSC)* (2020), pp. 1–6
5. G. Vijayakumar, M.G.S. Prasad, Model based predictive direct speed control with FCS of PMSM drive system, *IRET Transaction on Design and Control Systems (ITDCS)*, vol. 1, 1–6 Aug 2014
6. J. Rodriguez, P. Cortes, *Predictive Control of Power Converters and Electrical Drives*. The Atrium, Southern Gate, Chichester, West Sussex, PO19 8SQ, (United Kingdom, Wiley, 2012)
7. A.A. Ahmed, B.K. Koh, Y.I. Lee, A comparison of finite control set and continuous control set model predictive control schemes for speed control of induction motors. IEEE Trans. Ind. Inf. **14**(4), 1334–1346, Apr 2018
8. C. Xiong, H. Xu, T. Guan, P. Zhou, A constant switching frequency multiple-vector-based model predictive current control of five-phase PMSM with non sinusoidal back EMF. IEEE Trans. Ind. Electron. **67**(3), 1695–1707 Mar 2020
9. B. Ren, H. Chen, H. Zhao, W. Xu, MPC-based torque control of permanent magnet synchronous motor for electric vehicles via switching optimization. Control Theory Technol. **15**, 138–149 May 2017
10. L. Wang, S. Chai, D. Yoo, L. Gan, K. Ng, *PID and Predictive Control of Electrical Drives and Power Converters using MATLAB/Simulink* (Wiley, Singapore Pte. Ltd., 2015) 1 Fusionopolis Walk, #07–01 Solaris South Tower, Singapore 138628

# Comparative Study, Design and Performance Analysis of Grid-Connected Solar PV System of Two Different Places Using PV syst Software

**Azam Khan, Bilal Alam, Wajid Ali, Faisal Jamal, Amir Khan, and Nural Islam**

**Abstract** The Electricity is an important source of energy in daily life. Nowadays, we are depending upon electrical appliances. The increasing demand for electricity and degrading effects of non-renewable energy resources on the environment made the world shift focus toward the available renewable energy sources such as solar energy for the fulfillment of our energy needs. The most common way to generate electricity from the sun is by using a PV cell system that changes solar energy directly into electricity and efficiency of a photovoltaic system greatly relies upon the location of the place and type of PV modules used. The PV system is of greater importance in sites with a good amount of sunlight. In this, we collected and analyzed data from two grid-connected solar PV systems located at Maulana Azad Library, Aligarh, and Manipal University, Manipal. In this paper, we copy the grid-connected to a solar photovoltaic system using computer software PV syst v-7.0.10.17617. The total sum of energy produced by the solar-connected system and the different types of losses that can be seen in a given system is detailed and analyzed.

**Keywords** Solar photovoltaic · System simulation · Solar radiation · PV Syst

## 1 Introduction

The present energy situation isn't sustainable, as fossil fuel reserves are diminishing. In [1, 2] as of 2020, most experts predict that we only have around 40–80 years of non-renewable energy available, and this may not be enough to satisfy the increasing demand related to economic development and population increase. While the fossil

A. Khan · F. Jamal
Electrical Engineering Department, Z.H.C.E.T, Aligarh Muslim University, Aligarh, India

B. Alam (✉) · W. Ali · N. Islam
Department of Mechanical Engineering, Aligarh Muslim University, Aligarh, India
e-mail: amu.bilal.chs@gmail.com

A. Khan
Department of Mechanical Engineering, Delhi Technological University, Delhi, India

 
S. Patnaik et al. (eds.), *Smart and Sustainable Technologies: Rural and Tribal Development Using IoT and Cloud Computing*, Advances in Sustainability Science and Technology,
https://doi.org/10.1007/978-981-19-2277-0_26

fuels we consume constitute to 28% to the world's gas emissions released annually, fuel combustion causes pollution, which is dangerous for humans, plants, and animals.

The global solar market is burgeoning and it's predicted that the globe will have 1.5 trillion watts of installed solar PV capacity by 2030. Approximately 5500 trillion kWh/year of energy occurs in the Indian subcontinent. The majority of areas receive 4.5–7.5 kWh/sq./day. In [2–4] we can be successfully tied to generating great capacity in India. Solar also allows for distributed power acquisition and allows for faster power additions with shorter lead times. Off-grid distribution, as well as lower temperatures, will benefit the domestic energy vision and meet other energy needs like heating and cooling in both rural and urban areas. From the standpoint of energy security, solar energy seems the safest of all other available sources because it is widely available [3]. Theoretically, only a fraction of a completely different energy event (if taken successfully) could meet energy needs across the country. The National Solar Energy Centre has tested about 748 GW of solar energy by taking up 3% of the surface area of the wasteland and covered it with solar PV modules. Solar power has taken over the Indian National Action Plan for Climate Change and the National Solar Mission with the integration of vital equipment. The installed solar capacity of the solar system was 35,739 MW as at August 31, 2020. There are three main categories of solar PV and storage systems: grid-tied, grid/hybrid, and stand-alone/off-grid. In this paper, the grid connected to the design and analysis of the PV system is eliminated with the help of PV syst software [1, 5]. PV syst software can be a tool that enables the study, measurement, and data analysis of the complete PV systems [6, 7]. It works with grid-connected, stand-alone pumps as well as DC-grid (public transportation) PV systems, and it includes meteor and PV details, as well as standard solar power tools. This software is extremely useful for computer modeling and system analysis because it is very reliable and accurate.

This paper contains 4 sections in which Sect. 2 describes the methodology behind the whole process, Sect. 3 contains performance and analysis of the grid-connected system, and Sects. 4 and 5 contain the result and discussion of the whole system.

## 2 Grid-Tied PV System

A PV system, inverters, and other balance system components, like mounting system, charge controller, cables, power transformers, and necessary batteries, are included in the grid-connected bound PV system. The choice of components depends on the device requirements and the requirements of the customer, which are defined in [8, 9]. Figure 1 shows a schematic view of a photovoltaic system connected to the grid.

The DC current generated by GCPVS is converted into an AC current with the help of inverters. Solar array compatibility and optimized sizing are of vital significance in solar PV system performance and system design. The important factor of the inverter must be recognized in conjunction with the device design for power and frequency

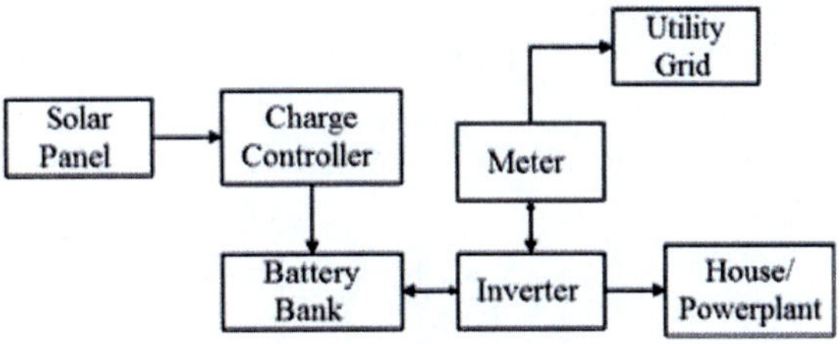


**Fig. 1** Grid-connected solar PV system

rating. In [10, 11] other connected grid components that are also very significant, such as switches, wire, connectors, and the assembly of the device mounting.

# 3 Methodology

In this paper, we are surveying two different cities of Uttar Pradesh and Karnataka. These cities are Aligarh and Manipal. The weather monitoring station at the Aligarh and Manipal latitude and longitude is shown in Table 1.

## *3.1 PV Syst Software*

PV software is generally used for estimating and simulating the cost, aging, and estimation of several inverters for PV panels. All over the world, PV syst is used for developing projects. The following steps are needed to obtain a final report.

### 3.1.1 Irradiation Data

The IR data was derived from the satellite data of Metronorm 7.3. The IR data obtained reflected hourly synthetic ambient temperature, beam, diffused and global IR for the whole year. The statistical equivalent of the deployed power plant GCPVS was measured and simulated using the metronor 7.3 satellite data file. Figure 2a and b show the synthetically generated meteo data of Aligarh and Manipal.

**Table 1** Latitude and longitude of Aligarh and Manipal

| | Longitude | Latitude |
|---|---|---|
| Aligarh | 27.8974 N | 78.0880 E |
| Manipal | 13.3525 N | 74.7928 E |

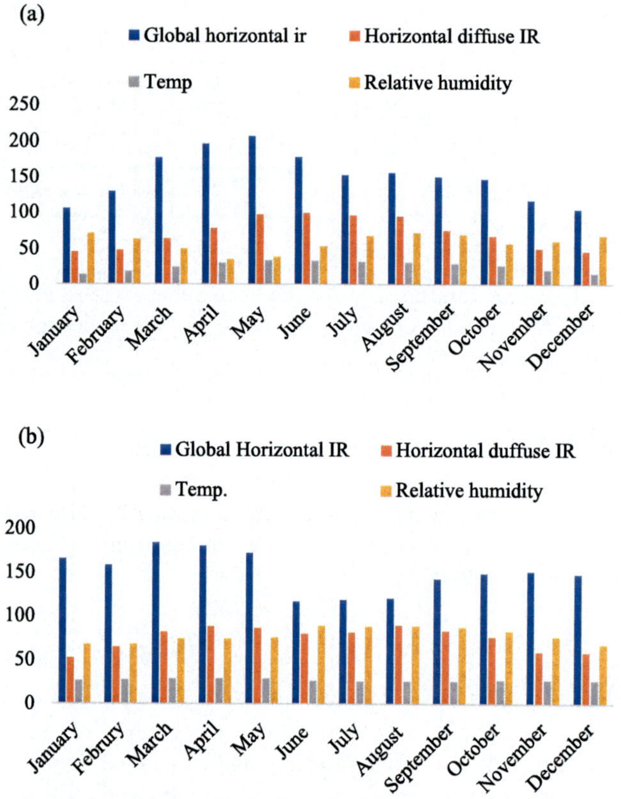


**Fig. 2** Synthetic generated meteo data **a** Aligarh **b** Manipal

### 3.1.2 Orientation and Horizon

In this paper, we simulate 123 kW and 110 KW Aligarh and Manipal systems respectively for a constant tilt optimization with reference to the latitude and longitude of that particular location. To improve efficiency and performance, the set tilt angle has been replaced by two axis variables. The orientation was followed in order to create an equivalent real-time simulation of the present system in PV syst. Figure 3a depicts the orientation with constant tilt for the Aligarh system, while Fig. 3b depicts the Manipal system.

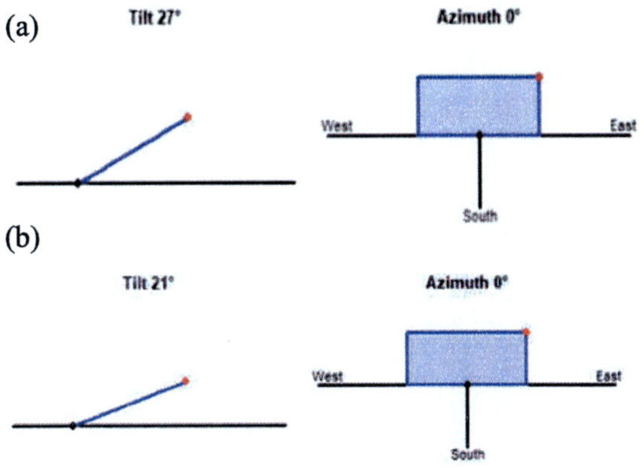


**Fig. 3** Tilt and Azimuth Angle **a** Aligarh **b** Manipal

**Table 2** Load demands

| | Aligarh | Manipal |
|---|---|---|
| Total wattage | 132.125 KW | 115.23 Kw |
| Total daily consumption | 1828.635 Kwh/day | 1554.29 Kwh/day |

### *3.2 Case Study*

In this proposed work, we design the solar PV rooftop grid-connected system for a load outlay of 1828.635 kWh/day. Table 3 shows the load data and daily energy consumption of the given location. And the load is shown in Table 2

## 4 Input Requirement

As each location absorbs varying levels of solar IR radiation, the architecture of the PV system is based on location. This is because of the orientation of precise place (longitude and latitude) with respect to the face of the sun.

### *4.1 Location*

The cities of Aligarh and Manipal are located in northern and southern India, respectively, at latitude and longitude, as shown in Table 1. Aligarh has flat terrain. These systems are designed for 132 and 110 KW of power.

**Table 3** Comparison of load data and daily energy consumption

| Load | QA | | WA | | TWA | | Time A | | Consumption A (Wh/day) | |
|---|---|---|---|---|---|---|---|---|---|---|
| | Aligarh | Manipal | Aligarh | Manipal | Aligarh | Manipal | Aligarh | Manipal | Aligarh | Manioal |
| Air purifier | 4 | 4 | 20 | 20 | 80 | 80 | 18 | 18 | 1440 | 1440 |
| Air condition machines (window/Ductable split) | 58 | 48 | 750 | 750 | 43,500 | 36,000 | 16 | 16 | 696,000 | 576,000 |
| Air cutter | 03 | 2 | 220 | 220 | 660 | 440 | 12 | 12 | 7920 | 5280 |
| Ceiling fan | 223 | 204 | 75 | 75 | 16,725 | 15,300 | 16 | 16 | 26,7600 | 244,800 |
| Pedestal fan | 27 | 27 | 55 | 55 | 1485 | 1485 | 15 | 15 | 22,275 | 22,275 |
| Ex. fan | 40 | 30 | 150 | 150 | 6000 | 4500 | 12 | 12 | 72,000 | 54,000 |
| Wall fan | 135 | 120 | 30 | 30 | 4050 | 3600 | 18 | 18 | 72,900 | 64,800 |
| Cabin fan | 31 | 31 | 75 | 75 | 2325 | 2325 | 8 | 8 | 18,600 | 600 |
| Desert cooler | 37 | 20 | 200 | 75 | 7400 | 4000 | 12 | 12 | 88,800 | 48,000 |
| Fridge | 02 | 02 | 250 | 200 | 500 | 500 | 24 | 24 | 12,000 | 1200 |
| Heat convector (Blower) | 04 | 03 | 2000 | 250 | 8000 | 6000 | 8 | 8 | 64,000 | 4800 |
| Electric water heater instant geyser | 02 | 02 | 4500 | 2000 | 9000 | 9000 | 8 | 8 | 72,000 | 72,000 |
| Inverter | 05 | 04 | 5000 | 4500 | 15,000 | 20,000 | 2 | 2 | 30,000 | 4000 |
| Mono block pump 2HP | 01 | 01 | 750 | 5000 | 750 | 750 | 2 | 2 | 1500 | 1500 |
| Stabilizer | 12 | 12 | 750 | 750 | 9000 | 9000 | 24 | 24 | 216,000 | 216,000 |
| Servo stabilizer | 08 | 08 | 450 | 750 | 3600 | 3600 | 24 | 24 | 86,400 | 86,400 |
| Water cooler | 05 | 05 | 750 | 450 | 3750 | 3750 | 24 | 24 | 90,000 | 9000 |
| Water purifier and aqua guard | 02 | 02 | 150 | 750 | 300 | 300 | 24 | 24 | 7200 | 7200 |

## 4.2 Tilt Angle

The angle of incident is known as the angle of slope at which the solar panel is placed or faces the sun. The location of the sun varies with respect to every time of day. The mounting angles of the panels often keep changing. The angle of tilt is often taken to be the same as the latitude of the location. Tilt angle and latitude angle are taken into consideration to optimize the sum of the solar energy on the plates. Based on the optimal value of tilt angle, we have found an annual IR yield. Figure 4 shows the tilt angle of the given system.

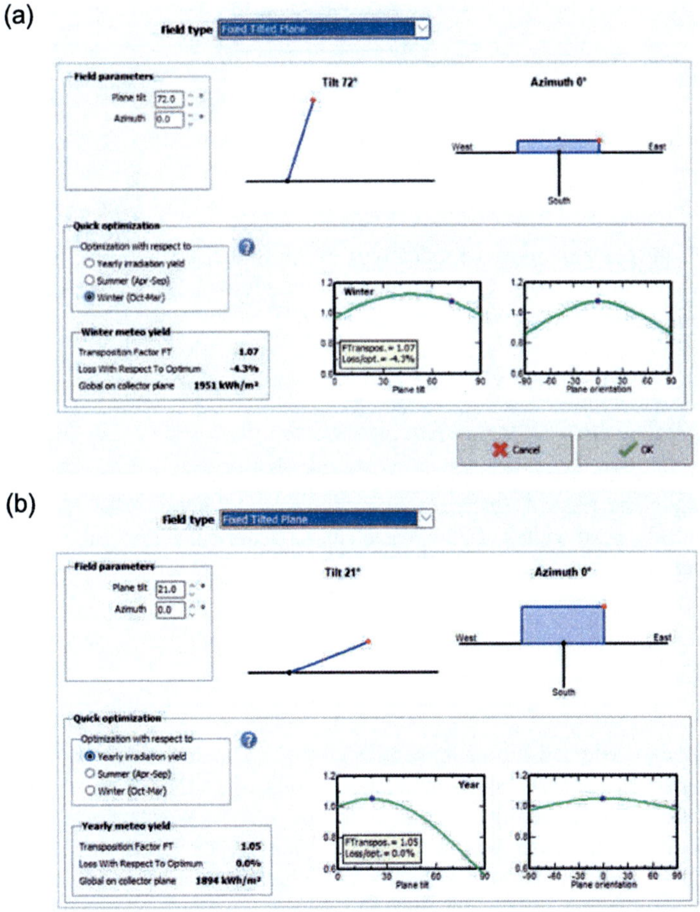


**Fig. 4** Orientation with fixed tilt for system **a** Aligarh **b** Manipal

**Table 4** Specification of Inverter and PV Array

| | For Aligarh | For Manipal |
|---|---|---|
| *A. PVArray characteristics* | | |
| 1. Manufacturer | Somera | Generic |
| 2. Model | VSM.72.350.05 | TSM-300-P14A |
| *B. Number of PV modules* | | |
| 1. In series | 18 odules | 19 modules |
| 2. In parallel | 19 strings | 19 strings |
| 3. Total Number of modules | 378 | 361 |
| Unit No.Power | 350 Wp | 300 Wp |
| *C. Inverter* | | |
| 1. Manufacture | Generic | Generic |
| 2. Model | TRIO-TM-50_0-400 | TRIO-27_6-TL-OUTD-S1-US |

## 4.3 Albedo

Albedo is the calculation of the reflectivity of the surface of the earth. It depends on the texture of the earth's position (latitude and longitude). The value fluctuates from 0 to 1. It is higher in snowy regions.

## 4.4 Module and Inverter Specification

For sizing of the 132-kW system at Aligarh, we used Somera VSM.72.350.05 poly crystalline silicon modules of 350 W each and a generic inverter. Table 4 shows the specifications of the inverter and PV panels.

## 4.5 Economics Analysis

The system is designed for a commercial load of 132 and 110 KW in the Maulana Azad Library at Aligarh and the innovation building of MIT, Manipal respectively. The tilt and the azimuth angle are shown in Fig. 3. A 1 kW solar rooftop plant will produce an average (taken for one year) 4.6 kWh of electricity per day (5.5 sunshine hours). Table 5 shows the MNRE solar rooftop data.

**Table 5** From MNRE solar rooftop calculator

| | For Aligarh | For Manipal |
|---|---|---|
| A. Size of power plant | 132.2 kw | 115 |
| B. Cost of the power plant | | |
| 1. MNRE current benchmark cost | 36,000 Rs/Kw | 36,000 Rs/Kw |
| 2. Without subsidy (Based on current MNRE benchmark) | 4,752,000 Rs | 4,140,000 Rs |
| 3. With subsidy 0 (Based on current MNRE benchmark) | 4,752,000 Rs | 4,140,000 Rs |
| C. Total electricity generation from the power plant | | |
| 1. Annually | 182,160 Kwh | 172,500 KWh |
| 2. Life time (25years) | 4,554,000 Kwh | 4,312,500 KWh |
| D. Financial savings. (tariff @5.5/KWh) | | |
| 1. Monthly | 83,490 Rs | 63,681 Rs |
| 2. Annually | 1,001,880 Rs | 764,175 Rs |
| 3. Life time (25years) | 25,047,000 Rs | 19,104,275 Rs |
| E. $CO_2$ emission mitigated is | 3734 tones | 3536 tones |
| F. The installation will be equivalent to planting | 5975 (Tree over the life time) | 5658 (Tree over the life time) |

## 5 Simulation and Result

PV syst is software that can provide all the details needed to design SAPV and GCPVS. It calculates PV modules, size of inverters, battery capacity, loss factor, energy calculation, PV module, hourly profile, P–V curve, and maximum power point, including the economic analysis. PV syst provides the proper location, tilt angle, and necessary data. Figure 6 shows the systematic loss of the solar grid-connected PV system. The losses of a solar grid-connected PV system are depicted in the loss diagram in step-by-step form. Several losses arise in solar PV arrays and modules, namely module mismatch loss, wiring loss, converter loss while in operation, loss in converter because of threshold power, loss in battery efficiency, loss in the inverter because of threshold power, charge and discharge current efficiency loss, PV loss due to temperature, unused energy. After rectifying the inverter, the input from the grid is calculated.

PR indicates the ratio between the output of the inverter and the output of the PV array. Figure 5a and b show the performance ratio. The required performance ratio is 29.3%, which means that in the case of Aligarh, 70.7% of the total energy produce by the solar power plant is lost, whereas in Manipal, 20.1% is lost. Figure 5c and d depicts the calculation of Normalized Production based on the simulation study. Some of the important factors are shown in the graphical representation, such as the unused energy, collection loss, system losses, and battery charging. Table 6 represents

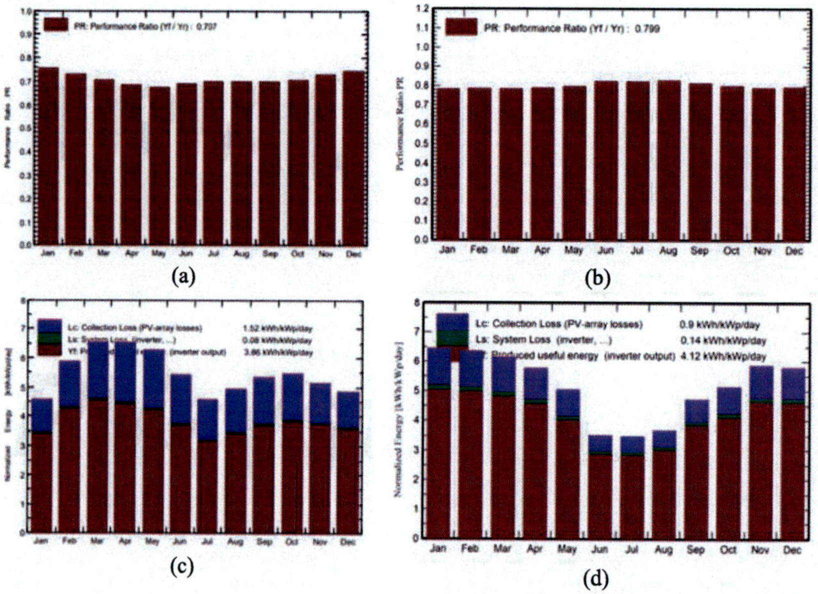


**Fig. 5** Performance ratio **a** Aligarh **b** Manipal, Normalized Production, **c** Aligarh **d** Maniapl

the final and main result for 132 and 115 KW solar power plant. This table shows the Global horizontal irradiation, Energy injected into grid, Performance Ratio, T.amb and etc.

## 6 Conclusion

We conclude the following in this paper for grid-connected Maulana Azad Library at Aligarh and Manipal University at Manipal using PV syst software. We surveyed these two cities, one from Uttar Pradesh and the other from Karnataka. The performance ratio of the solar PV system in Manipal is better than in Aligarh. The average annual performance ratio in Aligarh is 0.707, meanwhile for Manipal it is 0.799. Global solar irradiation is highest in Aligarh in May, while in the case of Manipal, this highest data occurs in March. In India, irradiation is generally received at 4.5–7.5 kWh/sq./day. The annual energy dispensed to grid is 186.61 Mwh.

This result indicates that this design and performance analysis of the grid-connected solar PV system plan will be profitable to the installer and put into use because of the operational advantage.

**Table 6** Balance and Main Result Comparison

| | | GlobHor KWh/m$^2$ | DiffHor KWh/m$^2$ | T_Amb °C | GlobInc KWh/M$^2$ | GlobEff KWh/m$^2$ | EArrayWh | E_GridWh | PR ratio |
|---|---|---|---|---|---|---|---|---|---|
| January | Aligarh | 105.5 | 48.8 | 13.73 | 142.6 | 139.9 | 14.52 | 14.25 | 0.755 |
| | Manipal | 165.7 | 52.41 | 26.49 | 199.9 | 195.9 | 17.55 | 16.98 | 0.784 |
| February | Aligarh | 129.2 | 48.0 | 17.93 | 165.1 | 161.9 | 16.22 | 15.90 | 0.728 |
| | Manipal | 158.6 | 64.75 | 27.32 | 186.4 | 174.2 | 15.66 | 15.16 | 0.787 |
| March | Aligarh | 176.7 | 65.8 | 24.08 | 202.6 | 198.1 | 19.21 | 18.82 | 0.702 |
| | Manipal | 184.0 | 82.34 | 28.34 | 190.8 | 186.4 | 15.78 | 16.24 | 0.786 |
| April | Aligarh | 191.1 | 82.2 | 30.09 | 196.0 | 191.2 | 18.09 | 17.70 | 0.683 |
| | Manipal | 180.2 | 88.50 | 28.96 | 173.5 | 268.9 | 15.37 | 14.86 | 0.798 |
| May | Aligarh | 206.6 | 99.0 | 33.62 | 195.3 | 190.0 | 17.84 | 17.45 | 0.675 |
| | Manipal | 172.2 | 86.48 | 29.02 | 156.5 | 151.6 | 13.99 | 13.52 | 0.798 |
| June | Aligarh | 177.5 | 105.5 | 32.92 | 162.7 | 157.8 | 15.11 | 14.80 | 0.688 |
| | Manipal | 117.1 | 80.14 | 26.07 | 105.8 | 102.1 | 9.78 | 9.44 | 0.824 |
| July | Aligarh | 152.7 | 101.5 | 31.90 | 141.7 | 137.2 | 13.37 | 13.10 | 0.699 |
| | Manipal | 119.2 | 81.48 | 26.04 | 108.2 | 104.3 | 9.99 | 9.64 | 0.823 |
| August | Aligarh | 156.0 | 92.9 | 30.86 | 153.1 | 148.7 | 14.42 | 14.12 | 0.697 |
| | Manipal | 121.0 | 89.66 | 25.81 | 113.8 | 110.1 | 10.58 | 10.23 | 0.815 |
| September | Aligarh | 150.1 | 80.2 | 29.40 | 160.3 | 156.3 | 15.12 | 14.81 | 0.698 |
| | Manipal | 143.1 | 83.59 | 25.94 | 141.7 | 137.7 | 12.92 | 12.50 | 0.815 |
| October | Aligarh | 143.1 | 71.4 | 26.40 | 169.7 | 166.3 | 16.18 | 15.85 | 0.706 |
| | Manipal | 149.4 | 76.84 | 29.89 | 159.7 | 155.8 | 14.31 | 13.84 | 0.800 |
| November | Aligarh | 116.0 | 52.0 | 20.47 | 154.8 | 151.7 | 15.22 | 14.94 | 0.729 |

(continued)

**Table 6** (continued)

| | | GlobHor KWh/m$^2$ | DiffHor KWh/m$^2$ | T_Amb °C | GlobInc KWh/M$^2$ | GlobEff KWh/m$^2$ | EArrayWh | E_GridWh | PR ratio |
|---|---|---|---|---|---|---|---|---|---|
| | Manipal | 151.3 | 59.44 | 27.05 | 176.3 | 172.6 | 15.60 | 15.09 | 0.790 |
| December | Aligarh | 106.3 | 45.1 | 15.55 | 150.4 | 147.6 | 15.15 | 14.87 | 0.747 |
| | Manipal | 148.2 | 58.51 | 26.79 | 180.0 | 176.0 | 15.99 | 15.47 | 0.794 |
| Yearly | Aligarh | 1810.8 | 892.5 | 25.63 | 1994.3 | 1946.6 | 190.45 | 186.61 | 0.707 |
| | Manipal | 1810.1 | 904.17 | 27.07 | 1884.0 | 1835.6 | 168.53 | 162.99 | 0.799 |

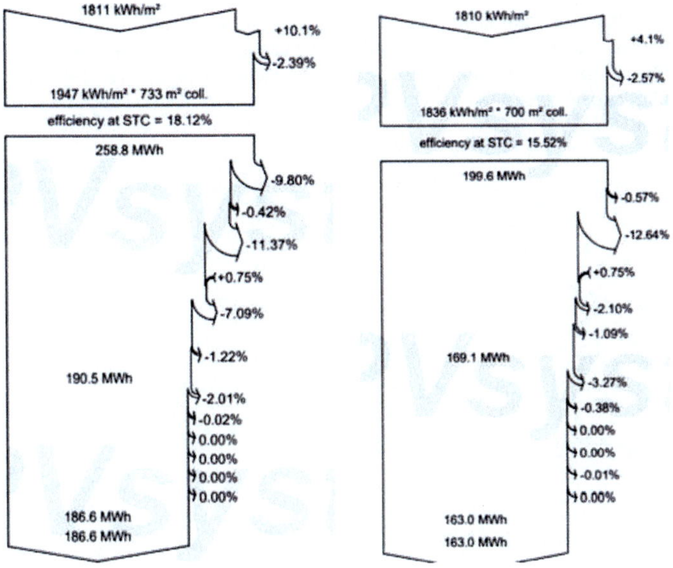


**Fig. 6** Loss diagram of the whole year **a** Aligarh **b** Manipal

# References

1. A. Chouder, S. Silvestre, N. Sadaoui, L. Rahmani, Modeling and simulation of a grid connected PV system based on the evaluation of main PV module parameters. Simul. Model. Pract. Theory **20**(1), 46–58 (2012). https://doi.org/10.1016/j.simpat.2011.08.011
2. E. Román, R. Alonso, P. Ibañez, S. Elorduizapatarietxe, D. Goitia, Intelligent PV module for grid-connected PV systems. IEEE Trans. Ind. Electron. **53**(4), 1066–1073 (2006). https://doi.org/10.1109/TIE.2006.878327
3. B. Marion et al., *Performance Parameters of Grid Connected*, vol. 1 (2005), pp. 1601–1606
4. D.H.W. Li, K.L. Cheung, T.N.T. Lam, W.W.H. Chan, A study of grid-connected photovoltaic (PV) system in Hong Kong. Appl. Energy **90**(1), 122–127 (2012). https://doi.org/10.1016/j.apenergy.2011.01.054
5. R. Kumar, C. S. Rajoria, A. Sharma, S. Suhag, Design and simulation of standalone solar PV system using PV syst software: a case study. Mater. Today Proc. (2020). https://doi.org/10.1016/j.matpr.2020.08.785
6. B. Alam, M. Nusrat, Z. Sarwer, M. Zaid, A general review of the recently proposed asymmetrical multilevel inverter topologies, in *Nnovations in Cyber Physical Systems. Lecture Notes in Electrical Engineering.* https://doi.org/10.1007/978-981-16-4149-7_61
7. J. Selvaraj, N.A. Rahim, Multilevel inverter for grid-connected PV system employing digital PI controller. IEEE Trans. Ind. Electron. **56**(1), 149–158 (2009). https://doi.org/10.1109/TIE.2008.928116
8. Y. Dong, J. Huang, M. Ding, H. Li, S. Zhang, Performance test and evaluation of photovoltaic system. IET Conf. Publ. **2015**(CP679), 1–4 (2015). https://doi.org/10.1049/cp.2015.0500
9. V. P. Sharma, A. Singh, J. Sharma, A. Raj, Design and simulation of dependence of manufacturing technology and tilt orientation for 100 kWp grid tied solar PV system at Jaipur, in *2016 International Conference on Recent Advances and Innovations in Engineering, 2016.* https://doi.org/10.1109/ICRAIE.2016.7939508

10. S. J. L., *P40_Renewables 2014 global status report* (2014)
11. N. Kumar, P. Yadav, S. S. Chandel, Comparative analysis of four different solar photovoltaic technologies. *International Conference Energy Economics and Environment—1st IEEE Uttar Pradesh Section Conference UPCON-ICEEE 2015* (2015). https://doi.org/10.1109/EnergyEconomics.2015.7235077

# Transparent Solar Cell: A Powerful Device of Upcoming Era

**Debani Prasad Mishra, Bipul Manoj Thakur, Himanshu Pandey, and Manas Malviya**

**Abstract** For almost any work in the present world, we need energy therefore it (energy) creates a major impact in the development of a country. In current scenario, due to excessive use and depletion of non-renewable energy resources, world is turning toward renewable sources for the production of energy. And one of the most abundant, never-ending sources of energy is sun. Harvesting solar energy is the easiest way of extracting clean energy. There are many researches going on solar cells. But due to limitation of land resources and many more different situations, researches are progressing toward transparent/semi-transparent solar cells. Many types of transparent solar cells are there like polymer solar cell (PSC), Near-Infrared transparent solar cell, thin film photovoltaics (TPVs), Perovskite solar cells. Transparent solar cells can be incorporated in the existing window panes where they can absorb and utilize unwanted light energy passing through the windows in buildings and automobiles. Such a productive use of architectural space can prove to be very economic for generation of the renewable energy. Here in this report, we will study about transparent solar cells and its future scope.

**Keywords** Solar energy · Solar cells · Semiconductors · Solar windows · Transparent photovoltaics

## 1 Introduction

The major role in the human as well as economic development is played by the energy. It may be considered a key factor also [1]. As in modern world the conventional energy sources are the major reasons for creating a huge amount of pollution, harvesting clean energy from the nature is the main goal of the researchers. The use of non-renewable energy sources is being considered to be the cause of climate change around the world. Non-Renewable energy sources are depleting day by day therefore researchers are focusing on renewable sources of energy. Among every renewable

D. P. Mishra (✉) · B. M. Thakur · H. Pandey · M. Malviya
International Institute of Information Technology, Bhubaneswar, India
e-mail: debani@iiit-bh.ac.in



S. Patnaik et al. (eds.), *Smart and Sustainable Technologies: Rural and Tribal Development Using IoT and Cloud Computing*, Advances in Sustainability Science and Technology,
https://doi.org/10.1007/978-981-19-2277-0_27

source of energy, solar energy is the most abundant available resource that can satisfy society's demand [2, 3]. Harvesting energy from sun is the most common way to generate a clean source of energy. Photovoltaic technologies are using solar energy most efficiently and abundantly. A Photovoltaic (P.V.) cell is a very useful device that coverts solar energy into electric energy with the help of different semiconductors available. This report is mainly interested in transparent solar cells, however to understand transparent solar cell we require to have a background knowledge of simple solar cell (silicon type). Transparent solar cells when used effectively (TSC) will not only produce clean energy but it will also be blessings in disguise as it will occupy places where it will serve several purposes simultaneously.

A study shows that the United States generates 3% of the total electricity [4] through the use of the solar cells. The area covered by the glass surface alone in the United States is 5–6 billion square meters. If this space is utilized properly with the help of transparent solar cell, then approximately 40% of the total electricity of the country will be produced by it.

Solar windows were set up at Qingdao Rocky Ltd. (Qingdao, China) and were tested. Their performances characteristics were slightly varying with different window models. Here the output from the window (size of 0.85 $m^2$) was 20 W, this was considered to be an estimated performance. Similarly another experiment was conducted in Perth, Australia. The results were measured using PROVA 200 A Solar Module Analyzer. The area was taken 0.85 $m^2$ and the power output came to be 25–26 W [5].

## 2 Discussion

### 2.1 *Working Principle of Transparent Solar Cells*

We can generally wonder that solar cell is a device that absorbs sunlight but in transparent solar cell it will not fully absorb the sunlight and if it does so its purpose will be defeated. But the transparent solar cell is developed in such a way that it will absorb the invisible portion of the spectrum like infrared rays and ultraviolet light while the visible light will pass through it [6].

Fig. 1 Shows how the visible part of the spectrum passes through cell and the infrared rays are responsible for the excitation of the electrons in the transparent solar cell. Here, VIS represents the visible portion of spectrum i.e., VIBGYOR. NIR represents, near infrared rays that will be absorbed by the cell

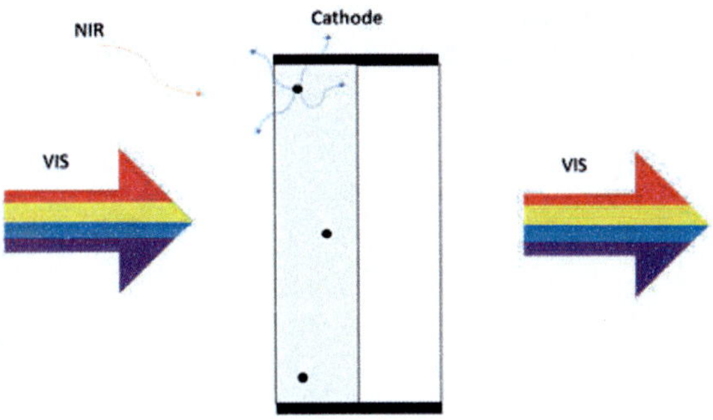


**Fig. 1** Schematic representation of transparent solar cell

## 2.2 *Different Types of Transparent Solar Cell*

Here we will see some different types of transparent solar cells that are efficient and widely used throughout the world in different applications of this technology.

**Thin Film Photovoltaics (TVPs)**

We observed that the thin film photovoltaics are very frequently used for transparent solar cells. The main focus is kept on transparency of the solar cell, therefore the materials which have high transparency are fabricated in this cell.

A thin film solar cell is made up of thin films of semiconductors settled on glass or plastic in various distinct ways [7]. The most widely developed thin film solar cells are CdS/CdTe and CdS/CDIS which have achieved astonishingly high efficiency of 20%.

Figure 2 shows the schematic representation of thin film transparent solar cell where different layers are used. The thin conducting oxide is generally made up of ZnO: Al, it is the main layer of the cell. The window layer and the absorbing layer is generally made up of semiconductors.

The thickness of this cell is ranging from few nanometers to micrometers. Due to very low thickness, this solar cell is very cost efficient [8], the lower the thickness will be, the good the cell is considered. In this type of solar cell, we can control the transparency by engineering the particle size, minimizing the film thickness, structure of the semiconductor oxide material to enhance the conductivity and the dye color, which will increase the variety to use different colors to absorb more light at different wavelengths [9].

**Polymer Solar Cell (PSC)**

A polymer solar cell is generally made by introducing semiconducting conjugated polymers as active (electron carrier) components in the generation of photocurrent and conversion of power within the photovoltaic cell that is responsible for conversion of solar energy into electrical energy [10–13].

These are organic solar cell made by fabricating transparent polymer materials with transparent conducting materials [14]. Conducting polymers are very good

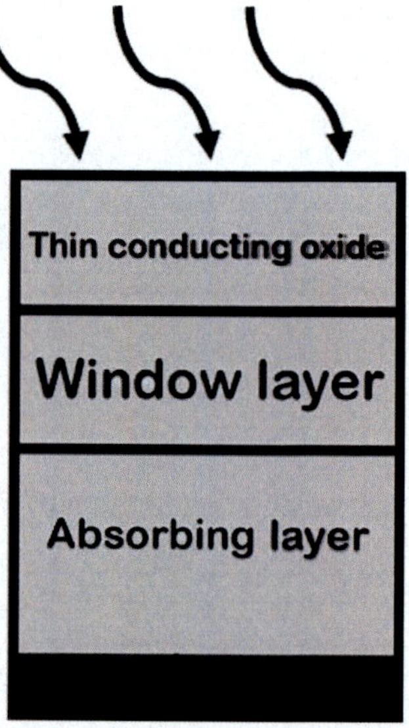


**Fig. 2** Schematic diagram of a thin film photovoltaic cell

charge carriers therefore they are helping in generating current easily. These cells absorb maximum ultraviolet and infrared rays and the visible light passes through it.

Figure 3 shows the typical structure of polymer solar cell with different layers namely transparent conducting polymer, transporting layer, photoactive layer, substrate. Each layer has a unique function.

The polymers that are generally used in this solar cells are PBDTT-DPP: $PC_{61}BM$, PDTP-DFBT: $PC_{71}BM$, PDPP4T-2F: $PC_{71}BM$, PBDTT-SeDPP: $PC_{61}BM$ etc. The power conversion efficiency of organic photovoltaic has a value approximately greater than 12%.

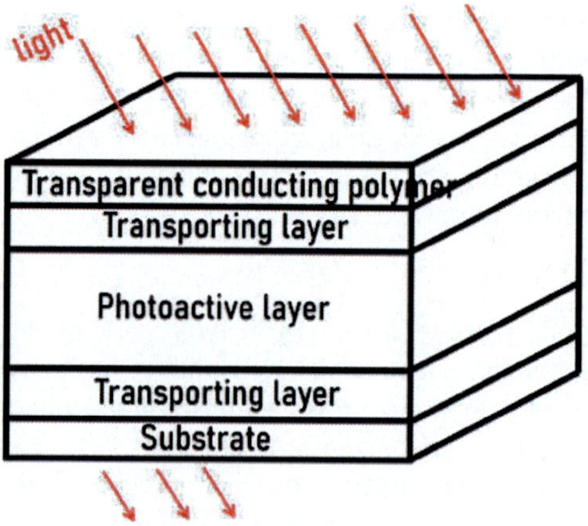


**Fig. 3** Schematic representation of PSC

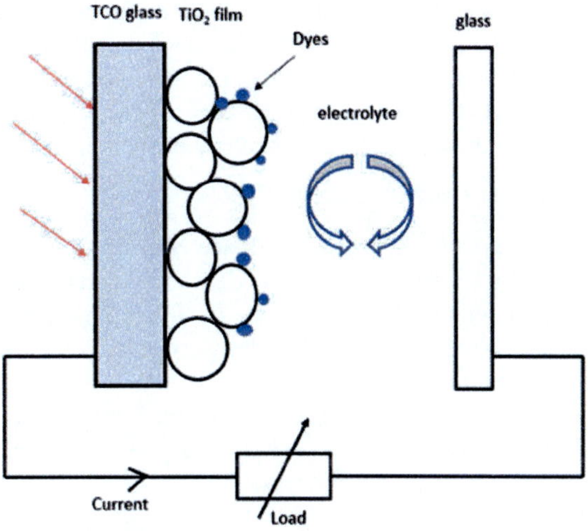


**Fig. 4** Below is the schematic of DSSC

**Dye-sensitized Solar Cells (DSSC)**

Figure 4 as given below shows the typical diagram of a Dye-sensitized solar cell where different dyes according to the needs of the technologies are used to generate photo electric current.

Other important elements are TCO glass, $TiO_2$ film and electrolyte.

Another transparent solar cell that has proved to be one of the most promising photovoltaic cells for integration in building is Dye-sensitized solar cell (DSSC) [15, 16]. A continuous work of many years led to the development of newest organic dyes of present world, which allowed us to identify various very efficient electron-rich units (such as triarylamine derivatives) and pulling/attracting electron-withdrawing units (such as carboxylic acid or cyanoacrylic acid) for the development and preparation of DSSC.

This solar cell follows the principle of photochemical mechanism which is very used technically [17]. It is easy to fabricate from the scratch, cheap transparent solar cell which can be used widely for different purposes. DSSC transparency is flexible as it has various degrees of transparency, this characteristic of it which makes it suitable for distinct types of window application in buildings. Its efficiency generally lies in between (8–12)%.

**Perovskite Cell**

Made by hybrid of organic–inorganic lead material for harvesting of light. Efficiency have increased up to 25% and have potential to gain higher efficiencies. Its low-cost design makes it more attractive to market. Normal perovskite attains a good efficiency and transparent perovskite cell achieves efficiency about 6–10% [18, 19].

Major drawback in this solar cell is that as we increase transparency, efficiency is decreased. Perovskites are organic materials which have good electrical properties which are required in solar cells. The design used in simulation has average visible transmittances of 29%. Here gold is added as electrode which results in efficiency of 6.5%. This model is sufficient for making a self-reliant energy building. An additional benefit of this layout is the absence of metal oxides in the cell which makes the manufacturing of cells easy and makes it compatible with flexible application.

**Quantum Dot Solar Cell (QDSC)**

Quantum Dots (QD) are man-made crystals (whose size lies in nanometers) which can carry electrons [20]. When QDs are integrated in semiconductors they are being used in solar cell technologies. In this report, we talk about semi-transparent or almost fully transparent QD sensitized solar cells which can be used for generation of electrical energy.

The characteristics QD film tells it has a strong light absorption close to ultraviolet (UV) region and nearly a peak in absorption in the near infrared (IR) region, the solar cell can thereby be made transparent as the visible region will be allowed to pass through it [20].

PbS QD solar cell is used generally for transparent solar cell because it is a perfect light absorber for solar cell application. It is also reported that some of this cell has 9% power conversion efficiency [21], which is high as compared with other solar cells.

## 3 Applications

**In Greenhouse Structures**

According to a report greenhouse cultivation, which has been taken up on about 150 acres in Nashik during the last one year, is more profitable than grape cultivation, say farmers and experts. "A farmer may produce around 2700 flower stems per acre in a day, and earn Rs 6750 by selling it". Greenhouse farming is profitable in itself and if we integrate the transparent solar cell in its structure, it will become the energy house also and the energy can be cultivated for different use. It will become twice beneficial than the older one as such large structures will produce a huge amount of energy.

**In Glazing Products**

The transparent solar cells are used as different types of windows, roofs, facades etc. Also, they have the ability to transmit daylight for the options of using natural lighting while producing electricity. By using this technology, the buildings will become the power house and they will produce huge amount of power on daily basis which will be good for humans as the energy generated is clean energy and the cost of electricity can be saved. There are many companies which are installing this technology in the buildings, they are namely Physee, Ubiquitious and many more.

**In Electronic Devices**

The transparent solar cells can be used in the devices such as mobiles phones, laptops, desktops, watches, clocks and many other electronic devices. It can be very useful as it will recharge the battery automatically anywhere and at any time. This will minimize the risk of switching off the device which will be very helpful. There will be minimum amount of energy present always in the device which can be handy at the time of emergency.

**In the Automobiles**

The transparent solar cells can be mounted as windshields, sunroof and windows in cars and other automobiles. The energy generated can be utilized for the secondary power consuming devices in the automobiles like radio, headlights, fan, wipers etc. This will enhance the life of primary battery of the automobiles as it has to bear less loads. This technology has tremendous applications once it is used widely.

**Scope of Transparent Solar Cell in India**

In a recent report the amount of energy generated by solar energy in India was 50.1 TWh or 3.6% of the total energy produced. This amount of energy produced through a renewable resource is less as compared to the other countries. India is a subtropical country therefore its landmass receives ample amount of sunlight throughout the year, if we utilize this energy then we can produce electricity very cost effectively.

By using transparent solar cell in different devices and equipment we will be able to collect and utilize this energy in different form. As this technology is being made cost effective this technology will be widely used in the country to meet the demands of such a huge population in an easy manner. In future this technology will leave a great impact on the lives of every people in India.

## 4 Conclusion

The transparent solar cell has very wide range of applications. The cost in transparent solar cell is little higher due to use of elements like silver, gold and its nitrates but its application can reap a lot of benefits. As the average visible transmission is increased it results in lesser efficiency. Solar cell having transparency 80%, efficiency near 1–2%. The feasible and economical solar cell which is achieved till now has resulted in 20%. Different types of transparent cells can have different application. A building can use cell having transparency 30%, cell phone has requirement of higher transparency about 70%. It has been tried to use transparent light absorbing material such as perovskite so that more light energy converts to electrical. Film methods with different types of deposition techniques are used. Transparent substrates LiF, ITO, glass, polymers are used. These methods resulted also in slimming the solar cell. Many solar cells are not commercially available. This will help in shortcoming solar field acquisition problem. More research will help to make self-sufficient energy building, vehicles, devices. Load on generation stations will be reduced. More independency from non-renewable source will be achieved, helping nature to recover, making earth green and fastening human growth.

## References


1. A.A.F. Husain et al., A review of transparent solar photovoltaic technologies. Renew. Sustain. Energy Rev. **94**, 779–791 (2018)
2. M. Nivas, R.K.P.R. Naidu, D.P. Mishra, S.R. Salkuti, Modeling and analysis of solar powered electric vehicle. Int. J. Power Electron. Drive Syst. (IJPEDS) **11**(3), Sep 2020. https://doi.org/10.11591/ijpeds.v11.i3.ppxx-ppyy
3. R.L. Andy Henion, Transparent solar technology represents "wave of the future." Michigan State Univ (2017)
4. https://www.google.com/search?q=total+energy+production+in+usa+by+solar+cell&oq=total+energy+production+in+usa+by+solar+cell&aqs=chrome..69i57j33i160.14754j0j7&sourceid=chrome&ie=UTF-8
5. Transparent Solar Panels: Reforming Future Energy Supply, By Solar Mag. 29 Feb 2020
6. S.S. Saswat, S. Patra, D.P. Mishra, S.R. Salkuti, R.N. Senapati, Harnessing wind and solar PV system to build hybrid power system. Int. J. Power Electron. Drive Syst. (IJPEDS) **12**(4), 2160–2168, Dec 2021. https://doi.org/10.11591/ijpeds.v12.i4.pp2160-2168
7. A. Orioli, An accurate one-diode model suited to represent the current-voltage characteristics of crystalline and thin-film photovoltaic modules. Renew. Energy **145**, 725–743 (2020)
8. N. Yan et al., Recent progress of thin-film photovoltaics for indoor application. Chin. Chem. Lett. **31**(3), 643–653 (2020)
9. S. Chakraborty, D.P. Mishra, Modeling and simulation of multi-junction TANDEM solar cell. Int. Conf. Intell. Sustain. Syst. (ICISS) **2019**, 140–145 (2019). https://doi.org/10.1109/ISS1.2019.8908124
10. K.-S. Chen, J.-F. Salinas, H.-L. Yip, L. Huo, J. Hou, A.K.-Y. Jen, Semi-transparent polymer solar cells with 6% PCE, 25% average visible transmittance and a color rendering index close to 100 for power generating window applications. Energy Environ. Sci. **5**, 9551 (2012)
11. C. Chen, L. Dou, R. Zhu, C. Chung, T. Song, Y.B. Zheng et al., Cells produced by solution processing. [120712084458005]. ACS Nano **6**(8), 7185–7190 (2012)
12. S.K. Jha, S. Roy, V.K. Singh, D.P. Mishra, Sun's position tracking by solar angles using MATLAB. Int. Conf. Renew. Energy Integr. Smart Grids Multi. Approach Technol. Model. Simul. (ICREISG) **2020**, 5–9 (2020). https://doi.org/10.1109/ICREISG49226.2020.9174533
13. S. Huang et al., Highly efficient inverted polymer solar cells by using solution processed MgO/ZnO composite interfacial layers. J. Colloid Interface Sci. **583**, 178–187 (2021)
14. B. Das, P.K. Panigrahi, S.R. Das, D.P. Mishra, S.R. Salkuti, Power quality improvement in a photovoltaic based microgrid integrated network using multilevel inverter. Int. J. Emerg. Electr. Power Syst. 000010151520210040 (2021). https://doi.org/10.1515/ijeeps-2021-0040
15. Q. Huaulmé et al., Photochromic dye-sensitized solar cells with light-driven adjustable optical transmission and power conversion efficiency. Nat. Energy **5**(6), 468–477 (2020)
16. M. Godfroy et al., Benzothiadiazole-based photosensitizers for efficient and stable dye-sensitized solar cells and 8.7% efficiency semi-transparent mini-modules. Sustain. Energy Fuels **5**(1), 144–153 (2021)
17. J. Zhang, S. Li, P. Yang, W. Que, W. Liu, Deposition of transparent $TiO_2$ nanotubefilms via electrophoretic technique for photovoltaic applications **58**, 785–790 (2015)
18. S.R. Das, P.K. Ray, D.P. Mishra, H. Das, Performance assessment of PV integrated model predictive controller based hybrid filter for power quality improvement. Int. J. Power Electron. (2020)

19. A. Salant, M. Shalom, I. Hod, A. Faust, A. Zaban, U. Banin, Quantum dot sensitized solar cells with improved efficiency prepared using electrophoretic deposition. ACS Nano **4**, 5962–5968 (2010)
20. E.S. Arinze et al., Color-tuned and transparent colloidal quantum dot solar cells via optimized multilayer interference. Opt. Expr. **25**(4), A101–A112 (2017)
21. Z. Ning, O. Voznyy, J. Pan, S. Hoogland, V. Adinolfi, J. Xu et al., Air-stable n-type colloidal quantum dot solids. Nat. Mater. **13**, 4–10 (2014)

# Machine Learning Techniques and Applications

# An Adaptive Firefly Optimization Algorithm for Human Gait Recognition

**P. Sankara Rao, Gupteswar Sahu, Priyadarsan Parida, and Srikanta Patnaik**

**Abstract** Gait recognition is a distinct biometric technique for identifying people at a distance based on their walking style. It is used extensively in video surveillance, medical, forensics, digital security, and defense applications. However, the effectiveness of gait recognition can be adversely affected by several covariates such as clothing and the carrying conditions of an individual. Furthermore, identifying a person from multiple view-angles is another significant and complex challenge in gait recognition. Therefore, an appropriate selection of gait features is a challenging task in human gait recognition. To select an optimal subset of gait features, we proposed a novel adaptive firefly algorithm (AFA) based on the flashing patterns and behavior of fireflies. In this paper, gait energy image (GEI) is used as a gait template to extract the essential features from the input gait data. This proposed algorithm is applied to these gait templates to choose the most relevant and informative gait features. Finally, these selected features are subjected to principal component analysis (PCA), and multi-class linear discriminant analysis (LDA), to improve gait recognition performance. Extensive experiments are done on the CASIA-B dataset have revealed that the suggested approach significantly improves the effectiveness of gait recognition than current state-of-the-art approaches.

**Keywords** Gait recognition · GEI · Feature selection · Firefly algorithm · PCA · And LDA

P. S. Rao · P. Parida (✉)
Department of Electronics and Communication Engineering, GIET University, Gunupur, Odisha 765022, India
e-mail: priyadarsanparida@giet.edu

P. S. Rao
e-mail: sankarrao.palla@giet.edu

P. S. Rao · G. Sahu
Department of Electronics and Communication Engineering, Raghu Engineering College Visakhapatnam, Visakhapatnam, Andhra Pradesh 531162, India

S. Patnaik
Department of Computer Science and Engineering, SOA University, Bhubaneswar, Odisha 751030, India

 
S. Patnaik et al. (eds.), *Smart and Sustainable Technologies: Rural and Tribal Development Using IoT and Cloud Computing*, Advances in Sustainability Science and Technology,
https://doi.org/10.1007/978-981-19-2277-0_28

# 1 Introduction

Human gait recognition is a behavioral biometric approach based on the person's walking style. Unlike other biometrics (fingerprint, iris, and face recognition), gait has distinct biometric features that can be used to identify people from a long distance even in uncontrolled situations (no cooperation from the subjects). Gait recognition has broad applications in numerous fields such as medical, sports, forensics, access control, and surveillance [1], etc. However, gait recognition suffers from different covariant conditions such as subject viewpoint, clothing, and carried objects, which may have an impact on the accuracy of gait recognition [2].

An important and challenging problem in gait recognition is extracting suitable features that will efficiently capture the gait characteristics. The selected features must be robust to covariant conditions and should provide good discrimination among the individuals. According to the literature review, most researchers have been using conventional methods (filter, wrapper, and embedded) to select significant features from an extensive feature dataset rather than introducing new techniques. These traditional methods are sometimes more prone to local optima during the search process and do not provide an optimal solution. Therefore, selecting an appropriate feature selection method is a critical task in gait recognition. Many authors have recently attempted to apply the meta-heuristics algorithms for the feature selection problems [3]. For example, Lin et al. combined particle swarm optimization (PSO) and support vector machine (SVM) to increase classification [4]. Neshatian et al. used a genetic algorithm (GA) to rank feature subsets in binary classification [5]. Faria et al. applied Ant Colony Optimization (ACO) technique in supply chain and concrete delivery application [6]. Xue et al. introduced a PSO with multiple feature selection goals [7]. Isaac et al. used GA to propose a template-based model-free approach for gait recognition [8]. Tafazzoli et al. extracted an optimal subset of gait features using GA [9]. Hema et al. suggested PSO algorithm to identify individuals using their gait [10]. Gao et al. proposed an improved human gait recognition algorithm based on an artificial bee colony (ABC) [11]. Aasha et al. suggested a gait-based human recognition using bat algorithm (BA) [12].

In this work, we propose a unique approach based on the firefly algorithm for the appropriate selection of feature subsets for human gait recognition. The paper shall be arranged as follows: Sect. 2 demonstrates the framework of the proposed approach. Section 3 shows the experimental results and Finally in Sect. 4 conclusion is presented.

# 2 Framework of the Proposed Approach

Figure 1 illustrates the block diagram of gait recognition using an adaptive firefly algorithm. In the first step, extract the GEI template from the gait video. After that given database is classified into two sets: Tuning set and Evaluation set. The tuning

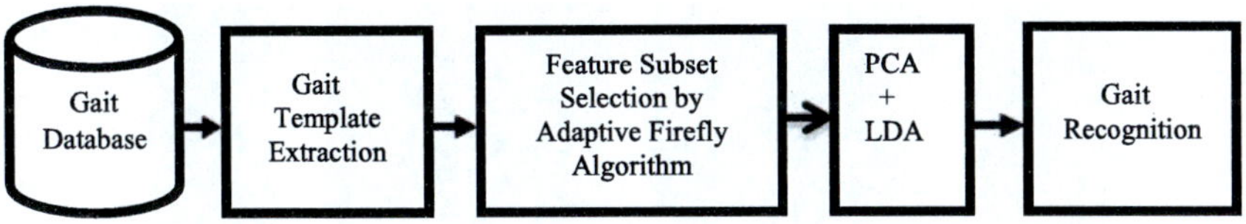


**Fig. 1** A proposed approach for gait recognition using an adaptive firefly algorithm

set is applied to the adaptive firefly algorithm to formulate the segments for optimal performance. These selected features are processed by PCA and multi-class LDA to get more accuracy. Finally, all the selected features are applied to evaluation set to identify the person and find the classification accuracy of the system using specificity, accuracy, sensitivity, precision, and percent of error.

## *2.1 Gait Template Extraction*

Gait templates have been the most widely measured representation for capturing temporal information in gait recognition approaches. GEI is the most effective and widely used approach for gait recognition. It is produced by averaging all silhouettes for a complete gait cycle [13]. Usually, the GEI template requires less computation time and storage space and is less susceptible to silhouette noise. We use GEI as our method's input and target image based on its success in gait recognition.

Let $S$ be the total number of silhouettes of a particular subject in a gait cycle, and each $k$th silhouette is denoted by $X(k)$ then the GEI of an individual sequence characterized (shown in Fig. 2) as:

$$\mathrm{GEI}_k = \frac{1}{S}\sum_{k=1}^{S} X(k) \tag{1}$$

Figure 3 represents the GEI of a person under various covariants. In this GEI, low-intensity pixels signify the dynamic parts of the body (suitable for recognition) and high-intensity pixels signify the static parts of the body (not ideal for recognition because it can be affected with covariate conditions).

**Fig. 2** GEI (the rightmost one) is generated by averaging all silhouettes (remaining images on the left side) in one gait cycle

Fig. 3 GEI of a subject under various covariants (*Normal, bag, and coat*)

### 2.2 Introduction to Standard Firefly Algorithm (FA)

In 2008 Yang [14] developed a firefly algorithm based on the flashing characteristics of fireflies: (i) Every firefly has the same sex so that a firefly despite its sex is attracted toward other fireflies. (ii) Attractiveness of fireflies decreases as the distance between them increases. (iii) The intensity or brightness of the firefly depends on the objective function.

Hence, for any two pair of fireflies $(m, n)$ updating formula at $X_m$, $X_n$ is:

$$X_m^{L+1} = X_m^L + \beta(X_m^L - X_n^L) + \alpha\varepsilon_m^L \tag{2}$$

where $\alpha$ is the controlling (randomization) parameter, $\beta = \beta_0 \exp(-\gamma d^2)$ represents the attractiveness with $\beta_0.d$ specifies the attractiveness at distance $(d = 0)$. $\gamma$ signifies the light absorption coefficient and $\varepsilon_m$ indicates the randomization parameter based on the uniform distribution.

The Cartesian distance can be calculated for any two pair of fireflies $m$ and $n$ at $X_m$ and $X_n$ is given by

$$dmn = \|X_m - X_n\| = \sqrt{\sum_{l}^{d} (X_{m,l} - X_{n,l})^2} \tag{3}$$

where $X_{m,l}$ is the component of the spatial coordinate $X_m$ of m the firefly.

In this paper, we choose $\beta_0 = 2$, $\gamma = 1$, and $\alpha \in [0, 1]$ [0, 1]. However, studies illustrate that it is slow to convergence for multimodal problems and quickly becomes trapped in the local optimum.

## 2.3 Adaptive Firefly Algorithm for Feature Selection

To overcome the disadvantages of the standard firefly algorithm [15] in our approach we considered the global optima concept in the movements of the firefly (instead of allowing a single firefly to influence or attract its neighbors).

**Strategy 1**: In this paper, a new updating formula was proposed to enhance the influence of the brightest firefly. (replace Eq. (2) with (4)).

$$X_{m+1} = X_m + \beta(X_n - X_m) + \beta_0 e^{-\gamma d^2_{m,gbest}}(X_{gbest} - X_m) + \alpha(t)\varepsilon_m \tag{4}$$

This implies that a firefly is attracted not only by brighter fireflies but also by the best solution discovered thus far $X_{gbest}$ (position of the global optima). In the above Eq. (4), the random number is represented by $\varepsilon_m(\varepsilon_m = \text{rand} - 0.5)$, $\beta(t)$ represents the attractiveness coefficient at time $t$, and $\alpha(t)$ is the randomness coefficient at time $t$.

Where the Cartesian distance between any two fireflies for global optima is given by:

$$d_{m,gbest} = \sqrt{(X_m - X_{gbest})^2 + (Y_m - Y_{gbest})^2} \tag{5}$$

**Strategy 2**: If a randomized parameter is upgraded based on the number of iterations, then the randomness coefficient equation is:

$$\alpha(L) = \frac{0.4}{1 + e^{0.06(L - L_{Max})}} \tag{6}$$

where $L_{Max}$ specifies the maximum iterations and $L$ states the current iterations.

**Strategy 3**: According to the FA's optimization mechanism analysis, learning and exchange between fireflies must take place within visual range. Suppose the firefly population is small and the range is large. In that case, it will be difficult for the firefly with a long range to gather information from other fireflies. It may swing back to its original position, obstructing information exchange between individuals. In this paper, the distance weight, denoted by $\gamma$, is defined as the exponential part of the spatial distance. The distance weight in the traditional firefly algorithm is a constant value, which significantly limits the algorithm's optimization capacity, so the distance weight study appears to be more important. The dynamic distance weighting mechanism is defined as:

$$\gamma(L) = \gamma_{Max} - (\gamma_{Max} - \gamma_{Min})\left(\frac{L}{L_{Max}}\right)^2 \tag{7}$$

where $\gamma_{\text{Max}}$ indicates the initial distance weight value, $(2 \leq \gamma_{\text{Max}} \leq 4)$; $\gamma_{\text{Min}}$ specifies the distance weight value when iterating to the maximum number of times, $(0.5 \leq \gamma_{\text{Min}} \leq 1)$; $L_{\text{Max}}$ specifies the maximum iterations, and $L$ designates the current iterations. Based on these adjustments $(\gamma)$, the proposed algorithm will not become trapped in the local extreme, and it increases the convergence speed, resulting in a better solution at an optimum will be achieved.

**Strategy 4**: To enhance the random movement of fireflies, in this paper levy flight concept is introduced. The updated formula has been modified as follows:

$$X_{m+1} = X_m + \beta(X_n - X_m) + \alpha \text{sign}(\varepsilon) \otimes \text{Levy} \tag{8}$$

where $\otimes$ indicates the component-wise multiplication of the Levy distribution's random vector and the sign vector.

Where arbitrary step length is given by Levy distribution function.

$$\text{Levy} \approx L(x) = x^{-t} 1 < t \leq 3 \tag{9}$$

These strategies were combined to design an adaptive firefly algorithm for selecting an optimal subset of gait features.

### *2.4 Pseudo-Code of an Adaptive Firefly Algorithm for Feature Selection*

**Input**: All parameters must be initialized.

Distinguish the dataset into two parts: (i) training, and (ii) testing.

While loop: Till $L \leq L_{\text{Max}}$ do.

From Eqs. (6) and (7) evaluate the value of $(\alpha, \gamma)$. Where $\alpha$ and $\gamma$ are the adaptive parameters.

Loop 1: For each $m = 1 \ldots N$ (do this for all fireflies).

Loop 2: For each $n = 1 \ldots N$ (do this for all fireflies).

Determine the distance $(d)$ between two fireflies $(X_m, X_n)$ from Eq. (5).

Substitute $\beta = \beta_0 \exp(-\gamma d^2)$ in Eq. (4)

If $(I_m < I_n)$ firefly 'm' moves toward firefly '$n$' in $d$-dimension via Levy flights (from Eq. (8)).

The Parameter values $(\alpha, \gamma)$ should be updated by (6) and (7).

As per Eq. (4), assess the new solution $X_{m+1}$.

end if.

end Loop 2.

end Loop 1.

Light intensity information $L_{X_{m+1}}$ is updated.

Rank the fireflies to determine global best position.

end while loop.

Determine the correctness of the selected features.

Find the best firefly's position (select feature subset);

Find the classification accuracies.

End.

**Output**: The best fireflies position (the selected feature subset).

### *2.5 Combination of PCA and LDA*

The proposed algorithm selects all the important gait features, which are further processed by a combination of PCA and LDA to extract and classify the regions. The use of PCA and LDA combination improves the classifier's correct classification rate (CCR) performance. PCA can reduce the dimension of a gait feature. Still, it does not consider class differences, resulting in the loss of some important information, and it gives a lower recognition rate. In contrast, LDA can solve this problem more effectively by improving the inter-class distance and reducing the data's within-class length. LDA can further reduce the dimension of a feature that has already been reduced by PCA [13]. Therefore, we combined PCA and LDA in this work to obtain more gait feature information.

## 3 Experimental Analysis

### *3.1 Parameter Setting*

The tuning parameters $\alpha$, $\beta_0$, $\gamma$, and $N$ used in the firefly algorithm have a significant impact on algorithm performance [16]. In order to attain the highest accuracy, we tested different values for these parameters with different combinations to find the optimized values. The key parameters used in the proposed algorithm are illustrated in Table 1.

**Table 1** Adaptive firefly optimization algorithm parameters and their values

| Parameter | Selected value | Meaning |
|---|---|---|
| $\alpha$ | 0.2 | Randomness parameter |
| $\beta_0$ | 2 | Light amplitude |
| $\gamma$ | 1 | Absorption coefficient |
| N | 25 | Number of fireflies (Swarm Size) |
| $L_{\text{Max}}$ | 25 | Maximum number of iterations |

## 3.2 *Performance Evaluation Metrics of Classification*

There are numerous metrics available for determining the effectiveness of the model. In this paper, the effectiveness was evaluated using accuracy, precision, sensitivity or recall, specificity, and percentage of error. These parameters are calculated from the confusion matrix [17].

The confusion matrix is primarily concerned with true positive (TP), false positive (FP), true negative (TN), and false negative (FN).

Where

TP Correctly identified components
FP Incorrectly identified components (Type-1 errors)
TN Correctly rejected components
FN Incorrectly rejected components (Type-2 errors)

**Precision**:

$$\text{Precision} = \frac{\text{TP}}{\text{TP} + \text{FP}}$$

**Sensitivity**:

$$\text{Sensitivity} = \frac{\text{TP}}{\text{TP} + \text{FN}}$$

**Specificity**:

$$\text{Specificity} = \frac{\text{TN}}{\text{TN} + \text{FP}}$$

**Accuracy or CCR**:

$$\text{Accuracy} = \frac{\text{TP} + \text{TN}}{\text{TP} + \text{TN} + \text{FP} + \text{FN}}$$

**Table 2** The proposed approach performance using GEI template under the normal walking condition on CASIA-B dataset

| View angle | Precision | Sensitivity | Specificity | Accuracy | % of error |
|---|---|---|---|---|---|
| 0° | 98.71 | 98.7 | 99.85 | 98.70 | 1.3 |
| 18° | 99.70 | 99.7 | 99.96 | 99.70 | 0.3 |
| 36° | 98.91 | 98.9 | 99.87 | 98.90 | 1.1 |
| 54° | 98.71 | 98.7 | 99.85 | 98.70 | 1.3 |
| 72° | 98.33 | 98.3 | 99.81 | 98.30 | 1.7 |
| 90° | 98.32 | 98.3 | 99.81 | 98.30 | 1.7 |
| 108° | 97.13 | 97.1 | 99.67 | 97.10 | 2.9 |
| 126° | 98.81 | 98.8 | 99.86 | 98.80 | 1.2 |
| 144° | 96.32 | 96.3 | 99.58 | 96.30 | 3.7 |
| 162° | 94.93 | 94.9 | 99.43 | 94.90 | 5.1 |
| $180^0$ | 98.61 | 98.6 | 99.84 | 98.60 | 1.4 |
| Mean % | 98.04 | 98.03 | 99.78 | 97.96 | 1.97 |

### *3.3 Results and Discussion*

The CASIA database comprises four datasets (A, B, C, and D). In our experiments, we consider all the conditions of the CASIA dataset-B (normal walking, bag, and coat) on the GEI template. The CASIA dataset (B group) contains 124 subjects with 11 view angles (0°, 18°, 36°, 54°, 72°, 90°, 108°, 126°, 144°, 162°, 180°). CASIA dataset-B is the most popular multi-view gait dataset [18].

As compared with the recent advances in gait literature, the proposed approach produced promising results in terms of precision, sensitivity, specificity, accuracy, and percent of error in gait recognition, as shown in Tables 2, 3 and 4.

Table 5 and Fig. 4 illustrate the performance comparison of the proposed algorithm with other existing gait recognition methods in terms of accuracy (%). The proposed approach on the GEI template attains an accuracy of 97.96%, 97.03%, and 96.57% for normal, bag, and coat conditions, respectively.

## 4 Conclusion

In this paper, we proposed gait recognition approach based on the adaptive firefly optimization algorithm on the GEI template for identifying individuals under various view angles and covariates such as clothes and carried conditions. The proposed system's performance was evaluated on the CASIA dataset (B group). The obtained results demonstrated that the suggested approach has a high recognition rate (CCR) compared to current advanced gait recognition approaches. Therefore, future research

**Table 3** The proposed approach performance using GEI template under-bag condition on CASIA-B dataset

| View angle | Precision | Sensitivity | Specificity | Accuracy | % of error |
|---|---|---|---|---|---|
| 0° | 98.27 | 98.2 | 99.15 | 97.72 | 2.27 |
| 18° | 98.89 | 98.64 | 98.46 | 99.26 | 0.73 |
| 36° | 97.56 | 98.6 | 99.73 | 98.43 | 1.56 |
| 54° | 97.98 | 97.36 | 98.65 | 97.59 | 2.4 |
| 72° | 98.16 | 97.12 | 99.37 | 97.39 | 2.6 |
| 90° | 95.74 | 95.7 | 99.52 | 95.7 | 4.3 |
| 108° | 95.93 | 96.23 | 99.01 | 96.71 | 3.28 |
| 126° | 97.72 | 98.24 | 98.6 | 97.69 | 2.3 |
| 144° | 95.39 | 95.94 | 99.32 | 95.05 | 4.94 |
| 162° | 94.34 | 94.82 | 98.92 | 93.79 | 6.20 |
| 180° | 98.01 | 96.72 | 97.88 | 98.02 | 1.97 |
| Mean % | 97.09 | 97.05 | 98.96 | 97.03 | 2.96 |

**Table 4** The proposed approach performance using GEI template under-coat condition on CASIA-B dataset

| View angle | Precision | Sensitivity | Specificity | Accuracy | % of error |
|---|---|---|---|---|---|
| 0° | 97.43 | 97.64 | 98.30 | 97.38 | 2.62 |
| 18° | 98.42 | 97.87 | 97.64 | 99.16 | 0.83 |
| 36° | 97.46 | 97.61 | 99.40 | 98.14 | 1.85 |
| 54° | 97.91 | 97.27 | 97.85 | 96.65 | 3.34 |
| 72° | 97.73 | 96.42 | 98.61 | 96.96 | 3.03 |
| 90° | 96.43 | 96.4 | 99.60 | 96.4 | 3.60 |
| 108° | 95.71 | 96.17 | 98.25 | 96.04 | 3.95 |
| 126° | 97.55 | 97.28 | 98.33 | 96.77 | 3.22 |
| 144° | 94.45 | 95.83 | 99.14 | 94.95 | 5.04 |
| 162° | 93.79 | 94.14 | 98.02 | 93.74 | 6.25 |
| 180° | 96.21 | 95.52 | 96.11 | 96.13 | 3.86 |
| Mean % | 96.64 | 96.56 | 98.3 | 96.57 | 3.42 |

will enhance the proposed concept by incorporating hybrid swarm intelligence techniques for the feature subset selection and the finest classification algorithms for human gait recognition.

**Table 5** Comparison of the proposed approach with current state-of-the-art approaches in terms of average accuracy (%) on CASIA-B dataset

| Year | Approaches/References | Normal walking (N) condition | Carrying a bag (B) condition | Wearing a coat (C) condition |
|---|---|---|---|---|
| 2016 | Wu et al. [19] | 93.90 | 72.37 | 53.98 |
| 2017 | Isaac et al. [8] | 95.05 | 94.79 | 93.42 |
| 2018 | Chtourou et al. [20] | 95.52 | 87.76 | 86.17 |
| 2019 | Zhang et al. [21] | 91.81 | 85.74 | 58.87 |
| 2019 | Chao et al. [22] | 94.96 | 87.23 | 70.35 |
| 2020 | Fan et al. [23] | 96.15 | 92.46 | 78.69 |
| 2020 | Hema et al. [10] | 96.22 | 96.10 | 95.01 |
| 2021 | Chao et al. [24] | 96.10 | 90.80 | 70.28 |
| Proposed method accuracy (%) | | 97.96 | 97.03 | 96.57 |

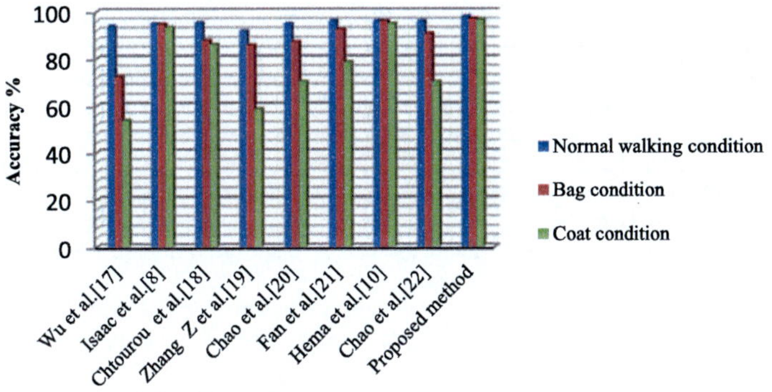


**Fig. 4** Performance comparison of the proposed approach on the GEI template with recent gait approaches on the CASIA dataset (B group) in terms of average % of accuracy

# References


1. J.P. Singh, S. Jain, S. Arora, U.P. Singh, A survey of behavioral biometric gait recognition: Current success and future perspectives. Arch. Comput. Methods Eng. **28**(1), 107–148 (2019). https://doi.org/10.1007/s11831-019-09375-3
2. A.O. Lishani, L. Boubchir, E. Khalifa, A. Bouridane, Human gait recognition using GEI-based local multi-scale feature descriptors. Multimedia Tools Appl. **78**(5), 5715–5730 (2019). https://doi.org/10.1007/s11042-018-5752-8
3. H. Xu, S. Yu, J. Chen, X. Zuo, An improved firefly algorithm for feature selection in classification. Wireless Pers. Commun. **102**, 2823–2834 (2018). https://doi.org/10.1007/s11277-018-5309-1
4. S.W. Lin, K.C. Ying, S.C.H. Chen, Z.J. Lee, Particle swarm optimization for parameter determination and feature selection of support vector machines. Expert Syst. Appl. **35**(4), 1817–1824 (2008). https://doi.org/10.1016/j.eswa.2007.08.088
5. K. Neshatian, M.J. Zhang, Genetic programming for feature subset ranking in binary classification problems, in *Proceedings of the 12th European Conference on Genetic Programming,*

*Lecture Notes in Computer Science*, vol. 5481, (2009), pp. 121–132. https://doi.org/10.1007/978-3-642-01181-8_11
6. J.M. Faria, C.A. Silva, J.M.C. Sousa, M. Surico, U. Kaymak, Distributed optimization using ant colony optimization in a concrete delivery supply chain, in *Proceedings of International Conference on Evolutionary Computation* (2006), pp. 16–21. https://doi.org/10.1109/CEC.2006.1688292
7. B. Xue, M.J. Zhang, W.N. Browne, Particle swarm optimization for feature selection in classification: A multi-objective approach. IEEE Trans. Cybern. **43**(6), 1656–1671 (2013). https://doi.org/10.1109/TSMCB.2012.2227469
8. E.R.H.P. Isaac, S. Elias, R. Srinivasan, K.S. Easwarakumar, View-invariant gait recognition through genetic template segmentation. IEEE Signal Process. Lett. **24**(8), 1188–1192 (2017). https://doi.org/10.1109/LSP.2017.2715179
9. F. Tafazzoli, G. Bebis, S. Louis, M. Hussain, Genetic feature selection for gait recognition. J. Electron. Imaging **24**(1), 1–24 (2015). https://doi.org/10.1117/1.JEI.24.1.013036
10. M. Hema, K. Babulu, N. Balaji, Individual gait recognition using particle swarm template segmentation. Int. J. Adv. Sci. Technol. **29**(6s), 2684–2697 (2020)
11. F. Gao, T. Tian, T. Yao, Q. Zhang, Human gait recognition based on multiple feature combination and parameter optimization algorithms. Comput. Intell. Neurosci. 1–14, (2021). https://doi.org/10.1155/2021/6693206
12. M. Aasha, S. Sivakumar, An efficient gait-based recognition using bat algorithm. Int. J. Comput. Appl. **132**(9), 41–45 (2015). https://doi.org/10.5120/ijca2015907548
13. H. Xue, Z. Hao, Gait recognition based on gait energy image and linear discriminant analysis, in *IEEE International Conference on Signal Processing, Communications and Computing* (2015), pp. 1–4. https://doi.org/10.1109/ICSPCC.2015.7338934
14. X.S. Yang, X. He, Firefly algorithm: recent advances and applications. Int. J. Swarm Intell. **1**(1), 36–50 (2013). https://doi.org/10.1504/IJSI.2013.055801
15. W.A. Khan, N.N. Hamadneh, S.L. Tilahun, J.M.T. Ngnotchouye, A review and comparative study of firefly algorithm and its modified versions. Optim. Algorithms Methods Appl. 281–312, (2016). https://doi.org/10.5772/62472
16. M.O. Yuan-bin, M.A. Yan-zhui, Z. Qiao-yan, Optimal choice of parameters for firefly algorithm, in *Fourth International Conference on Digital Manufacturing & Automation* (2013), pp. 887–892. https://doi.org/10.1109/ICDMA.2013.210
17. R.P. Sankara, G. Sahu, P. Parida, A contemporary survey on human gait recognition. J. Inf. Assur. Secur. **15**(3), 94–106 (2020)
18. S. Yu, T. Daoliang, T. Tieniu, A framework for evaluating the effect of view angle, clothing, and carrying condition on gait recognition, in *18th International Conference on Pattern Recognition*, vol. 4 (2006), p. 441–444. https://doi.org/10.1109/ICPR.2006.67
19. Z. Wu, Y. Huang, L. Wang, X. Wang, T. Tan, A comprehensive study on cross-view gait-based human identification with deep CNNs. IEEE Trans. Pattern Anal Mach. Intell. **39**(2), 209–226 (2016). https://doi.org/10.1109/TPAMI.2016.2545669
20. I. Chtourou, E. Fendri, M. Hammami, Walking direction estimation for gait-based applications. Procedia Comput. Sci. **126**, 759–767 (2018). https://doi.org/10.1016/j.procs.2018.08.010
21. Z. Zhang, L. Tran, F. Liu, X. Liu, On learning disentangled representations for gait recognition. IEEE Trans. Pattern Anal. Mach. Intell. 1–15, (2019). https://doi.org/10.1109/TPAMI.2020.2998790
22. H. Chao, Y. He, J. Zhang, J. Feng, GaitSet: regarding gait as a set for cross-view gait recognition. Proc. AAAI Conf. Artif. Intell. **33**, 8126–8133 (2019). https://doi.org/10.1609/aaai.v33i01.33018126
23. C. Fan, Y. Peng, C. Cao, X. Liu, S. Hou, J. Chi, Y. Huang, Q. Li, Z. He, GaitPart: Temporal part-based model for gait recognition. Proc. IEEE/CVF Conf. Comput. Vision Pattern Recognit. (2020). https://doi.org/10.1109/CVPR42600.2020.01423
24. H. Chao, K. Wang, Y. He, J. Zhang, J. Feng, GaitSet: Cross-view gait recognition through utilizing gait as a deep set. IEEE Trans. Pattern Anal. Mach. Intell. 1–12, (2021). https://doi.org/10.1109/TPAMI.2021.3057879

# Job Prediction Astrology for Using Classification Techniques in Machine Learning

**Snehlata Barde and Sangeeta Tiwari**

**Abstract** Astrology is a traditional science technique used for prediction where astrologers collect the information from the person who is interested to know his future, about himself, and many solutions of problems that are related to his life by the Date of birth, Time of birth, and place of birth. Astrologer prepared the horoscope that indicates all the planet positions and different significance of twelve houses. For example, the tenth house indicates the job related information like it's profession, position in the workplace etc. In this paper we overcome the problem of conventional tools of astrology to produce machine learning-based classification techniques and predict the chance of getting the Job of a person. For that we collected data from 200 people. By the classification technique we calculate the result definitely it will increase the accuracy rate.

**Keywords** Classification technique · Artificial intelligence · Astrology · Weka tool

## 1 Introduction

Astrology is identified as a medium of looking into the future. From ancient period to till date everyone who visits astrologers knows about the future. Astrologers, by examining the planetary position, talk about the various milestones of an individual's life. Understanding the combination of planetary position will tell the important phases of life [1]. For prediction the astrologers use the horoscope of individuals. Horoscopes produce extremely accurate results—certainly job predictions.

Now it is the time to change our all-traditional theories to scientific theories for improving the credibility in astrology, many challenges are to be happening. Astrology is a traditional science which is widely used in India. Astrology combines knowledge of geography, astronomy, mathematics, and societal changes. A good astrologer combines all knowledge to predict [2, 3].

S. Barde (✉) · S. Tiwari
MSIT, MATS University, Raipur, CG, India
e-mail: v.snehabarde@gmail.com



S. Patnaik et al. (eds.), *Smart and Sustainable Technologies: Rural and Tribal Development Using IoT and Cloud Computing*, Advances in Sustainability Science and Technology,
https://doi.org/10.1007/978-981-19-2277-0_29

As Biometrics is used for person recognition where its physiological and behavioral characteristics are used for identification that indicates the person's characteristics but person's kundali talk about person's nature and future [4].

In this research paper we work on scientific-based classification techniques of artificial intelligence that help to predict the possibilities of job by the 10th house of horoscope of the person and calculate the accuracy percentage of classification techniques.

## 2 Related Work

There are many researchers who have made their contribution. Hall et al. (2009) introduced a new technique based on data mining that is used for prediction of health astrology. For that author collected 100 data set for the research and applied the different classification algorithm for calculating the result and analyze the report and predict about the health of a person. WEKA is a well-organized software tool used for machine learning calculation. For preprocessing used some different tool for data editing, clustering, association rule, classification regression, and visualization [5].

Kelly et al. (1997) presented a preconception gender assessment using data mining techniques based on implementation of natural laws and favoring factors. Their main objective of a system is to access more accurate and safer the possible preconception gender on the base of natural law and favoring factor. They apply three data classification mining techniques on our dataset. They used a rapid miner tool to generate results. They achieved more accuracy by the Naïve Bayes on their dataset [6].

McRitchie et al. (2011) introduced a smart health prediction using data mining. The proposed system used Naïve Bayes classification on prepared historical dataset for extracting the hidden knowledge. They use association rule mining to generate the final report. Suggested methodology works on prepared historical data that provided accurate results, which helped patients to get diagnosis instantly [7].

Neelam et al. (2013) studied the position of planets and stars at the time of birth of a person that predict the result of becoming a doctor in future and studied how much the data mining technique contributes in the healthcare field [8].

Mohanty et al. (2021) analyzed UCI kid data collection. Complete and missing data are analyzed using Mean Standard Deviation standardization, dimension reduction using Diffusion Mapping, and lastly ASD classification using Deep Neural Network Prediction and Classification (DNNPC) model. Various performance parameters of DNNPC classifier model are validated. The performance of DNN models can be improved by adding more layers.

Mohanty et al. (2021) proposed method used Landmark Isomap to reduce dimension and then enhanced Deep Neural Network prediction with classification (iDNNPC) to detect ASD class. Lengthy and costly ASD clinical diagnosis method to conduct out clinical research, ASD datasets are stored at authenticated sites like Kaggle and UCI Machine Learning (ML) repository. An ASD Test smart phone

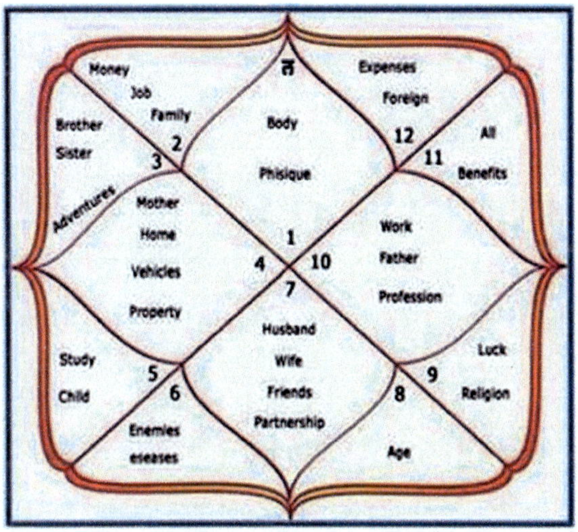


**Fig. 1** Horoscope

app collected data from adults, adolescents, children, and toddlers. The performance parameters supported the classifier model's success.

Keerthi Kiran et al. (2013) give a complete literature analysis on vehicle detection and classification techniques and address open issues in this research field. It also covers several vehicle datasets used to evaluate various approaches.

## 3 Methodology

### *3.1 Data Set (HOROSCOPE)*

Astrologers prepare janm kundaly of a person by using planets position in twelve houses, mahadasha, Antardasha, and Nakshtra, Dasa, that explore about the person's future. Indian Vedic astrology inculcates so many literatures which include verses for prediction of the scenario; some of them are Bhrigu Samhita, VrihatJatak and special notes of the rishis. Figure 1 shows the horoscope image.

### *3.2 Feature Extraction*

According to the person's horoscope, the 10th house is determine his/her profession it is predicted more accurate result when we know that which planet is placed in house? What is the rashi of the house? And how many type of different yog present in kundali [9]. We arranged the data in the tabular form that indicates the responsible

**Table 1** Data prepared for result

| Attribute | Place of planet in house | Zodiac Sign | Profession |
|---|---|---|---|
| Sun | 10 | 1–12 | Govt Job |
| Mars | 10 | 1–12 | Army/Engineer |
| Mercury | 2 | 1–12 | Teacher |
| Jupiter | 9 | 1–12 | Professor/Teacher |
| Saturn | 10 | 1–12 | Police/Army/Engineer |
| Saturn | 10 | 7 | Judge |
| Rahu and mass relation ship | 10 | 1–12 | Police/Army |
| Ketu + Mass | 1 | 1–12 | Police/Army |

planets, position of planets in horoscope, zodiac sign, and possibilities of job. Table 1 shows the details.

These are some possibilities of jobs that are indicated by the person's horoscope. There are many more possibilities that indicate the profession of job.

Now we are moving to the scientific calculation for this we apply some computer technique that work on the data and calculate the result that enhances the accuracy of prediction and prove the authenticity of astrology that indicates the planet position, zodiac sign, and other combination.

## 4 Data Processing Techniques

There are various classification techniques are available in Artificial intelligence which is used to generate the accurate result that makes easy to prediction. We perform the data on three classification techniques for calculating the accuracy result [10].

### *4.1 Logistic Regression*

Logistic regression is a very simple and most prevalent statistic modal that estimates the relation between dependent or independent variables and produces a single variable as an output. It is mostly used in machine learning for prediction. Machine Learning divided into two category Supervised Learning and Un supervised. Logistic regression is a part of classification techniques of Supervised Learning.

The output of a result in logistic regression model represents the discrete value of a dependent variable in categories of true/ false, pass/fail, 1/0. These are indicator variable where 1 indicates the linear function value means the output depends on the input value when input is changing the output will automatically change. Linear relationship between predictors and one responsible dependent variable shown in a linear form of a straight line as:

$$Y = B_0 + B_1x_1 + B_2x_2 + B_3x_3 + \ldots\ldots + B_nx_n \quad (1)$$

When $Y = 1$ then the linear relation between the variables is

$$\log\left(\frac{p}{1-p}\right) = \mathrm{B_0 + B_1x_1 + B_2x_2} \quad (2)$$

The Probability is shown by the equation (3)

$$P = \frac{(\mathrm{B_0 + B_1x_1})}{1 + e(B_0 + B_1X)} \quad (3)$$

## 4.2 *Naïve Bayes Classification*

Naïve Bayes is another supervised classification technique based on bayes method. It is suitable for prediction by checking the probability of a variable. It is a combination of two word Naïve indicates the individual properties of an independent variables and Bayes is based on theorem or rule.

There are many research related applications like sentimental analysis, spam filtration, use this classification method.

Suppose there are '*n*' classes: $B_1, B_2, B_3\ldots\ldots B_\mathrm{n}$ Let Naïve class is conditionally independent then

$$\mathrm{P(X|Bi)} = \pi\mathrm{P(X_k|Bi)}$$
$$\mathrm{k} = 1 \quad (4)$$

$$\mathrm{P(X|Bi) = P(X_1|Bi) * P(X_2|Bi) * \ldots\ldots\ldots * P(X_n|Bi)} \quad (5)$$

## 4.3 *DTNB*

DNTB is also a simple classification technique where decision table is represented in the form of conditional probability. In this classification attributes are divided into two different disjoint set one works on decision table and second works on Naive Bayes. At every step, the forward selection method is used for selecting the attributes that are modeled by Naïve Bayes and the decision table finds the remainder.

The Decision Table and Naïve Bayes classification can be combined to produce class probability estimates. Let us $X_{\top}$ be the set of attributes for Decision Table and $X^{\perp}$ be the set of attributes for Naive Bayes, the overall class probability estimates are calculated as

$$P(Y/X) = A * PDT(Y/XT) * PNB\big(Y/X^{\perp}\big)/P(Y)$$

where $PDT(Y/XT)$ and $PNB(Y/X^{\perp})$ indicate the class probability estimates which are generated by the Decision Table and Naive Bayes respectively.

## 5 Experiments and Analysis

Dataset stored the information of Person's DOB, TOB, and POB as a record and prepared the horoscope chart for looking the position of planet, combination of planet, significance of tenth house, and other relevant information. All content is arranged in column wise in a table. The Process of the scientific method is shown in (Fig. 2).

We loaded this dataset in ARFF format tabular data in the WEKA, which is the most suitable software for machine learning, weka has many classification and analysis methods that prepare the result and provide the accurate prediction.

We have used supervised learning classification method which is most popular and useful for prediction. That gives the scientific result that improves the confidence

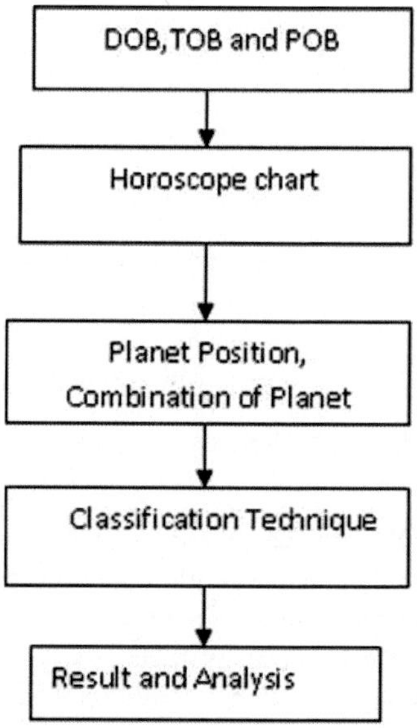


**Fig. 2** Process of scientific method using classification technique

**Table 2** Result of classification technique

| Size of data set | Classification technique | Accuracy (%) |
|---|---|---|
| 200 | Logistic regression | 78.56 |
| 200 | Naïve Bayes | 81.25 |
| 200 | DTNB | 92.72 |

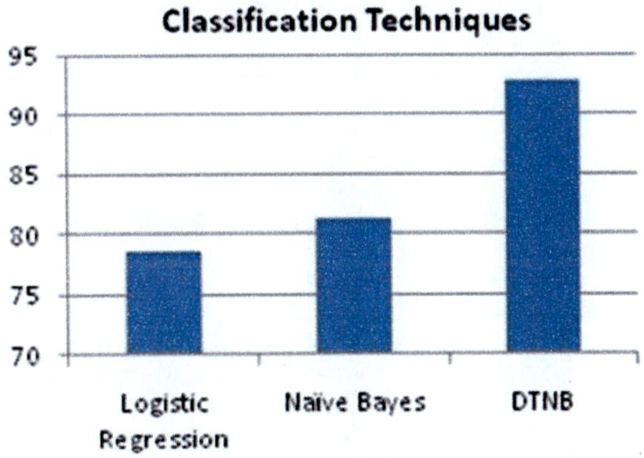


**Fig. 3** Graphical representation of accuracy (logistic regression, Naive Bayes and DTNB)

and beliefs in the astrology field. The logistic regression, naïve bayes, and DTNB technique applied on the dataset and result is calculated in weka generates the Table 2 and graph2.

The result calculated in the term of accuracy percentage wise by the logistic regression 78.56%, Naïve Bayes 81.25% and the better results produced by DTNB with an accuracy 92.72 that are indicated in graph (Fig. 3).

## 6 Conclusion

We have collected limited data of a person those are working in a government job, private, Business or searching a job. Total information of 200 persons which one 100 are doing the different job prepared the training dataset and remaining 100 persons data is applied as a testing dataset. By the help of online available software Astrosage Kundali prepared the horoscope chart of individuals. We calculate the result on logistic regression, Naïve Bayes and DTNB classification techniques: with the accuracy in percentage. After getting the result we found that DTNB is better than logistic regression and naïve bayes, 92.72%.is higher accuracy than 78.56%, 81.25%.

## References


1. N. Marbell, Profile self-selection: A test of astrological effect on human personality. Synapse Foundation monograph. Republished (1986–87) NCGR J. (winter), 29–44 (1981)
2. Z.P. Dobyns, Resultsof the VernonClark experiment." astrology Now, 1 (65, April), 81. Ertel, Suitbert. (2009). appraisal of Shawn Carlson's renowned astrology tests. J. Sci. Explor. **23**(2), 125–137 (1976)
3. M. Baigent, *From the omens of Babylon: Astrology and ancient mesopotamia* (Penguin Arkana, London, 1994)
4. S. Barde, PCA based multimodal biometrics using ear and face modalities. Int. J. Inf. Technol. Comput. Sci. (IJITCS) **6**(5), 43–49, Apr 2014
5. M. Hall, E. Frank, G. Holmes, B. Pfahringer, P. Reutemann, I.H. Witten, The WEKA data mining software. ACM SIGKDD Explor. Newsl. (2009)
6. I.W. Kelly, Modern astrology: ACritique. https://www.researchgate.net/publication/275866394
7. K. McRitchie, Support for astrology from the carlson double-blind experiment. ISAR Int. Astrologer Aug 2011
8. N. Chaplot, P. Dhyani, O.P. Rishi, A review on machine learning concepts for prediction based application. Int. J. Computat. Sci. Eng. Technol. Mar 2013
9. O.P. Rishi, N. Chaplot, Predictive role of case-based reasoning for astrological predictions about profession: System modeling approach, in *International Conference on Communication and Computational Intelligence*, Dec 2010
10. S. Barde, Machine learning make possible to astrological prediction for government job using classification techniques, in *2021 Asian Conference on Innovation in Technology, ASIANCON* (2021)

# Developed Face and Fingerprint-Based Multimodal Biometrics System to Enhance the Accuracy by SVM

**Snehlata Barde and Kishor Kumar Singh**

**Abstract** Biometrics system is known for the identification of a person by its physiological and behavioural characteristics but it has some limitation that overcome by the multimodal biometrics system by combining the modalities that helps to improve authentication. In this paper, we work on self-created database of Face and Fingerprint Region. For feature extraction we applied LBP (Local Binary pattern) with PCA for face and minutiae extraction for figure print. We develop an adaptive trait matching algorithm for effective classification. This scheme effectively converts high-dimensional dense traits into a much more compact representation. We used fusion for feature encoding through the Multi-SVM classifier.

**Keywords** Multimodal biometrics · LBP · PCA multi-SVM

## 1 Introduction

Unimodal biometric systems are easily spoofable and duplicated. An example of how easily phony or rubber fingerprints may fool fingerprint identification systems. The use of unique biometrics can result in interclass similarities. Face recognition may not operate effectively for identical twins due to the camera's inability to identify two subjects. Because of note shortcoming of unimodal biometric systems is that no single technology may be suitable for all types of applications. Furthermore, the use of only one biometrics trait has a high potential for copying, and distortion is also possible due to poor quality, incomplete acquisition conditions due to underdeveloped biometrics traits. Most biometric systems are separated into three main phases, which are as follows:

- Enrols as a first-time user when registering for the biometric system.
- This information is saved in the second step.
- After that, a comparison is made.

S. Barde (✉) · K. K. Singh
MSIT, MATS University, Raipur, CG, India
e-mail: v.snehabarde@gmail.com



S. Patnaik et al. (eds.), *Smart and Sustainable Technologies: Rural and Tribal Development Using IoT and Cloud Computing*, Advances in Sustainability Science and Technology,
https://doi.org/10.1007/978-981-19-2277-0_30

In some ways, the multimodal biometric system is a progressively evolved pattern resulting from the evolution of the Unimodal system. In which an attempt was made, which is still ongoing, to overcome the various flaws encountered when using the Unimodal system. Whether it is the accuracy of the results or the security of the database. Currently, apart from the robust security of the database during the process of identification and verification, we use two or more biometric traits in Multimodal biometric system to achieve the accurate result more safely and effectively. Different sorts of algorithms are employed for feature extraction depending on the nature of the attributes.

Achieving fusion in multimodal biometric systems is accomplished by comparing two or more biometric traits against two or more different algorithms, which are then combined to make a conclusion. It turns out that this procedure is extremely effective in cases where the identities of thousands of people need to be verified all at once, such as in a civil identification context on a massive scale. The procedure of fusing the obtained results is then completed. The attributes collected from the biometrics are also integrated at several levels during the fusion process. Finally, the process is finished by merging the findings of various classifiers.

## *1.1 Face Recognition*

For identification, face recognition relies on human facial features. Face recognition, as the most extensively used and convenient biometric technology, has gained acceptability due to its security, low cost, convenience of use, and social use. It is a fantastic option in both the civil and commercial sectors. Facial recognition, one of numerous ways of "biometric" person identification, evaluates the physical aspects of an individual's body in order to effectively discriminate all but one people. Many changes occur in the face over time; for example, a person's age may vary after the previous photograph taken with a camera. Aside from that, the individual can conceal his or her face to prevent being photographed. As a result, fixing such difficulties is a difficult process. However, it is stated that such issues can be overcome with the use of various technologies, which may reveal the best approach of identifying persons. Face recognition is a simple biometric method we use to recognize one another. Identification of one or so more persons in videos or photos is done by comparing processes.

## *1.2 Fingerprint Recognition*

Fingerprint recognition matching is the technique of recognizing a person's finger print. In general, the finger print is a commonly accepted biometric feature for identification. Fingerprint recognition is commonly used to differentiate people. The minutiae detection is crucial in identifying the optimum correlation spot of the input image to the train dataset. At first, the image is pre-processed and the pre-processed image is further segmented and the region of interest is detected from the image. Then the minutiae is detected from the ROI area. At last the validation purpose is carried out to valid the error correction and the matching point of the finger print. The thresholding algorithm is used to segment the region of interest.

In this article, we integrated face and fingerprint modalities and analysed their efficiency. The paper's key components are as follows.

### *1.3 Fusion*

Fusion plays the main role in multimodal biometrics. According to the research being done, there are primarily five types of fusion utilized in multimodal biometrics. When two or more images from various sensors are combined, this is referred to as sensor level fusion; when two or more features extracted using separate modalities are combined, this is referred to as feature level fusion. In the case of matching score level fusion, the scores collected from various matchers are combined; in the case of decision level fusion, the accept/reject decisions acquired from numerous systems are integrated. Rank level fusion is a way of combining more than two identification results to improve identity efficiency. To fuse the biometrics matching scores from several biometric traits, rank level fusion can be employed [1].

## 2 Literature Review

Jain et al. (2005) studied and examined the various score normalization techniques and fusion rules based on face, finger, and Hand geometry traits. They proved that prior to merging scores, normalizing them enhances the recognition rate of a multi-modal biometric model that includes face, fingerprint, and hand geometry features for identity verification and optimized normalization [2]. Barde et al. (2014) introduced a multimodal biometrics system based on facial and ear traits. In this case, a PCA-based neural network classifier was utilized for extracting features. They discovered overall performance of recognition accuracy, FAR, and FRR, as well as improved reliability [3]. Zhou et al. (2007) presented a multi-rout detection-based identity authentication system using face and finger print. In this, they presented a novel notion of optimal two-dimensional face to boost performance. They fuse both traits by SVM with normalization new method [4]. Barde et al. (2018) conducted research on three modalities: the face, the ear, and the foot, and used fusion to determine the results in terms of FAR, FRR, EER, and Matching score, they used the PCA classifier. Afterward, the results of normalization were pooled, and then the sum rule fusion was used to obtain the maximum matching score possible [5]. Darwish et al. (2010) presented a system based on a large dataset of 2000 face and fingerprint pairs that uses a consistent local binary pattern for the face and minutiae extraction for the fingerprint. To boost performance, matching score level fusion is used in conjunction with a product rule [6].

Telgad et al. (2014) developed a system using face and fingerprint at score level fusion. PCA and the minutiae extraction approach were used to extract the facial and fingerprint features, respectively. The Gabor filter is used in fingerprints, and PCA is used in face dimension reduction [5]. Rattani et al. (2007) proposed system using face and finger traits fusion at feature level. They also evaluated the inherent benefit of fusion performed at the feature extraction level vs. the matching scores level [7]. Keerthi Kiran et al. (2021) introduced a texture-based vehicle recognition system. Local features are obtained individually utilizing the LBP, MBP, GDP, and LAP descriptors. Experimenting with different colour spaces to see how they affect vehicle classification efficiency. It divides grey-level car photos into five classes using the CompCars dataset [8]. Keerthi Kiran et al. (2021) proposed to improve the vehicle recognition rate by combining sophisticated improvement techniques with current pre-trained networks. A new model is proposed to evaluate the use of augmentation approaches with CNN networks for car classification. The proposed method improves original CNN networks by combining several improvement strategies [9].

## 3 Methodology

Face and fingerprint recognition are used in this paper to demonstrate an appropriate solution. Face picture recognition is accomplished via the use of local binary pattern texture, and fingerprint recognition is accomplished through the use of minutiae extraction. Figure 1 shows the flow of proposed work.

We applied LBP method for feature extraction of face and Minutiae extraction for fingerprint feature extraction. And then for optimization used PCA and get optimized data as PCA fitness value. Feature fusion was applied and get the interface component. Using multi-SVM classification has been performed and got the matched and unmatched data. Figure 2 shows the flow diagram of our proposed work.

**Fig. 1** Proposed flow of work

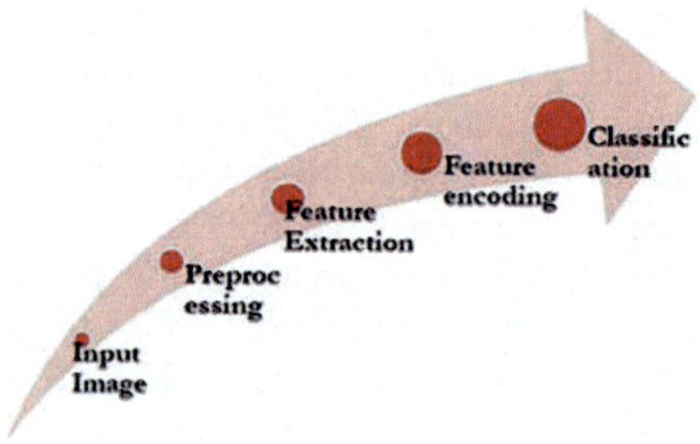

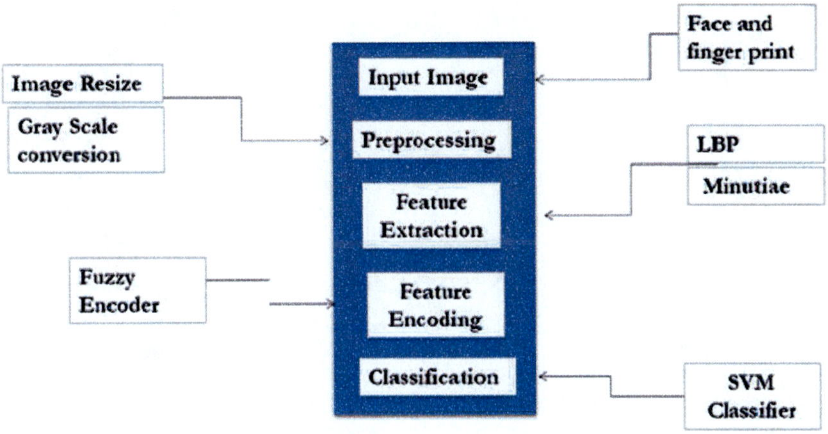


**Fig. 2** Flow diagram of proposed framework

## *3.1 Face*

Face recognition is now proposed by the system. It has two phases. Encoding high dimensional dense characteristics into compact representations by maximizing in correlation is first conceived. Second, the system builds an adaptive feature matching method. This system proposes a maximum correlation compression encoding technique. This technique effectively compresses high-dimensional dense features. Based on this feature descriptor, the Adaptive Matching Framework is suggested, which enhances performance.

Using the LBP, we can evaluate the relationship between neighbourhood pixels, and then encode that relationship into a binary word. This enables for the recognition of patterns and characteristics whilst remaining impervious to variations in intensity [10]. The current submission includes pixel-wise approach as well as filtering-based approach, both of which result in significantly reduced run-times. These solutions produce the same results, despite the fact that they execute at various speeds.

The local binary pattern technique is a major texture analysis breakthrough. The LBP's main features are its adaptability to changing light and its ease of use in demanding real-time situations. Due to its good results, LBP texture and concept have a considerably larger role in image processing [10]. The local binary pattern (LBP) operator is a non-parametric operator used to characterize the regional spatial pattern of a picture. An ordered set of regular patterns at a given pixel point. To obtain a region's LBP texture descriptor, first give a label to each pixel in the region by multiplying the value of the $5 \times 5$ neighbourhoods by the value of the centre pixel. The label histogram is used as a texture descriptor [11]. The code of the LBP value of a pixel $(x_0, y_0)$ is indicated in Eqs. 1 and 2. Figure 3 shown the process of calculating the threshold value.

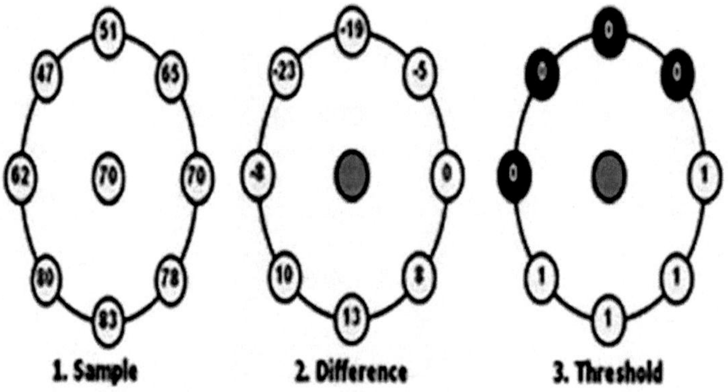


**Fig. 3** Process of calculate the threshold value

$$\mathrm{LBP}_{\mathrm{p,q}} = \sum_{\mathrm{p=0}}^{\mathrm{p-1}} \mathrm{s(xp - xp)^2p} \tag{1}$$

$$s(x) = \begin{cases} 1, & \text{if } x \geq 0 \\ 0 & \text{otherwise} \end{cases} \tag{2}$$

## 3.2 *Finger*

At first, the image is pre-processed and the pre-processed image is further segmented and the region of interest is detected from the image. Then the minutiae is detected from the ROI area. At last the validation purpose is carried out to valid the error correction and the matching point of the finger print. The thresholding algorithm is used to segment the region of interest. Then the minutiae detection and the centroid detection play a vital role in the recognition process. The features of the input/query image are extracted and then the feature is compared with the already trained image in the dataset.

## 3.3 *Principal Component Analysis*

For optimization we have used PCA; Principal Component Analysis (PCA) is used in person recognition algorithms (PCA). Finding the primary directions (also known as the eigenvectors) of the covariance matrix of a multidimensional data set is the core notion behind decorrelation, which is used to emphasize differences and similarities between two or more variables. To put the biometric system through its paces, photos of people's faces were taken from the training set of photographs [12].

## 3.4 SVM Classifier

In machine learning, Support-vector machines are suitable classification systems that evaluate data for classification and regression. An SVM training technique produces a model that allocates new instances to one of two categories, creating it a non-probabilistic binary linear classifier. SVM links training instances to spatial positions to maximize the distance between two categories. It predicts which section new instances belong to depending on which area of the margin they occur [14].

# 4 Result and Experiments

For this we have worked on self-created database of 100 persons shown in Fig. 4. We get face image in colour (RGB mode) from high quality camera and to obtain the fingerprint we used optical fingerprint sensor.

In this work, after getting input of face images pre-processing has been done with resizing shown in Fig. 5. After that we convert RGB image to grayscale, cropping and then converting the grayscale image to binary image BW (Black and White), by replacing all pixels in the input image with luminance greater than level with the value 1 (white) and replacing all other shown in Fig. 6.

**Fig. 4** Database images of face and fingerprint

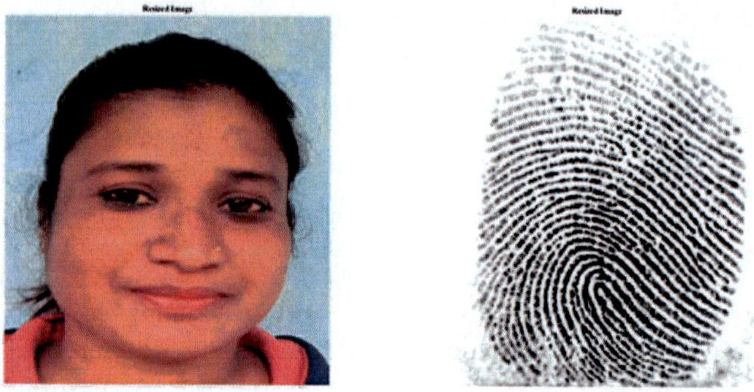

**Fig. 5** Resized images

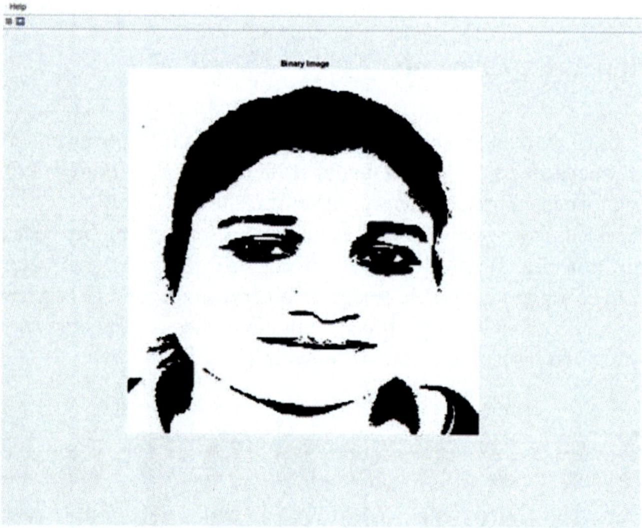

**Fig. 6** Binary images

After that we use the Viola-Jones algorithm to detect people's faces, noses, eyes, mouth, or upper body result is shown in Fig. 7. For the finger print input images after the pre-processing work convert it in binary image. After that on this binary images Morphological operations applied to get enhanced thinned binary image shown in Fig. 8a which is used to minutiae extraction. In minutiae point extraction termination and bifurcation pixel discrimination computed and get the centroid image shown in Fig. 8b and calculated the centroid of each Ridges image shown in Fig. 9. The contour of each ridge's images shown in Fig. 10.

After that we have calculated the match score of face and finger modalities of each person and then applied the sum rule method for the fusion and calculated the score

**Fig. 7** Detected face

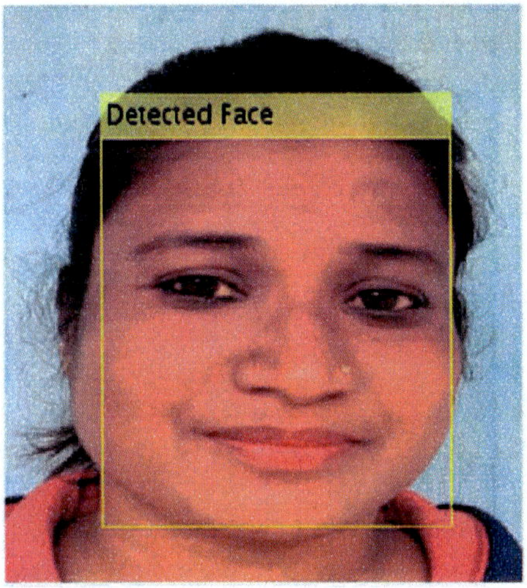


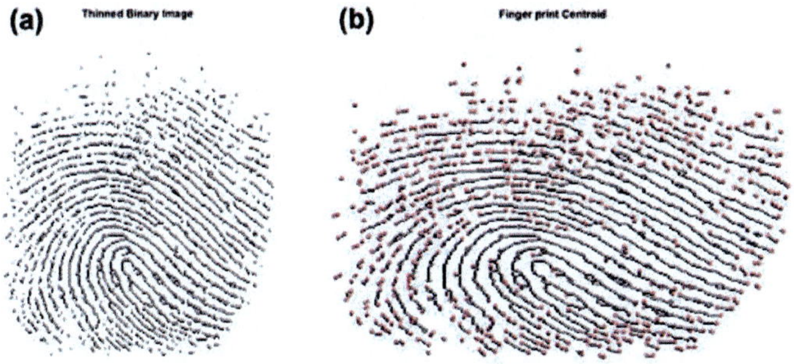


**Fig. 8** **a** Thinned binary image. **b** Centroid image

after fusion and find out the value of FAR, FRR that helps to find out the accuracy in percentage. Table 1 shows the feature of match score. This matching data are used to find out the genuine person or imposters and calculate the value of FAR and FRR that provide the accuracy of method. Figure 11 shows the data of plotting of matching score.

Firstly, face and finger algorithms are tested individually and individual weight for face is found to be 92% and for finger 96% as shown in Table 2. However, in order to increase the accuracy of the biometric system as a whole the individual results are combined at matching score level. At second level of experiment the matching scores from the individual traits are combined and final weight is calculated. The overall performance of the system has increased showing weight for face and finger of 98.24% with FAR of 1.06 and FRR of 0.93 respectively.

**Fig. 9** Centroid of each ridges image

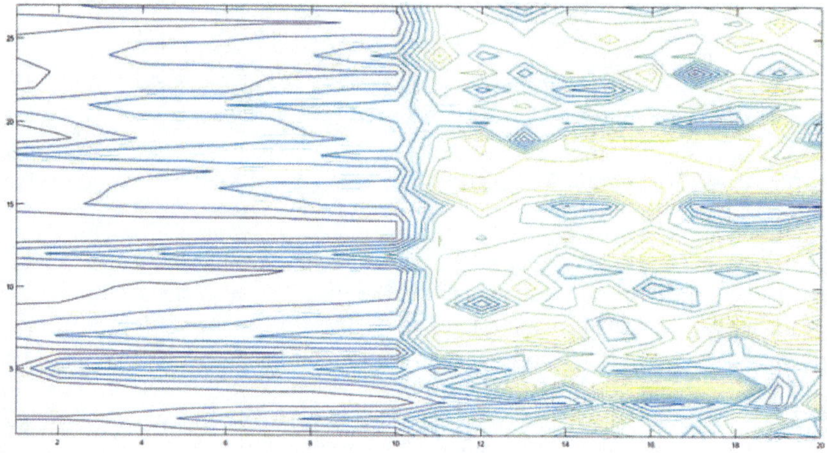

**Fig. 10** Contour of each ridges image

**Table 1** Feature match score

| Image of a person | Face match score | Finger match score | Image of a person | Face match score | Finger match score |
|---|---|---|---|---|---|
| P1 | 240.57 | 147.52 | P38 | 257.10 | 153.74 |
| P6 | 251.32 | 150.23 | P60 | 253.51 | 151.65 |
| P8 | 247.97 | 152.07 | P71 | 251.85 | 154.51 |
| P10 | 246.58 | 148.52 | P82 | 256.30 | 153.69 |
| P12 | 253.87 | 151.45 | P87 | 257.60 | 148.78 |
| P15 | 241.69 | 152.52 | P94 | 251.64 | 155.21 |
| P27 | 255.58 | 154.40 | P99 | 245.50 | 152.57 |

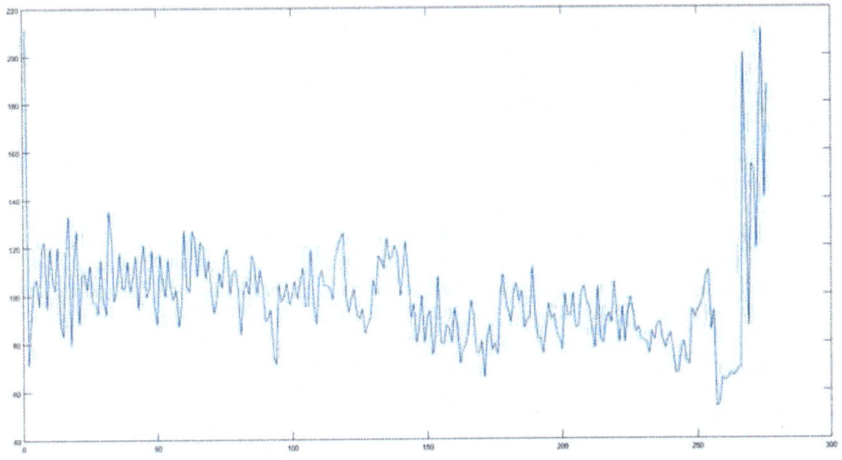

**Fig. 11** Plotting of feature fuse image

**Table 2** Face, finger, and combined accuracy

| Traits | Approach | Match score % | FRR % | FAR % |
|---|---|---|---|---|
| Face | PCA | 92 | 10 | 98 |
| Finger | PCA | 94 | 11 | 99 |
| Face–finger | SVM | 98 | 10 | 93 |

## 5 Conclusion

The system proposes a face and fingerprint recognition approach and its adaptive matching. It has two advantages: it can account for non-linear variations without using kernel methods. The system can enhance accuracy by picking a small subset of training samples for adaptive face and figure matching are 92 and 94% since those examples usually better reflect the characteristics of a given testing sample. The proposed new approach SVM performs in various face and figure shown 98% accuracy to recognition scenarios.

## References


1. V. Keerthi Kiran, S. Dashand, P. Parida, Vehicle recognition using extensions of pattern descriptors, in *IOP Conferences Series: Materials Science and Engineering*, vol. 1166(1), Jul 2021, p. 012046
2. A. Jain, K. Nandakumar, A. Ross, Score normalization in multimodal biometric systems. Pattern Recognit. **38**, 2270–2285 (2005)
3. S. Barde, A.S. Zadgaonkar, G.R. Sinha, PCA based multimodal biometrics using ear and face modalities. I.J. Inf. Technol. Comput. Sci. **05**, 43–49 (2014)
4. J. Zhou, G. Su, C. Jiang, Y. Deng, C. Li, A face and fingerprint identity authentication system based on multi-route detection. Neurocomputing **70**, 922–931 (2007), Elsevier B.V.
5. S. Barde, Fusion based multimodal biometrics system. Int. J. Adv. Technol. Eng. Sci. **06**(05), 112–120 (2018)
6. A.A. Darwish, W.M. Zaki, O.M. Saad, G. Schaefer, Human authentication using face and fingerprint biometrics, in *2010 Second International Conference on Computational Intelligence, Communication Systems and Networks* (2010), pp. 274–278
7. S. Aleem, P. Yang, S. Masood, P. Li, B. Sheng, An accurate multi-modal biometric identification system for person identification via fusion of face and finger print, in *Special Issue on Smart Computing and Cyber Technology for Cyberization © Springer part of Springer Nature 2019*
8. V. Keerthi Kiran, S. Dashand, P. Parida, Improvement on deep features through various enhancement techniques for vehicles classification. Sens. Imaging **22**(1), 41, Dec 2021
9. D. Huang, C. Shan, M. Ardabilian, Y. Wang, L. Chen, Local binary patterns and its application to facial image analysis: a survey. IEEE Trans. Syst. Man Cybern. Part C Appl. Rev. **41**(6), 765–781 (2011)
10. D. Huang, Y. Wang, Y. Wang, A robust method for near infrared face recognition based on extended local binary pattern. In: International Symposium on Visual Computing, pp. 437–446. Springer (2007)
11. S. Barde, Person identification using face and foot modalities, in Int. J. Adv. Eng. Res. Stud. **IV/II**, 160–163 (2015), E-ISSN2249–8974
12. G.W. Mwaura, W. Mwangi, C. Otieno, Multimodal biometric system:- fusion of face and fingerprint biometrics at match score fusion level. Int. J. Sci. Technol. Res. **6**(04), Apr 2017
13. A. Rattani, D.R. Kisku, M. Bicego, M. Tistarelli, *Feature Level Fusion of Face and Fingerprint Biometrics* May, 978-1-4244 (2007)
14. L.K. Singh, M. Khanna, H. Garg, Multimodal biometric based on fusion of ridge features with minutiae features and face features. Int. J. Inf. Syst. Model. Des. **11**(1), 37–57, (January–March 2020). https://doi.org/10.4018/IJISMD.2020010103

# A Survey on Deep Learning Techniques for Anomaly Detection in Human Activity Recognition

**R. N. L. S. Kalpana, D. Nageshwar Rao, and Ajit Kumar Patro**

**Abstract** Continuous monitoring of health in elderly case has gained more importance nowadays to estimate the medical emergency remotely. Wearable sensors devices like smart watches etc. are used for this purpose. Wearable sensor devices give a human-friendly environment for extracting physiological signals data used for estimation of a current situation. Human activity recognition (HAR) is a famous application of the wearable sensor devices (Towards automatic feature extraction for activity recognition from wearable sensors: a deep learning approach. 2017 IEEE 17th International Conference on Data Mining Workshops.(ICDMW). IEEE, 693–702). Rapid growth in use of sensors in application of HAR is influenced by the many challenges. Challenges like feature extraction, annotation scarcity, class imbalance are some of challenges in HAR. Using no of sensors in human body may cause the domination, one signal by another strong signal leads to wrong identification of activity or malfunction of sensor is not identified which fails in estimation of medical emergency. Deep learning methods have gained importance because of various methods to address the challenges in HAR. For HAR sensor-based data many open source datasets are available like WISDM, UCI-HAR. Deep learning has emerged many advanced techniques to identify the anomaly in sensor signal as well as identification of current health condition of patient with suitable architecture. This paper contains an overview of different machine learning-based algorithms for human activity recognition applications. Data sets from open source are taken for consideration for activities like walking, running, climb up, down and sub activities are also determined for generalization of results. Various Deep learning methods for HAR challenge addressing are summarized with merits and demerits, generalization of activities to give overview of current situation( HAR-Net: Fusing Deep Representation and Hand- Crafted Features for Human Activity Recognition. In International Conference On Signal And Information Processing,Networking And Computers.

R. N. L. S. Kalpana (✉) · D. Nageshwar Rao
Department of ECE TKR College of Engineering and Technology, Hyderabad, India
e-mail: rns.kalapana@giet.edu

A. K. Patro
Department of ECE, GIET University, Gunupur, Odisha, India

 
S. Patnaik et al. (eds.), *Smart and Sustainable Technologies: Rural and Tribal Development Using IoT and Cloud Computing*, Advances in Sustainability Science and Technology,
https://doi.org/10.1007/978-981-19-2277-0_31

Springer, 32–40). This paper brief knowledge on the commonly used DLL techniques and algorithms for fault diagnosis also covered. In this paper we propose an improved DLL architecture with top performing technique for future purpose.

**Keywords** Deep learning · Human activity recognition · Physiological signal

## 1 Introduction

Biomedical signal processing is analysing the signal received from the human body employing different sensors, extracting information that is hidden in the original signal for analysing the health condition of the subject. It consists of filtering, preprocessing, amplifying, and transforming from one format to another based on the requirement. Human activity recognition is predicting human gesture and activity based on the biomedical signal by means of sensor Biomedical signal processing is playing a vital role in human activity recognition (HAR) [1, 2] (Fig. 1).

Basic signal processing framework begins with extracting signal from sensor and filtering using windowing technique by batch process or frame by frame technique [3]. Data set can be prepared for anomaly detection from raw data into identified samples using temporal window. Temporal window can be generated by non-overlapping, overlapping, leave, and trail out techniques. Anomaly is generally outlier rare event and doesn't conform to the normal behaviour. For example, abnormal reading of blood pressure or temperature of patient whilst the patient is in normal physical condition.

Fusion of machine learning in fault detection of sensors as well as HAR have gained importance in now a day. Machine learning algorithms are 3 types based on the data and learning style used: supervised, unsupervised, semi-supervised [3]. Supervised learning algorithms are linear regression, logistic regression, support vector machine, Naive Bayes, k-nearest neighborhood. Unsupervised algorithms are clustering, vector quantization [4]. Supervised learning algorithms are trained using the label data set ('correct', true, false) as minimum samples. In training process algorithms make appropriate predictions based on labelled sampled data for appropriate prediction, improve its estimation using reiterations to reach the desired level of accuracy'. Cost function is the error between the estimated value and real value. Linear regression, logistic regression, support vector machine etc. comes under supervised learning. In unsupervised learning no labelled data is available with training dataset. Inference needs to be maintained in between training data set and model hidden structure. In clustering technique finding the pattern in collection of data which is called

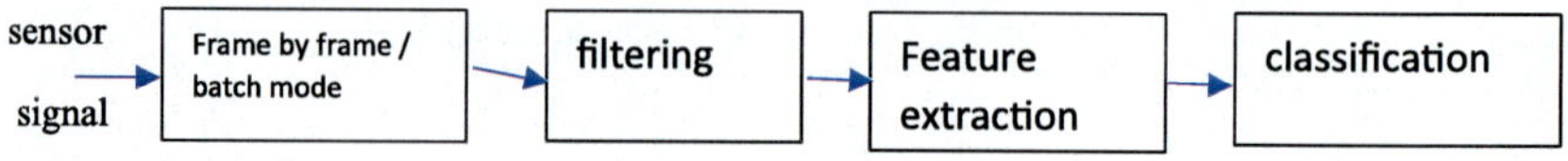


**Fig. 1** Basic signal processing frame work

as clustering. K no of clusters can be prepared with given data. Distance between clusters to be calculated for classification. Previous researchers used support vector machine, hidden Markov Model, random forest, KNN, Naive Bayes. Deep learning came into fame because of better results. Deep learning is a branch in machine learning that has gained popularity recently, capable of learning multiple levels of abstraction using several layers.

In elderly cases of emergency, the patient biomedical signal needs to be collected precisely failing which causes the medical emergency. It requires the expert analysis of the signals from time to time which is difficult and time-consuming. So, we need one technique which analyses the signals, and any emergency will notify correctly along with the data of malfunctioning of the sensor. Malfunctioning of one sensor causes the wrong signal, generating noise that dominating another sensor signals also needs to be analysed to avoid the anomaly. In recent years deep learning got high prosperity in abstracting sensitive data in many areas like vision computing, language processing speech processing. Early works in HAR [1–3] find effectiveness in deep learning methods in HAR allows use is wide variety with little confront with lack of theoretical support.

In this paper fault detection methods of the sensors using various deep learning methods are covered. Deep learning methods in fault detection methods of the sensor are nowadays growing rapidly. Deep learning methods used in HAR human activity recognition predict human gestures through the received signals.

**Sensor modality**:

The performance of HAR is depend on sensor system. Sensor system is divided into 4 types [4]. Wearable sensors, ambient sensor, object sensor, and other sensors.

Wearable sensors: Accelerometer is working based on the rate of change of velocity. It is used to measure the velocity of patient with which he is moving. Gyroscope is used to get info of the orientation of body. It specifies the body is standing position, laying on bed or floor. Magnetometer is combination of gyroscope and accelerometer. It Works on the magnetic field at location. ECG is biometric tool for recognition of activity from electrical signals generated by heart. ECG signals generate univariate time signal.

Ambient sensors are embedded in environment like Wi-Fi-based activity recognition system uses received signal strength to assess (RSS) HAR signal, channel state information (CSI) for coarse information like walking and jogging. RFID technology used by object sensors uses electromagnetic fields to track the objects or location of patient. Tags may be active or passive. Tags relay on low power source, which can be generated by pressure points like ankle.

Other sensors modalities like audio sensor are used in HAR for tacking for fine grained activities. But they are exploited in case of chewing and pressure sensors can be paced in diverse locations. Energy harvest is beauty of pressure sensor.

## 2 Deep Learning

The NN consists of number of hidden layers. These networks are called feedforward networks [5]. Deep neural network which uses deep learning techniques for learning. Deep NN is available in various architectures ex: DBN, autoencoder, CNN, RNN each having their own specialty can be used on different applications. CNN: convolution neural network—A network layer which uses convolution operation is called as convolution layer. It can be applied to one-dimensional or two-dimensional data. It consists of two parts. Convolution pooling layers and second part is fully connected layer predicts the target data based on training data. The output is featured map. RNN: recurrent neural network. It is a type of DL architecture to learn sequences. It will convert the sequences into single output. It can handle various length sequences. LSTM is type of RNN. LSTM can extract features from the sensors. LSTM is type of RNN which is used for the single dimensional and continuous data. It is used for long-term dependencies like anomaly detection, speech processing because of its memory cell. In LSTM input gate, output gate, and cell unit will connect like a chain like structure. It involves feedback connection for better classification process. DBN network is having one hidden layer and one visible layer. Each unit in layer is fully connected to other layer but not connected to the same level. Energy function will model the joint configuration between visible and hidden layers.

Autoencoder is another type of DL architecture to closely construct the output close to input. Autoencoder is having two parts encoders, decoder. Input and output having same no of nodes. It has special feature of bottleneck to learn the essential features only. Encoder learns the representation of input and decoder reconstructs the input closely based other learned representation. Denoising and dimensionality reduction is main advantage of autoencoder leads to computational efficiency. Autoender is constructed by (2n-1) DBN network.

There are two types of human activity recognition techniques, video-based and sensor-based activity recognition techniques. [4] Video-based HAR is having problems with security as well as privacy. Wearable sensors are free of privacy problems. Examples of wearable senor are a barometer, accelerometer, etc. There are freely available datasets on the internet like M-health, USCHAD, WISDIM, OPPORTUNITY. For any kind of application of deep learning techniques, applications for data analysis DATA first need to be pre-processed, transferred into identical samples using windowing techniques. A temporal window can be generated with 3 techniques full non-overlapping, semi non-overlapping window, leave on trial methods. The author worked on handcrafted methods using classification methods of machine learning. He used training and testing data for extracting main features. A new hybrid approach with 3 dense neural network layers has proposed two parts feature extraction and NN architecture. He conducted three types of experiments based on validation techniques. The proposed technique has more generalization ability in activity recognition.

Long-term monitoring of individuals in a normal environment has made a route to the development of many wearable sensor networks for a wide range of medical applications [5]. DL techniques are useful for continuous monitoring data of a patient. Wearable technology combines electronics, fashion, biomedical, textile.

Data science is the science combination of data analytics machine learning artificial intelligence through which the machine will learn automatically like a human being. Deep learning solves machine learning problems. The basic unit of the NN is perceptron. Perceptron consists of input and output input combines the no of inputs and using activation function it will decide the output state. The design of the activation function will decide the strength of the network. Perceptron consists of net function and activation function. The output of the activation function is an output of perceptron.

## 3 Methods and Implementation

Deep learning starts its activities with the acquisition of data from sensors and taking data pre-processing, denoising removing noise from data normalizing data, synching of all data samples after that last step is feature extraction for training and testing of data (Fig. 2).

The author compared several DL techniques possible for wearable senor devices and for different applications found that DL having good generalization capability.

In WSN the network sensor's performance is based on the environment and complexity of the network [6]. High reliability is very much required in health care issues. The author worked on 3 aspects of the WSN collection of network fault information including network status info, fault detection, and diagnosis process. Advantages and disadvantages of each method are related to technical and algorithm. According to his theory fault detection could be in 2 ways centralized and distribute. Centralized detection suffers from heavy-duty problems whereas distribution is suitable for fault detection which tells about network state information and locating faulty nodes. The distributed method reduces the number of exchange messages, latency is maximum. Support vector machines and fuzzy logic are having less complexity in building models for fault detection. SVM is having excellent performance in classification but suffers from an overfitting problem. And fuzzy logic having learning ability is week. For fault diagnosis methods NN type of sensor network is having good performance when compared to fuzzy as well as SVM.

The series data continuously generated from sensors (accelerometer, gyroscope, etc.) is required for mobile sensing and for tracking of the location of the sensor as

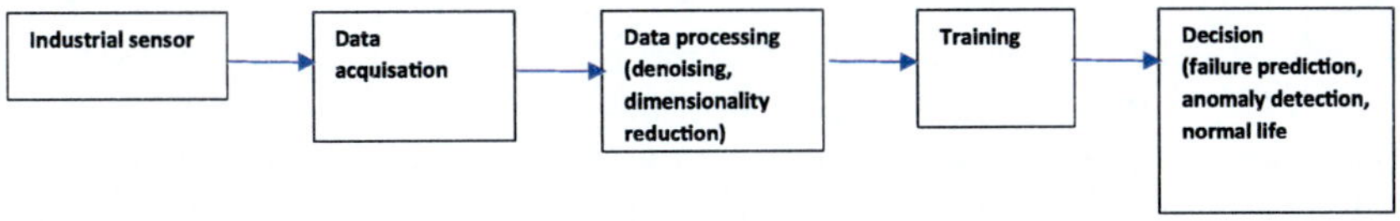


**Fig. 2** Visualization of data analysis pipeline

well as for activity recognition technique. The author proposed a deep learning technique called DEEP SENSE [7] which is an integration of CNN, RNN for detecting noise in the signal and used for regression as well as classification. Regression is used in the case of continuous data because of diverse sensor noise patterns and heterogeneous human behaviour (used in car tracking case) and classification for logical or categorical data. He collected the raw sensor data and computed appropriate features from raw data with processing techniques then used with deep sense technique for activity recognition. Deep sense extracts the distinct feature carries local–global and temporal features. He applied this technique in-car tracking data and compared it with state of art techniques with good results. Deep sense can withstand with time-dependent noise and bias. Whereas new users can't be generalized in the state of art technique n less tracking. The disadvantage is time-consuming. In this paper, the author acquired signals from the sensor using batch mode or frame by frame mode.

The speed of the DLL model in recognizing the activity is also a major role. The study on improving the speed of the CNN network test is time-consuming in classification. The author has taken non-linear units into account applied to the VGG-16 MODEL [8]. A solution was suggested for non-linear optimization problems without the need of STOCHASTIC GRADIENT PROBLEM (SGD) [9], widely used in the VGG-16 model for image net classification. The speed has been increased 4 times than the regular one at cost of accuracy. Basic accelerating models are focus on the decomposed one or few layers and this paper concentrates on optimization apart from the decomposition of layers. The data reconstructed techniques are stochastic gradient process also BACK propagation and singular value decomposer.

A lot of surveys on the fusion of wearable sensor devices, mobile sensors, and deep learning techniques [10] for feature learning. This paper provides a depth summary of deep learning techniques for human activity recognition (HAR). He stated that automatic feature extraction can be done easily with advanced deep learning techniques in image classification and speech classification and health applications. Using DLL techniques inner class similarities, as well as intra-class variation, can be classified accurately at cost of more data for training, compared to convention machine learning techniques. With a large amount of sensor data set the problem of the overfitting and high computationally intensive system can be avoided. The basic DLL techniques can be classified as restricted Boltzmann machine, autoencoder, sparse coding convolution neural networks, and recurrent neural networks. These techniques are used for the analysis of patient conditions in different situations like whilst playing patient emotions, daily activity training. RBM technique can be used for the condition of Alzheimer's patient, autoencoder for spars coding for converting raw data into Harwich Extraction of salient features and converting to vector with these techniques reduce the no of parameters requirement is less, the complexity of the system, used for high dimensional vectors for complex activity recognition.

The main aim of the classification algorithm is to find a parameter that reduces the error in the reconstructed signal. Preparing and fine-tuning is an important role in this case. Out of all classification models like the hidden Markov model, support vector machine, multinominal regression, a multinominal regression model is the best model for the classification of non-linear data. Data sets for HUMAN activity recognition using deep learning WISDM, PAMAP2, MHEALTH, SKODA were used. The open-source framework for users is TensorFlow, PyTorch.

The system of HAR is based on the DLL technique recurrent neural network, in which colliding bodies optimization is used for the feature extraction [11]. This process has 3 parts: data collection followed by feature selection and recognition. The change in features is measured like covariance and mean. The accrue changes in the future will be managed by an algorithm for the proper estimation of human activity. HAR and WISDM datasets were used. RNN is classifier. CBO RNN is compared with convolution classifier (CNN, SVM, RNN) shown best results in human activity recognition.

The author worked on anomaly detection using a famous DLL technique called DBN in electromotor [12]. In this paper, the author proposes two techniques, the tailer method consisting of two sections teacher and tester methods. The teacher method specifies how to test for faults and the test will work on the available data set for testing. The tailer method is based on the DBN of the autoencoder. It is a general encoder. The signal will be regenerated from the bottleneck by layers which is similar to the real signal. This method showed the best results when compared with SVM in the validation case and testing case 98%. Training time increases with the size of the data.

Multi-modality sensor data is used for HAR and will make an accurate classification [13]. Problems such as class imbalance, annotation scarcity, feature extraction, heterogeneity of sensor data, concurrent activity, computational cost, and privacy are the issues of the HAR referred to in this paper. Concept drift (heterogeneity with time inactivity recognition), the concept of evolution (new activities in action), and the open set problem are introduced in this paper. Interpretable deep learning method which decides the important part of the input data, attention mechanism is discussed.

Future challenges like new activities identifications, predictions of future activities, and standardization in state of art techniques are stated in this paper [14]. Reinforcement learning is used in the finding of faults in flying sensors to improve the efficiency of the network in multi-UAV when GPS fails.

The time series representation via temporal and contextual [15] contrasting method will follow 3 steps. Raw time series data is differentiated using strong and weak augmentation with no correlated data using unsupervised learning. In the second novel temporal contrasting module leaves temporal representation using tough cross vie prediction, the third approach is further discriminating module using max similarity amongst different contexts and min similarity between the context of diff samples. Active stake third one.

The study on sensor types applications of the different classifiers and their limitations are tabulated. From this research the KNN, NN, decision tree using automatic weight procedure are studied [16] and high-resolution remote sensing having limitations that there are lack of association with data mining and causes conflicts. CNN, SVM was used with Vanilla CNN architecture for GOT saving classifier. Having problem of no redundancy for sensor data. SVM was used with non-linear mapping decision tree K-fold for fault detection suffers from overfitting. The author proposed an EKNN, EELM, ERLM technique amongst ERELM paper covered about max faults in WSN fed to classifier for abnormal observation. ERLEM is having a high TPR (true positive rate) [17]. The work in the survey on contemporary methods for outlier detection is used in activity monitoring network performance, fault detection, structure defect detection, medical condition monitoring, Time-series monitoring cases. The author applied 3 approaches: in one technique No prior knowledge, in second one it analogues to supervised learning with labelled data, in the third approach, there will be the normal or abnormal condition of data. If the new data comes, it will first consider as normal one if it comes under normal boundary otherwise it will come to an abnormal case. These operations can be done by selecting a suitable classifier for the dataset. The author suggested 3 approaches like classification approach and clustering approach, and novelty approach.

## 4 Discussions

Before applying to any applications, these conditions must check [16]:

1. Data acquisition should be standard way suitable for the next stages
2. Normalization and pre-processing of data for the DL APPLICATIONS
3. Developing suitable security model, network model
4. Using expert knowledge and for generalization category.

The summary of the results of the 20 papers in DL methods suitable for WBAN (health care applications) are tabulated compared results with accuracy.

| S. No. | Data | Application | Network | References | Accuracy |
|---|---|---|---|---|---|
| 1 | Vibration signals of electromotor | Anomaly detection motor imagination | DBN, CNN, AUTOENCODER | [3] | 75 |
| 2 | EMG | Limb motion classification | DBN | [3] | 77 |
| 3 | ECG | Abnormal ECG recognition<br>Diabetes detection | DNN<br>CNN | [4] | 66<br>95 |
| 4 | Mobile senor data | Car tracking, activity recognition | DEEP SENSE (CNN + RNN) | [5] | 90 |
| 5 | Image data | Classification | CNN | [6] | X0.3% |

(continued)

(continued)

| S. No. | Data | Application | Network | References | Accuracy |
|---|---|---|---|---|---|
| 6 | MHELTH | Inner class similarities, intra-class variations, playing | RBM, AUTOENCODER | [7] | 80 |
| 7 | WISDM MHEALTH | Classifier | CBO-CNN | [8] | 90 |
| 8 | Electromotor dataset | Anomaly detection | DBN, autoencoder | [9] | 98 |
| 9 | Multimodality sensor data | Class imbalance, annotation scarcity | RNN | [10] | 90 |
| 10 | Flying sensor data | Future activities | Reinforcement learning | [11] | 90 |
| 11 | Time series data | Anomaly detection | Vanilla CNN | [12] | 89 |
| 12 | Time series data | Anomaly detection | LSTM—AUTOENCODER | [18] | 98 |

In WBAN anomaly detection network has to be chosen based on the effectiveness and accuracy. CNN has to be chosen in 1D, 2D data for sensory data and image signal in structural damage detection [18].

RNN is chosen in case of sequence learning in time series forecasting. The anomaly detection is based on the forecasting and comparing the actual signal is suitable with RNN.

Author applied DLL techniques PCA in combination with DNN applied in the field of the neuro developmental disorder like Autism spectrum disorder (ASD) got clinically approved results in identifying ASD in toddlers, children, adults with 0.004% of error.

With detail study of the many technique in DLL one technique is proposed to identify the anomaly in sensor signal and HAR is given below. In this ideology free data sets (HAR, WISDM) have been chosen. Data need to ne normalized to avoid noise. Optimal feature selection has to be done. Classification of data with application of LSTM and RNN will give satisfied results.

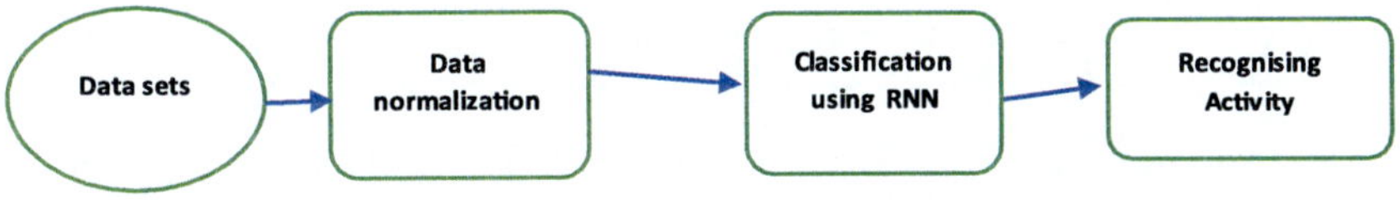

# 5 Conclusion

Deep learning techniques having high accuracy with great complexity of computation for critical patient health care conditions having good generalization capabilities than any other NN techniques. Generalization capability is improved with the DLL Techniques based on the sensor signal DLL technique used for sensor working conditions for sequential data. A wider set of activities and sub-activities can be predicted along with sensor working conditions that can be analysed using DLL techniques for future use. The proposed work will have good classification capability with LSTM and RNN techniques including the anomaly detection.

# References


1. Towards automatic feature extraction for activity recognition from wearable sensors: a deep learning approach, in *2017 IEEE 17th International Conference on Data Mining Workshops* (ICDMW) (IEEE), pp. 693–702
2. HAR-Net, Fusing *deep representation and hand- crafted features for human activity recognition,* in *International Conference on Signal and Information Processing, Networking and Computers* (Springer, Berlin), pp. 32–40
3. Recognizing daily and sports activities in two open source machine learning environments using body-worn sensor units. Comput. J. **57**(11), 1649–1667 (2014)
4. A Brief Survey of Machine Learning Methods and their Sensor and IoT Applications, In *2017 8th International Conference on Information, Intelligence, Systems and Applications (IISA),* https://doi.org/10.1109/IISA.2017.8316459
5. Human activity recognition using wearable sensors: review, challenges, evaluation benchmark Xiv:2101.01665v2 6 Jan 2021
6. Deep learning approaches for human activity recognition using wearable technology. 2018, https://doi.org/10.5937/mp69-18039, e-ISSN: 2466-5525
7. A review on fault diagnosis in wireless sensor networks, in *IOP Conference Series: Earth and Environmental Science* **428**, 012070 (2020), IOP Publishing. https://doi.org/10.1088/1755-1315/428/1/012070
8. Deep sense: a unified deep learning framework for time-series mobile sensing data, in *Processing 2017 International World Wide Web Conference Committee (IW3C2),* April 3–7 (2017)
9. X. Zhang, J. Zou, K. Hey, J. Sun, Accelerating very deep convolutional networks for classification and detection. 2015 IEEE, 0162–8828 (c)
10. Review Deep learning algorithms for human activity recognition using mobile and wearable sensor networks: State of the art and research challenges. Expert Syst. Appl. **105**, 233–261 (2018)
11. Automated human activity recognition by colliding bodies optimization-based optimmal feature selection with recurrent neural network (RNN) Oct 2020. https://doi.org/10.20944/preprints202010.0367.v1
12. Automated fault detection using deep belief networks for the quality inspection of electromotors. Technisches Messen. **81**(5), 255–263 (2014). https://doi.org/10.1515/teme-2014-1006
13. Deep learning for sensor-based human activity recognition: Overview, challenges and opportunities. arXiv:2001.07416v1, 21 Jan 2020
14. Detection of faults in flying wireless sensor networks using adaptive reinforcement learning. Int. J. Recent Technol. Eng. (IJRTE) **8**(1S4), June 2019. ISSN: 2277-3878

15. Time-series representation learning via temporal and contextual contrasting: [cs. LG] 26 Jun 2021, arXiv:2106.14112v1
16. Machine learning algorithms and fault detection for improved belief function based decision fusion in wireless sensor networks. Sensors **19**, 1334 (2019). https://doi.org/10.3390/s19061334
17. A survey of outlier detection methodologies. Artif. Intel. Rev. **22**, 85–126, (2004). Kluwer Academic Publishers
18. Deep learning methods for sensor based predictive maintenance and future perspectives for electrochemical sensors. J. Electrochem. Soc. **167**, 037552 (2020)

# Evaluation of Different Paradigms of Machine Learning Classification for Detection of Breast Carcinoma

**Nibedita Pati, Millee Panigrahi, and Krishna Chandra Patra**

**Abstract** In women, Breast Cancer (BC) is a common problem worldwide, and it is usually diagnosed at the last stages. BC is considered the main reason for death in women in all developed and underdeveloped countries. Therefore, its early detection and diagnosis are required, which will help to predict the disease in the early state. As a result, it will increase the treatment efficiency. To solve the BC problem Machine Learning (ML) techniques are used because it provides high classification accuracy. This paper presents a comparison between different paradigms of Machine Learning algorithms. In this paper, these algorithms are implemented to classify Wisconsin Breast Cancer Dataset (WBCD) as benign and malignant. The result shows that Random Forest (RF) algorithm provides good accuracy of 99.1% with the least computation time of 0.829 s. This helps the medical practitioners to obtain a more accurate detection of cancerous cells with least computational complexity and hardware requirement.

**Keywords** Breast cancer · Machine learning · Wisconsin breast cancer dataset (WBCD) · Random forest

## 1 Introduction

The breast, lung and colorectal organs of a woman are mostly affected by cancer. It is the most common reason for death in women due to the detection in the last stages. The machine learning (ML) approach is defined as analyzing the data, extracting the information from data, and setting the key features. ML is a tool that designs

N. Pati (✉)
Sambalpur University Institute of Information Technology, Sambalpur, Odisha, India
e-mail: nibedita.tech2007@gmail.com

M. Panigrahi
Trident Academy of Technology, Bhubaneswar, Odisha, India

K. C. Patra
Sambalpur University, Sambalpur, Odisha, India



S. Patnaik et al. (eds.), *Smart and Sustainable Technologies: Rural and Tribal Development Using IoT and Cloud Computing*, Advances in Sustainability Science and Technology,
https://doi.org/10.1007/978-981-19-2277-0_32

a model, i.e. computation model for classification and prediction of the dataset. From the researcher's point of view, ML is the best approach that will help for early detection and type of the disease [1]. In 2019, in the US alone, 26,800 new cases of BC diagnosed in women, and approximately 41,760 women died due to this life-threatening disease. Two types of BC are named as Ductal Carcinoma in situ (DCIS) and Lobular carcinoma in situ (LCIC). DCIS refers to the cell line to milk ducts, but they are generally not spread into the surrounding breast tissue. DCIS is non-invasive or pre-invasive BC. LCIS is termed as Lobular carcinoma in situ, which is abnormal cells in the milk glands in the breast [2]. Machine Learning approaches in medical areas are now developing because of their accuracy in classifying and predicting cancer disease. As a result, it helps to downsize the cost of medication and encourage improving the clinical experiments. ML consists of using a modern and filtered information control tool that helps to find out the relationship between the dataset [3]. It is usually benign but indicates the developing risk of BC. BC has no symptoms, but the extent of BC and how much it will affect the breast cancer cell can be calculated at diagnosis [4].

This paper has been categorically divided into six sections. Section 1 consists of the introduction to the article. Section 2 gives an idea about the work done by different reputed authors on the BC topic. Section 3 is based on the discussion of different types of Machine Learning Algorithms. Section 4 describes the workflow of the proposed model. Section 5 describes the experimental result and analysis. Section 6 describes the conclusion and references.

## 2 Related Work

This research referred few important works by the researchers.

Ojha and Goel have compared various classifier algorithms and obtained an accuracy of 81% in DT (Decision Tree) and Support Vector Machine (SVM) [5]. Tafish and El-Halees suggested a model that solved the disease's complexity by using different classification techniques on BC data. The classification gives an accuracy of 77.63% [6]. Abdel-Zaher and Eldeib proposed a CAD scheme for the detection of Breast Cancer using the unsupervised path of the Deep Belief Network and then the Supervised back propagation approach. They have used (WBCD) dataset, and the classifier computes an accuracy of 99.68% [7]. Amrane et al. proposed a model which consists of two different classifiers used for the detection of Breast Cancer, i.e. Naive Bayes (NB) and K nearest neighbor (KNN) where the accuracy of KNN is remarkable at 97.5% [8]. Bazazeh and Shubair compared the most three popular ML methods which are used for Breast Cancer, such as Support Vector Machine (SVM), Random Forest (RF), Naive Bayes (NB), on Wisconsin data set. The performance of classifiers is compared by using parameters such as accuracy, recall, precision and area of ROC [9]. Sivapriya et al. proposed a model using the Random Forest algorithm, which achieved the highest accuracy, 99.06%, with the least error rate. They processed four ML algorithms SVM, Logistic Regression, Naive Bayes and Random Forest [10].

**Table 1** Contains the comparison table with the accuracy obtained by different authors

| Ref. No. | Authors | Classifier compared | Accuracy |
|---|---|---|---|
| [5] | Ojha and Goel | DT and SVM | 81% (SVM) |
| [6] | Tafish and El-Halees | SVM, ANN and k-NN | 77.63 (SVM) |
| [7] | Abdel-Zaher and Eldeib | ANN | 99.68 |
| [8, 9] | Amrane et al. and Bazazeh and Shubair | NB and k-NN | 97.15 (kNN) |
| [10] | Sivapriya et al. | NB, SVM, Random Forest and LR | 99.06% (random forest) |
| [11] | Milan Islam et al. | SVM and kNN | 98.57 (SVM) |
| [12] | Sharma et al. | NB, Random Forest and k-NN | 95.90 (kNN) |
| [13] | Akbugday | NB, SVM and k-NN | 96.85 (k-NN) |
| [14] | Sharifmoghadam and Jazayeriy | AdaBoost extreme learning machine | 97.14 |

Milan Islam et al. proposed a model by using two supervised machine learning techniques, i.e. SVM, KNN for the prediction of Breast [11]. Sharma et al. concluded a comparison between KNN, NB and Random forest algorithm on Breast Cancer data set. Wisconsin data is used for testing purposes, and it is supposed KNN is the most effective, with an average accuracy of 94% [12]. Akbugday has investigated three algorithms, i.e. KNN (K Nearest Neighbors), Naïve Bayes (NB) and Support Vector Machine (SVM), by using different parameter results have been generated and provide accuracy of 96.85% [13]. Sharifmoghadam and Jazayeriy proposed a model where the AdaBoost-based algorithm adapts ELM in ensemble learning. The result of the proposed model is of accuracy 97.14% [14] (Table 1).

## 3 Methods and Materials

The used dataset is publicly available in CSV format which is obtained from UCI Machine Learning Repository. This database is created by Dr. William H. Walberg. It consists of 569 instances and 32 attributes. In the process of diagnosis this paper is going to predict if the cancer is M = Malignant or B = Benign. No missing value is obtained in the dataset [15]. This dataset has no missing value. Out of 569, 357 are labeled as Benign (B) and 212 as Malignant (M). The proposed model is simulated in MATLAB 2019 environment and executed in INTEL Core i5 (10th Generation) Processor with 2.27 GHz speed and 8 GB RAM.

## 4 Work Flow of the Proposed Model

The below flow diagram describe about the proposed model in Fig. 1.

Data preprocessing of the dataset is done by WEKA software tool. The CSV file is uploaded in WEKA environment. After pre processing data splitting is done i.e. for data training 75% of data and 25% data is used for testing. Performance is evaluated by using different Machine Learning Algorithms.

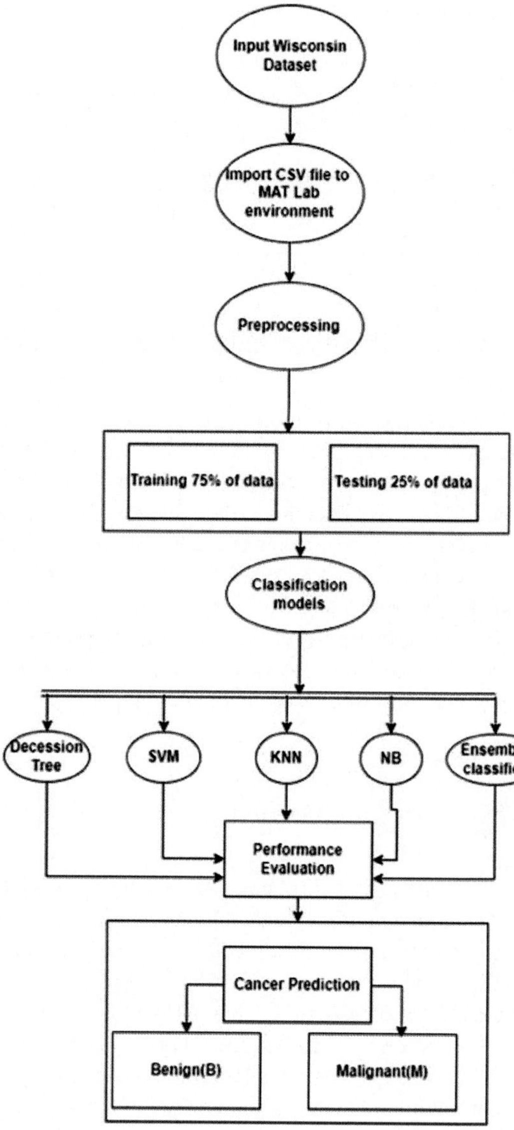


**Fig. 1** Flow of work of the proposed model

## 5 Result and Discussion

The results obtained from different ML algorithm is summarized in Table2.

Table 3 Contains the comparison table with accuracy obtained by different authors.

**Table 2** Result classification of breast cancer for different classifiers

| Classification model | TPR (%) | FNR (%) | PPV (%) | FDR (%) | Accuracy (%) | AUC | Computation time in (s) |
|---|---|---|---|---|---|---|---|
| *Decision tree (DT)* | | | | | | | |
| Fine | 97 | 03 | 88 | 12 | 92.1 | 0.97 | 4.01 |
| Medium | 97 | 03 | 88 | 12 | 92.1 | 0.97 | 2.65 |
| Coarse | 55 | 45 | 94 | 06 | 76.2 | 0.77 | 2.537 |
| *K nearest neighbor (KNN)* | | | | | | | |
| Fine | 100 | 0 | 97 | 3 | 98.4 | 0.98 | 1.343 |
| Medium | 74 | 26 | 68 | 32 | 69.8 | 0.75 | 1.759 |
| Coarse | 0 | 100 | 0 | 0 | 50.8 | 0.50 | 1.39 |
| Cosine | 77 | 23 | 62 | 38 | 65.1 | 0.66 | 2.489 |
| Cubic | 68 | 32 | 62 | 38 | 63.5 | 0.70 | 1.93 |
| **Weighted KNN** | **100** | **0** | **97** | **03** | **98.4** | **1.00** | **1.27** |
| *Naive Bayes (NB)* | | | | | | | |
| Gaussian | 29 | 71 | 64 | 36 | 57.1 | 0.64 | 1.247 |
| Kernel | 35 | 65 | 79 | 21 | 63.5 | 0.75 | 2.558 |
| *Support vector machine (SVM)* | | | | | | | |
| Linear | 55 | 45 | 68 | 32 | 65.1 | 0.69 | 1.365 |
| Quadratic | 90 | 10 | 97 | 03 | 93.7 | 0.94 | 1.306 |
| **Cubic SVM** | **100** | **0** | **97** | **03** | **98.4** | **1.00** | **1.237** |
| Fine Gaussian | 90 | 10 | 93 | 7 | 92.1 | 0.99 | 1.639 |
| Medium Gaussian | 65 | 35 | 65 | 35 | 65.1 | 0.78 | 1.074 |
| Coarse SVM | 10 | 90 | 75 | 25 | 54 | 0.64 | 1.365 |
| *Ensemble classifier* | | | | | | | |
| Adaboost | 97 | 3 | 100 | 0 | 98.4 | 0.98 | 2.166 |
| **Random forest** | **100** | **0** | **97** | **03** | **99.1** | **1.00** | **0.829** |
| RUS boosted | 97 | 03 | 88 | 12 | 92.1 | 0.97 | 2.016 |

[a] The best results are in bold font

Here *TPR* true positive rate of sensitivity, *FNR* false negative rate, *PPV* positive predictive value, *FDR* false discovery rate, *AUC* area under curve

From the table mentioned above we have obtained the result as **Weighted KNN, Cubic SVM, Random Forest** of accuracy **98.4%, 98.4%** and **99.1%** respectively with computation time of **1.27 s, 1.237 s and 0.829** s. Among these above **Random forest** provides accuracy of **99.1%**, with computation time of **0.829** s. Figure 2 describes the accuracy of different ML algorithms

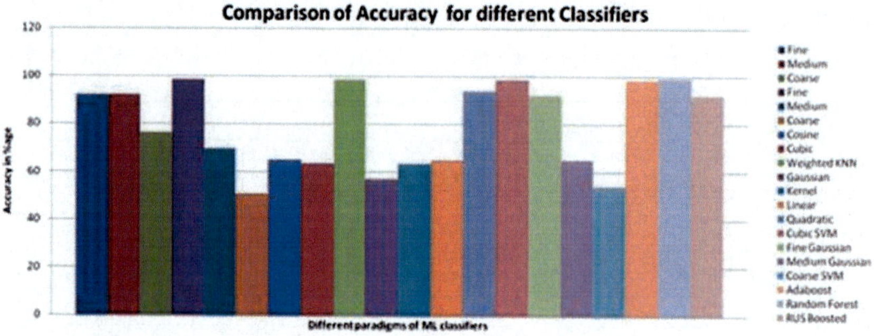


**Fig. 2** Accuracy for different comparative methods on WBCD data

**Table 3** Comparison of result with existing conventional models

| Author | Model used | Accuracy (%) |
|---|---|---|
| **Proposed method** | **Random forest** | **99.1** |
| Gupta and Gupta [16] | K-NN | 94.4 |
| | SVM | 93.9 |
| | DT | 96.9 |
| Baneriee et al. [17] | Bagging | 95.6 |
| | Boosting | 93.45 |
| | Voting | 87.23 |
| Sharma et al. [18] | Random forest | 94.74 |
| | Naïve Bayes | 94.47 |
| | KNN | 95.9 |
| Hazra et al. [19] | SVM | 98.5 |
| | NB | 97.3 |
| | Ensemble | 97.3 |
| Islam et al. [20] | SVM | 97.14 |
| | KNN | 97.14 |
| | RF | 95.71 |
| | ANN | 98.57 |
| | LR | 95.71 |

## 6 Conclusion

Among all cancer type diseases for women, Breast Cancer is the most frequent disease. So, if it is found at the early stage then it will help to save the lives of thousands of human beings. For identification of Breast Cancer we have used Wisconsin Breast Cancer Data using different Machine Learning Technique. Machine Learning helps as to classify and predict the cell whether Benign or Malignant. In this paper different

ML algorithms have been used on WBCD data. These algorithms have compared in order to find the best classifier in terms of accuracy and computation time to construct the model. Therefore it is concluded from the proposed model that Random forest provides the best result of accuracy 99.1% and less computation time of 0.829 s.

## References


1. S. Laghmati, A. Tmiri, B. Cherradi, Machine learning based system for prediction of breast cancer severity, in *International Conference on Wireless Networks and Mobile Communications (WINCOM)* (IEEE, 2019), pp. 1–5
2. https://www.cancer.org/content/dam/cancer-org/research/cancer-facts-and-statistics/breast-cancer-facts-and-figures/breast-cancer-facts-and-figures-2019-2020.pdf
3. A. Kumar, M. Poonkodi, Comparative study of different machine learning models for breast cancer diagnosis, in *Innovations in Soft Computing and Information Technology* (Springer, Singapore, 2019), pp. 17–25
4. N.G. Maity, S. Das, Machine learning for improved diagnosis and prognosis in healthcare, in *IEEE Aerospace Conference*, pp. 1–9 (2017)
5. U. Ojha, S. Goel, A study on prediction of breast cancer recurrence using data mining techniques, in *7th International Conference on Cloud Computing, Data Science and Engineering—Confluence* (2017)
6. M.H. Tafish, A.M. El-Halees, Breast cancer severity degree predication using data mining techniques in the Gaza strip, in *International Conference on Promising Electronic Technologies (ICPET)* (Deir El-Balah, 2018), pp. 124–128
7. A.M. Abdel-Zaher, A.M. Eldeib, Breast cancer classification using deep belief networks. Expert Syst. Appl. **S0957–4174**(15), 00710–00711 (2015)
8. M. Amrane, S. Oukid, I. Gagaoua, T. Ensari, Breast cancer classification using machine learning, in *Electric Electronics, Computer Science, Biomedical Engineerings' Meeting (EBBT)* (IEEE, 2018), pp. 1–4
9. D. Bazazeh, R. Shubair, Comparative study of machine learning algorithms for breast cancer detection and diagnosis, in *5th International Conference on Electronic Devices, Systems and Applications (ICEDSA)*, pp. 1–4 (2016)
10. J. Sivapriya, V. Aravind Kumar, S. Siddarth Sai, S. Sriram, Breast cancer prediction using machine learning **8**(4) (2019) ISSN: 2277-3878
11. Md. Milon Islam, H. Iqbal, Md. Rezwanul Haque, Md. Kamrul Hasan, Prediction of breast cancer using support vector machine and k-nearest neighbors, in *Humanitarian Technology Conference (R10-HTC)* (IEEE, 2017), pp. 226–229
12. S. Sharma, A. Aggarwal, T. Choudhury, Breast cancer detection using machine learning algorithms, in *International Conference on Computational Techniques, Electronics and Mechanical Systems (CTEMS)* (IEEE, 2018)
13. B. Akbugday, Classification of breast cancer data using machine learning algorithms, in *Medical Technologies Congress (TIPTEKNO)* (Izmir, Turkey, 2019), pp. 1–4
14. M. Sharifmoghadam, H. Jazayeriy, Breast cancer classification using adaboost—extreme learning machine, in *5th Iranian Conference on Signal Processing and Intelligent Systems ICSPIS* (2019)
15. Expert Systems with Applications **38**(7), 9014–9022. UCI breast cancer Wisconsin (original) dataset. https://archive.ics.uci.edu/ml/datasets/Breast+Cancer+Wisconsin+%28Original%29
16. M. Gupta, B. Gupta, A comparative study of breast cancer diagnosis using supervised machine learning techniques, in *Second International Conference on Computing Methodologies and Communication (ICCMC)* (IEEE, 2018), pp. 997–1002

17. C. Baneriee, S. Paul, M. Ghoshal, A comparative study of different ensemble learning techniques using Wisconsin breast cancer dataset, in *International Conference on Computer, Electrical and Communication Engineering (ICCECE)* (IEEE, 2017), pp. 1–6
18. S. Sharma, A. Aggarwal, T. Choudhury, Breast cancer detection using machine learning algorithms, in *2018 International Conference on Computational Techniques, Electronics and Mechanical Systems (CTEMS)* (IEEE, 2018), pp. 114–118
19. A. Hazra, S.K. Mandal, A. Gupta, Study and analysis of breast cancer cell detection using Naïve Bayes, SVM and ensemble algorithms. Int. J. Comput. Appl. **145**(2), 39–45 (2016)
20. M.M. Islam, H. Iqbal, M.R. Haque, M.K. Hasan, Prediction of breast cancer using support vector machine and K-nearest neighbors, in *IEEE Region 10 Humanitarian Technology Conference (R10-HTC)* (IEEE, 2017), pp. 226–229

# Enhanced Feature Matching Algorithm for Medical Image Registration

**Minu Samantaray, Millee Panigrahi, and Krishna Chandra Patra**

**Abstract** In medical imaging technology, image registration plays a vital role. Image registration is a process of transforming the test image to reference image based on different parameters and features. In traditional feature extraction methods the feature points are unevenly distributed for which the redundant features are more. This results to time consuming process and low registration accuracy. Hence it is a challenging task to match the images with higher accuracy. In the proposed method enhanced and robust image registration algorithm is based on SURF and BRISK feature extraction methods with upgraded GMS and RANSAC algorithm. In this method the keypoints of the medical images are detected through SURF algorithm then features are extracted by BRISK algorithm and false match keypoints are removed through GMS algorithm and improved RANSAC algorithm is used. Experimental results show the efficiency and effectiveness in terms of runtime, accuracy compared to traditional feature extraction matching algorithms which can be implemented in real time for occasion with the requirements of high accuracy.

**Keywords** Image registration · SURF · FAST · BRISK · GMS · RANSAC

## 1 Introduction

Image registration is the process of warping a test image with a reference image. This process can be carried out using two different categories—first one is area based image registration and second is feature based image registration [1]. Area based image registration uses the intensity based information that helps to match

M. Samantaray (✉)
Sambalpur University Institute of Information Technology, Cuttack, Odisha, India
e-mail: minu1485@gmail.com

M. Panigrahi
Trident Academy of Technology, Bhubaneswar, Odisha, India

K. C. Patra
Sambalpur University, Burla, Odisha, India



S. Patnaik et al. (eds.), *Smart and Sustainable Technologies: Rural and Tribal Development Using IoT and Cloud Computing*, Advances in Sustainability Science and Technology,
https://doi.org/10.1007/978-981-19-2277-0_33

the regions whereas feature based image registration uses the features which are initially extracted from the images and are matched by using similarity measures and calculating their spatial relationship. However feature based method gives better predictability in terms of rotation and illumination. The basic steps of feature based registration method includes—Feature (Keypoint) Detection, Feature (Keypoint) Description, Matching of extracted features and image transformation [2] as shown in Fig. 1.

The detection of features are estimated using different algorithms such as Fast Fourier Transform, Gabor Wavelets, Root Mean Square Error (RMSE) and other segmentation methods. The second step involves feature descriptors where matching features between two images are estimated. It includes different feature descriptors algorithms such as Scale Invariant Feature Transform (SIFT), Speeded Up Robust Features (SURF), BRISK, FAST, HARRIS [3]. In the aspect of feature matching, there are different algorithms endorsed to get rid of false matching such as RANSAC, MSAC, GMS, LMedS.

Based on demand in terms of speed, real time and more accuracy in image registration, this work proposes an enhanced image registration method that takes

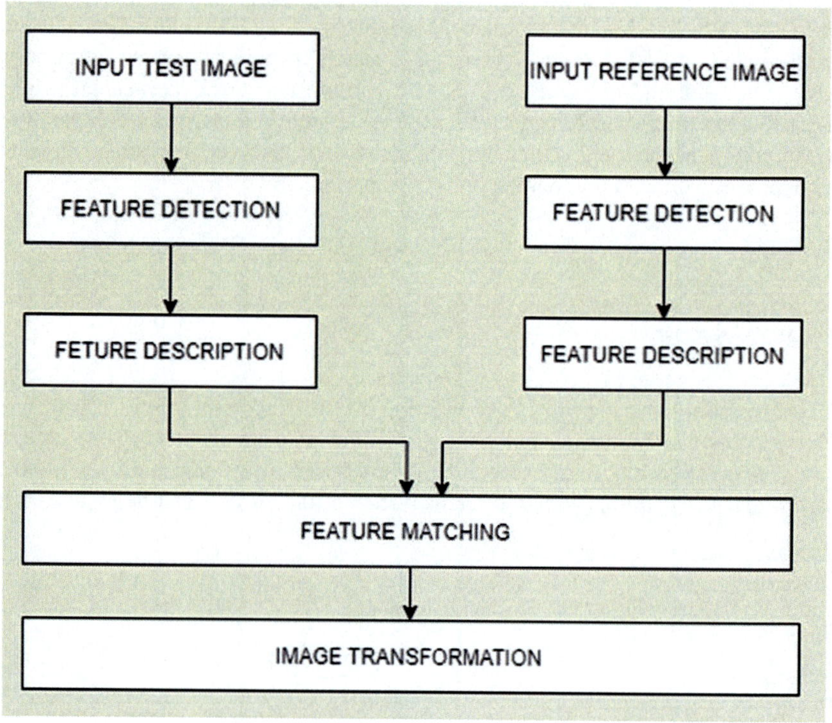


**Fig. 1** General work flow of image registration process

the combined SURF, BRISK and FAST algorithm with upgraded matching algorithm. The paper is organized as follows. Section 2 describes overview of SURF, FAST, BRISK algorithm, GMS and RANSAC algorithms. Section 3 discusses about enhanced image registration process. Section 4 discusses about the experimental results. The conclusion and future work is described in Sect. 5.

## 2 Feature Descriptor and Feature Matching

SURF—The Speeded Up Robust Feature is a descriptor in fast and robust algorithm [4]. It uses the summed read table i.e. integral image and the determinant of $H$-matrix where $H$ stands for Hessian. Hessian matrix has better accuracy and computing time. The determinant of $H$-matrix is calculated based on interest points, convolution with Gaussian kernel followed by second order derivative [5]. The $H$-matrix at point $X = (x, y)$ and scale $\mu$ is

$$H(x,\mu) = \begin{bmatrix} L_{xx}(x,\mu) & L_{xy}(x,\mu) \\ L_{xy}(x,\mu) & L_{yy}(x,\mu) \end{bmatrix} \tag{1}$$

$L_{xx}(x, \mu)$ = Convolution of Gaussian second order derivative in $x$ point and same for $L_{xy}(x, \mu) L_{yy}(x, \mu)$.

FAST—(Features from Accelerated Segment Test) is an algorithm used to extract features through corner detection [6]. It selects a pixel '$p$' in an image as point of interest and intensity is defined as $I_{\mathrm{p}}$. Then appropriate threshold value '$t$' is set. A circle of radius 3 consisting of 16 pixels around the test pixel '$p$' is set. Then corner is detected if a set of '$n$' neighboring pixels in the selected circle are higher in intensities from $I_{\mathrm{p}} + t$ or $I_{\mathrm{p}} - t$ where $n = 12$. For faster process, three or four pixel values are tested in four different directions of circle which are above or below $I_{\mathrm{p}} + t$ and then for all 16 pixels. This algorithm has high performance speed.

BRISK—Binary Robust Invariant Scalable Keypoints is a corner detecting feature extraction method that uses AGAST (Adaptive and Generic Corner detection based on Accelerated Segment Test) and FAST Algorithm [7]. AGAST algorithm is used to detect interest point by arranging a point pair line segments in four concentric rings that consists of 60 point pairs. Then scale-space keypoint detection mechanism is applied by computing non-maximal suppression on each intra octave and octave subpixel maximum in patch. BRISK is generally based on feature orientation where rotation invariance is achieved for each feature which further results a brightness test and generates a binary descriptor.

RANSAC [8] algorithm selects randomly a pair of 4 points from the input sampling set and transforms the model. It calculates the squared residual error for all the matching points and considers the consensus set which are within the range of threshold value. These steps are repeated till the maximum likelihood is estimated [9].

GMS [10] algorithm considers smoothness of true matching by taking multiple features. It sets the threshold value through statistical process which will distinguish the correct matched features from false matched features.

Combining the advantage of GMS algorithm which is robust in matching keypoints and high speed and RANSAC algorithm is based on repetitive process that leads to stability; an upgraded matching algorithm is introduced named as GMS-RANSAC algorithm.

## 3 Enhanced Image Registration Process

To achieve better performance in image registration, this paper proposes an enhanced image registration methodology as follows:

1. Feature extraction using SURF and BRISK
2. Removal of false matching keypoints using GMS
3. Calculation of transformation matrix after removal of false matching keypoints through RANSAC algorithm.

## 4 Experimental Results and Analysis

The experiment in this paper is completed by choosing medical dataset randomly. The feasibility of the model has been tested. Brain CT image, Liver CT image and Breast mammogram image are the reference images considered for the experiment as shown in Fig. 2a–c respectively. The test images are generated by rotating each image by an angle of 30° shown in Fig. 2d. The reference image and the test image are matched using SURF and BRISK algorithm. These two algorithms are combined for better feature matching. As shown in Fig. 2e the matched points are selected between the original image and test image where red lines indicate the points of original image and green lines indicate the points for test image. SURF has fewer features matching cost as compared to BRISK whereas BRISK is high for up scaling of images. The computational time for SURF is less with better efficiency. Therefore both the algorithms are concatenated to achieve better result as shown in Table 1. To achieve greater matching accuracy with lowest error rate RANSAC and GMS algorithms are combined. A comparison is made with other approaches and as seen in Table 2 the proposed method gives better registration matching accuracy of 98.3%.

The proposed work is simulated in MATLAB 2020 environment and executed in INTEL Core i5 (10th generation) Processor with 2.27 GHz speed and 8 GB RAM.

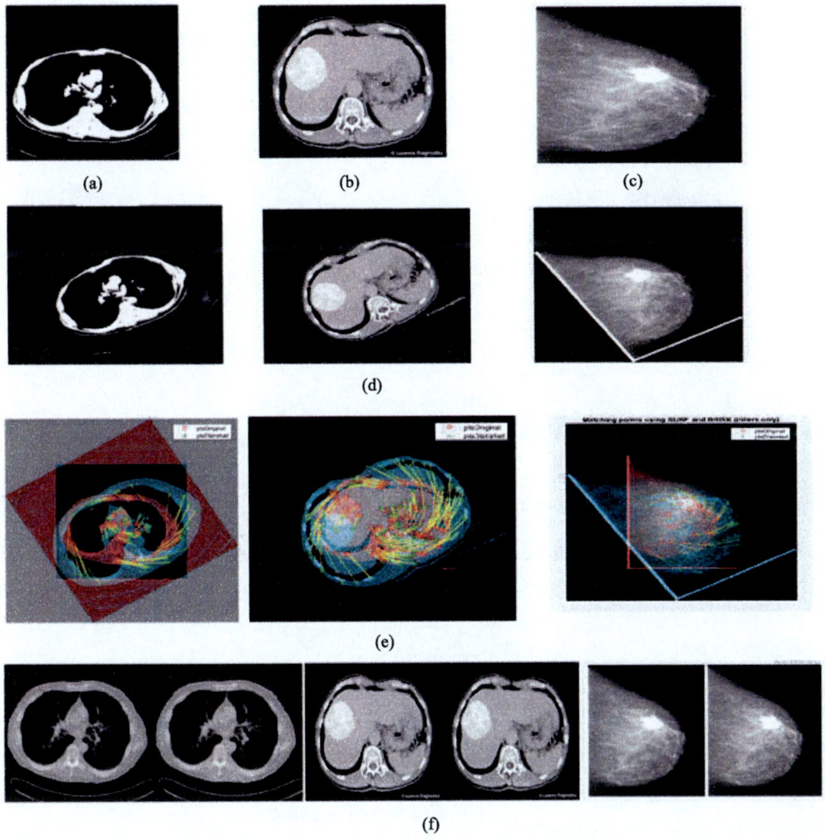


**Fig. 2** Experimental outputs of different biomedical images. **a** Brain CT image, **b** liver CT image, **c** breast mammogram image, **d** rotated images at angle 30°, **e** image matching using SURF and BRISK algorithm, **f** image registration

**Table 1** Evaluation parameters of different images using proposed algorithm

| Input images | Runtime | RMSE | Accuracy (%) |
|---|---|---|---|
| Image (a) | 8.6573 | 0.8124 | 96.25 |
| Image (b) | 9.2814 | 0.9271 | 97.31 |
| Image (c) | 8.9423 | 0.8296 | 98.37 |

**Table 2** Performance comparison of different algorithms

| Author | Algorithm used | Accuracy (%) |
|---|---|---|
| Cen et al. [1] | BRISK + GMS | 95.7 |
| Abuzneid and Mahmood [2] | SURF + RANSAC | 96 |
| Lan et al. [3] | ORB + GMS + RANSAC | 97.2 |
| **Proposed method** | **SURF + BRISK + GMS + RANSAC** | **98.3** |

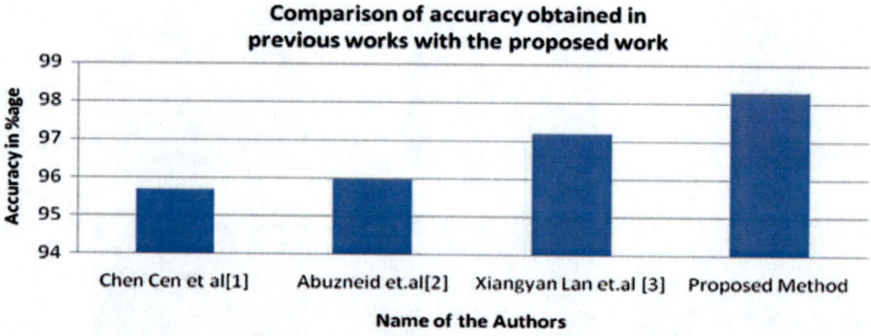


**Fig. 3** Comparison of accuracy obtained in previous works with the proposed work

### *4.1 Evaluation Parameters*

It includes Runtime, RMSE (Root Mean Square Error) and Accuracy.

- Runtime evaluates the efficiency of algorithm. Faster runtime better the efficiency.
- RMSE is defined as the square root of the mean of the square of all of the errors.

$$\text{RMSE} = \sqrt{\frac{1}{N}\sum_{k=1}^{N} f(x_k - y_k) - f(x'_k - y'_k)} \quad (2)$$

- Accuracy measures the total number of correct matches out of total matches.

$$\text{Accuracy} = \frac{\text{No. of correct matches}}{\text{Total matches}} \times 100 \quad (3)$$

From the graph it clearly observed that the proposed methodology for medical image registration has been successful to achieve the highest accuracy (Fig. 3).

## 5 Conclusion

This paper describes the enhanced image registration algorithm which combines SURF and BRISK feature extraction methods with upgraded GMS and RANSAC algorithm. This increases the efficiency in terms of runtime and improved accuracy compared to the traditional matching algorithms. It has got great worth in different aspects of applications such as video stitching, image mosaicing, diagnosis of medical diseases and other real time where high accuracy is prior requirement.

## References


1. C. Cen, R. Li, X. Xu, Fast robust matching algorithm based on BRISK and GMS. J. Phys. Conf. Ser. **1237**(2), 022070 (2019). IOP Publishing
2. M. Abuzneid, A. Mahmood, Image registration based on a minimized cost function and SURF algorithm, in *International Conference Image Analysis and Recognition* (Springer, Cham, 2017), pp. 321–329
3. X. Lan, B. Guo, Z. Huang, S. Zhang, An improved UAV aerial image mosaic algorithm based on GMS-RANSAC, in *2020 IEEE 5th International Conference on Signal and Image Processing (ICSIP)* (IEEE, 2020), pp. 148–152
4. E. Paul, A.S. Ajeena Beegom, Mining images for image annotation using SURF detection technique, in *2015 International Conference on Control Communication & Computing India (ICCC)* (IEEE, 2015), pp. 724–728
5. H. Bay, T. Tuytelaars, L. Van Gool, Surf: speeded up robust features, in *European Conference on Computer Vision* (Springer, Berlin, Heidelberg, 2006), pp. 404–417
6. E. Sabanovic, D. Matuzevicius, Experimental investigation of feature descriptors for retinal image registration, in *2017 5th IEEE Workshop on Advances in Information, Electronic and Electrical Engineering (AIEEE)* (IEEE, 2017), pp. 1–4
7. Z. Li, D.R. Selviah, Comparison of image alignment algorithms, in *Proceedings Paper, London Communications Symposium* (University College London, 2011)
8. R. Karani, T.K. Sarode, Image registration using discrete cosine transform and normalized cross correlation, in *IJCA Proceedings on International Conference and Workshop on Emerging Trends in Technology* (2012), pp. 28–34
9. Q.H. Tran, T.J. Chin, G. Carneiro, M.S. Brown, D. Suter, In defence of RANSAC for outlier rejection in deformable registration, in *European Conference on Computer Vision* (Springer, Berlin, Heidelberg, 2012), pp. 274–287
10. J. Bian, W.Y. Lin, Y. Matsushita, S.K. Yeung, T.D. Nguyen, M.M. Cheng, GMS: grid-based motion statistics for fast, ultra-robust feature correspondence, in *Proceedings of the IEEE Conference on Computer Vision and Pattern Recognition* (2017), pp. 4181–4190

# Smart Technology for Smart Living

# A Strategical Method of Proper Resizing and Reusing of Construction Formwork Materials

**Bholanath Mukhopadhyay, Rajesh Bose, and Sandip Roy**

**Abstract** This study can provide effective information about resizing and reusing of construction materials. Many advantages and disadvantages are present in resizing and reuse of construction material, which are included in this study. Variability of building is maintained by reusable plastics. Reuse of time timber formwork has few effective site management issues. Effective cost-control ability is also noticed in resized and reused construction formwork materials. Cost-effectiveness of plywood use is treated as an effective benefit of resizing and reusing process. Construction waste is also controlled successfully through this reusing and resizing process.

**Keywords** Resizing · Reusing · Construction industry · Timber · Plywood

## 1 Introduction

In this paper, we will discuss the proper resizing and reusing of the construction formwork materials. Formwork materials are the materials that are used for creating a mold for pouring the concrete during the construction. Various types of formwork are used in the construction of the project such as reusable plastic formwork, steel formwork, and timber formwork. Context of the paper contains description for advantages of resizing and reusing construction formwork materials, its evaluation, and cost advantages with illustrated diagram. The study shows that about 70% of plastic and other waste materials could be controlled with the use of construction formwork. Reuse of form materials helps to enhance the quality of the project as well as also affect the cost of the project. In this paper, we will discuss the evaluation of

B. Mukhopadhyay (✉) · R. Bose · S. Roy
Brainware University, Kolkata, India
e-mail: bnmukhopadhyay@gmail.com

R. Bose
e-mail: drrb.cs@brainwareuniversity.ac.in

S. Roy
e-mail: hod.css@brainwareuniversity.ac.in

 
S. Patnaik et al. (eds.), *Smart and Sustainable Technologies: Rural and Tribal Development Using IoT and Cloud Computing*, Advances in Sustainability Science and Technology,
https://doi.org/10.1007/978-981-19-2277-0_34

formwork such as plastic, wood, and timber. In the paper, we will find the timber is the most common formwork that is used by contractors. The advantage of resumes and resizing of formwork will also be discussed in the paper.

## 2 Advantage of Resizing and Reuse of Construction Formwork Materials

The reusable plastic formwork used in the construction can provide a simple structure by building high variable. The reuse of the plastic formwork in the construction is suitable for the robust and lightweight panel. In the mass housing schemes reusing of plastic, formworks help in providing quality base materials. According to Gálvez-Martos et al. [3], reusing the form materials at the construction sites enhances the efficiency of each project as well as affects the cost of the project. Uni-Span formwork solutions are the formwork designs by the engineers that are used in the energy and industrial sector for enhancing the quality of the project [6]. Resizing the formwork in the construction helps to enhance the quality of the new project that gets constructed. Resizing the formwork also decreases the labor requirements and total cost of the project. Resizing the formwork in the construction can decrease the construction wastage as well as helps in maintaining sustainability.

## 3 Evaluation of Construction Formwork Materials

Evaluation of the framework materials enhances the quality of the formwork materials that are used in the construction. Evaluation of the formwork identifies the quality and value of the materials. In this segment, evaluation is referred to the assessment of the formwork materials such as plastic, wood steel that helps us to measure the improvement in their performance while using this formwork in the construction. Evaluation of these formwork increases the durability of these materials as construction homework. Wood is the formwork that is used as formwork and it is composed of the elements of the fibers and fibrous materials that occur between the bark and the pith. Timber is the form of wood that is used as formwork materials in construction. Plastic is the formwork that is used in the construction and consists of synthetic and semi-synthetic organic compounds that can be molded into solid objects. In the construction of roads, buildings, and bridges the plastic works are done through the plastic formwork. These formworks are used to form the structural shape of the concrete such as beams, slabs, and columns (Fig. 1).

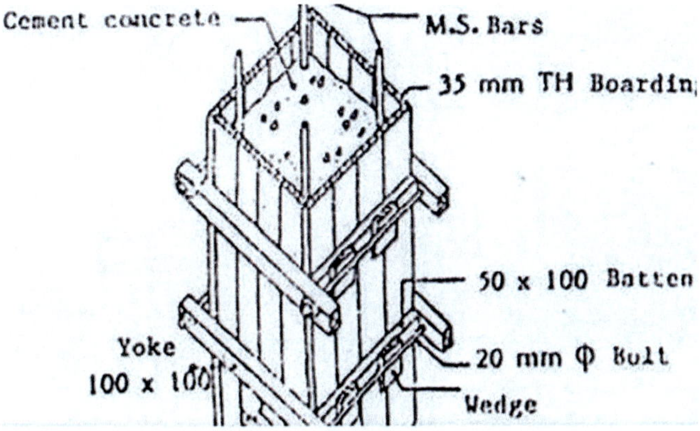


**Fig. 1** Timber formwork used in building. *Source* Özay et al. [7, p. 890]

## 4 Understanding Resize and Reusing the Formwork Materials in the Construction with Diagram

The given figures identified the various factors for the use of timber formwork. This framework helps the timber formwork to reuse their timber formwork in the construction. It was found that the efficiency of workmen, working attitude of workmen, and formwork stripping process have a great impact on the reuse of timber formwork (Fig. 2).

## 5 Why I Am Doing Resize and Reuse Formwork Materials from One Construction Project to Another Construction Project

Resize and reuse of the formwork materials from one construction project to another construction project have several advantages. The formwork materials that have been used at the one project can be reused for the other project that will save the cost of the project as well as enhance the quality of the other project. The formwork materials help the constructor or the contractor to cast these materials in the other project to enhance the structural quality of wall, floors, and roofs of the building. Resizing and reusing these materials also reduce the cost of another project [1].

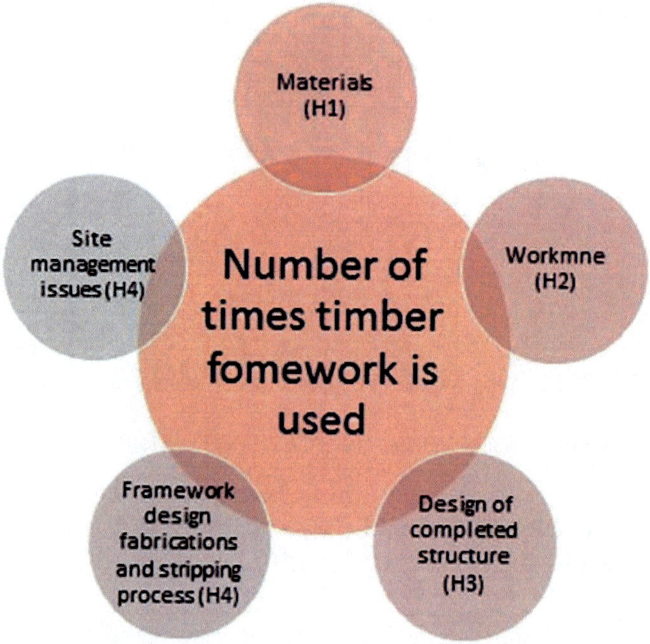


**Fig. 2** Reusing and resizing of timber homework

## 6 Formwork Materials of the Construction Industry

Timber, plywood and steel are mostly used formwork materials of the construction industry. The growth and development of the construction industry depend on these formwork materials. In addition, it is noticed that concrete structure is maintained properly with the involvement. It is important to collect proper information about the usability of these formwork materials. In addition, these formworks can play a crucial role in maintaining the support of concrete element properly. Removable and permanent factors of the construction industry are maintained with the implementation of these formwork components. Timber is an effective formwork material of the construction industry, this formwork is easily dismantled. It is easy to replace timber easily [5].

Plywood is also an effective formwork material; low cost is treated as an effective advantage of plywood, which is crucial for the construction industry. The smooth finish of the construction industry is maintained properly by plywood. Strength and durability are two effective advantages of this formwork material. Steel is treated as a common and important formwork component of the construction project. Domínguez et al. [2] stated that reusability is treated as an effective advantage of these formwork materials. However, the cost of steel is very high. The usability of steel is decreased due to these disadvantages.

## 7 Cost Control Advantages Using Resizing and Reusing

Lower expense is maintained properly with the involvement of resizing and reusing processes. On the other hand, the operational efficiency of the construction industry is maintained with the involvement of operational efficiency and streamlines technology. The cost of the construction project is maintained properly with the help of these resizing and reusing aspects. With the help of this factor, the construction industry can maintain its involvement in the competitive market. On the other hand, resizing and reusing processes can play a crucial role in maintaining execution cost. Along with this, it is noticed that resizing and reusing factors of the construction industry can play a crucial role in maintaining inventory management of the construction industry [4]. Development of construction project depends on this factor properly.

## 8 Conclusion

Based on this study, it is concluded that reusing and resizing of construction material can maintain cost effectiveness of construction industry. Reason behind resizing and reusing of construction formwork material is concluded through the context of the study. The study shows concluded that near about 70% of plastic could be replaced with the effective use of construction formwork. Result of the study that shows the reason for using construction formwork is best option to reduce the usage of permanent materials. Implementation of proper construction formwork can maintain removal and permanent materials of construction industry. Execution cost of construction industry is controlled successfully with the implementation of resizing and reusing processes. Plastic waste of construction project is reduced successfully through reusing and resizing process.

## References

1. M.C. Anjoletto Filho, S.M.B. Serra, Comparison between collective protective systems in Brazil: safety platforms and safety net type V. SN Appl. Sci. **2**(12), 1–12 (2020)
2. I. Domínguez, S. Ward, J.G. Mendoza, C.I. Rincón, E.R. Oviedo-Ocaña, End-user cost-benefit prioritization for selecting rainwater harvesting and greywater reuse in social housing. Water **9**(7), 516 (2017)
3. J.L. Gálvez-Martos, D. Styles, H. Schoenberger, B. Zeschmar-Lahl, Construction and demolition waste best management practice in Europe. Resour. Conserv. Recycl. **136**, 166–178 (2018)
4. V. Hernandez-Chover, Á. Bellver-Domingo, F. Hernández-Sancho, The influence of oversizing on maintenance cost in wastewater treatment plants. Process Saf. Environ. Prot. **147**, 734–741 (2021)
5. O. Kontovourkis, P. Konatzii, Environmental and cost assessment of customized modular wall components production based on an adaptive formwork casting mechanism: an experimental study. J. Clean. Prod. **286**, 125380 (2021)

6. G. López Garulo, Architecture and ceramics, hand in hand. Bachelor's thesis, Universitat Politècnica de Catalunya, 2020
7. E. Özay, G. Delaney, N. Traverse, A. Pries, Making Bibelot: casting material research within cultural frameworks. Front. Archit. Res. **8**(2), 121–135 (2019)

# Design and Development of a Di-Wheel Multipurpose Robot for Smart Agriculture Application

**Swagat Kumar Samantaray and Shasanka Sekhar Rout**

**Abstract** This paper presents an innovative development of a multi-functional mobile robot for various agriculture applications. This is archive by design of a mobile robot called Di-Mobile Wheel. The Di-Mobile Wheel Robot is prototype design for exploiting and applying in agriculture application for precision farming. The Di-Mobile wheel robots balancing robots current research may provide the future locomotion for agricultural application. The concept of inverted pendulum principle is utilized to provide suitable control and provide an autonomous navigation with great balance. The Di-Mobile Wheel robot operates safely in unstructured farming environments. A Kalman Filter algorithm is used for providing optimal state by prediction from previous and current state of measurement. The auto-calibration includes automatic load calculation and set point. The design of the whole system has different phases, such as robot assembly, and its circuit, mounting of different sensors, implementation of algorithms through programming. The Di-Mobile Wheel robot goes through some testing and those results have been reported.

**Keywords** Di-wheel robot · Kalman filter algorithm · PID controller · Smart agriculture

## 1 Introduction

Agriculture plays an important role toward the growth in the society. And current computer revolution and latest developments in the field of electronics provided an opportunity to shift from conventional way of farming to the smart farming techniques by use of robots and other appropriate technologies which helps the farmers to improve their skills and livelihood. The improved data acquisition and processing

S. K. Samantaray (✉)
VIT Bhopal University, Kotri Kalan, Bhopal, India
e-mail: swagatkumarsamantaray@vitbhopal.ac.in

S. S. Rout
GIET University, Gunupur, India



S. Patnaik et al. (eds.), *Smart and Sustainable Technologies: Rural and Tribal Development Using IoT and Cloud Computing*, Advances in Sustainability Science and Technology,
https://doi.org/10.1007/978-981-19-2277-0_35

power of system have increased the research in the direction of autonomous robots and its applications especially in agriculture [1]. The affordable autonomous robots will be improving the farming techniques and farmers will be befitted by using such technology [2–4]. Agricultural robots are getting more attention as earlier because of its advantages over manual method of farming [5]. This paper, it presented a hardware and software co-design for a Di-wheel robot which has the ability to use in the agriculture field for various task such as seed sowing, soil parameter monitoring, plant watering etc.

## 2 Di-Wheeled Robot

The Di-wheel robot is a one type of symmetrical shape with several layers with various components installed and integrated. The robot consists of controller, sensors, printed circuit board for different circuitry. The locomotion is provided through two wheels and motors mounted on the robot chassis. The sensors on the robot provides the environmental condition around it. Accelerometers and gyroscopes provides the stability and provides feedback to control system. So the information corresponding to position and direction is also recorded with suitable feedback. The accuracy improved by implementing Kalman filter provides a suitable approximation with data from different sensors. The computational power of controller allows the robot to take suitable feedback from sensors and is able to balance itself. The controlling of robot can be done using hardware or software. A hardware-based system is designed and appropriate feedback strategy has been applied with software-based control. The linear systems are implemented with the state-space control and Proportional Integral and Derivative (PID) controller [6–8]. By incorporating PID controllers the motor driver signals to the motor can be controlled. Recent development in the field of non-liner control have fuzzy and neural networks control approach.

### 2.1 *Inverted Pendulum Theory*

For a Di-wheel robot the reliability and its control parameters need to be analyzed through some mathematical model. In the Di-wheel robot the inverted pendulum principle is implemented. The friction coefficient is neglected as the robot will work on the different uneven surface and also it will simplify the circuit requirements [9]. Here the movements of wheels are taken as unison to provide stability. The main principle is to keep both the wheels in such a way that it makes balance. As the robot moves forward then to provide stability the wheel need to synchronize and move forward and return back to the beneath of the robot chassis mass. The system dynamics are associated with the mathematical problem and other dynamics. The mass of the cart and the point mass at the end of the rod are denoted by '$M$' and '$m$'. The length of the rod denoted as '$l$' as shown in Fig. 1 [10].

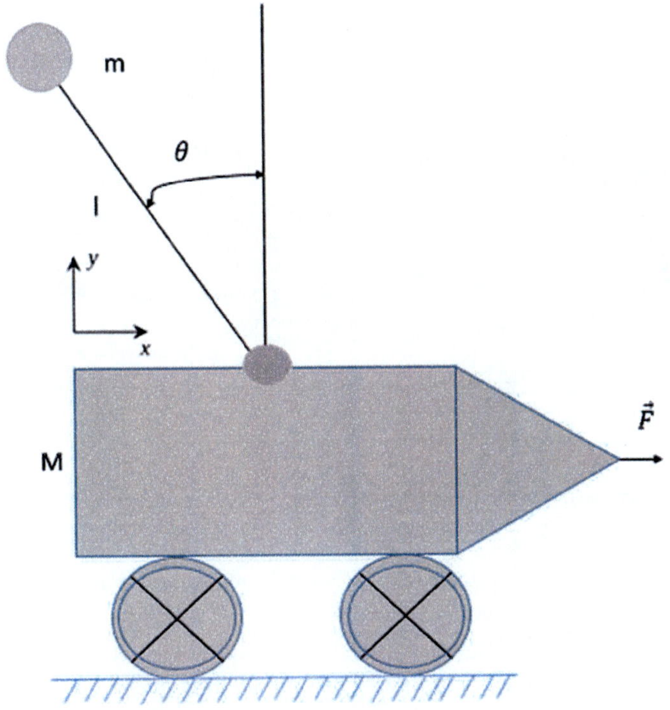


**Fig. 1** Inverted pendulum principle

## 3 Design Concept

The interfacing of different electronic components with Arduino used in this Di-mobile wheel robot is shown in Fig. 2. The interface of the inertia measurement unit (IMU) with Arduino controller and stepper mother driver has been done to get data from sensors and control the working of robot. The Bluetooth module (HC05) is also connected for remotely controlling the robot, through mobile or remote. The system is powered by 12 V power supply.

The Di-Mobile Wheel robot sectored in different layers of symmetrical shape, where the components such as controller, printed circuitry board, sensors and power supply are kept in layered structure. The summery of competes is listed in Table 1. It possesses a two-layered mechanical structure which includes chassis layer and the control layer. The chassis layer consists of the drive mechanism and the battery power source. Each modular wheel is connected with a DC motor and a harmonic reducer along the rotational axis of the wheel. The Control layer consists of controller, different sensors, motor drivers and implemented PID algorithm.

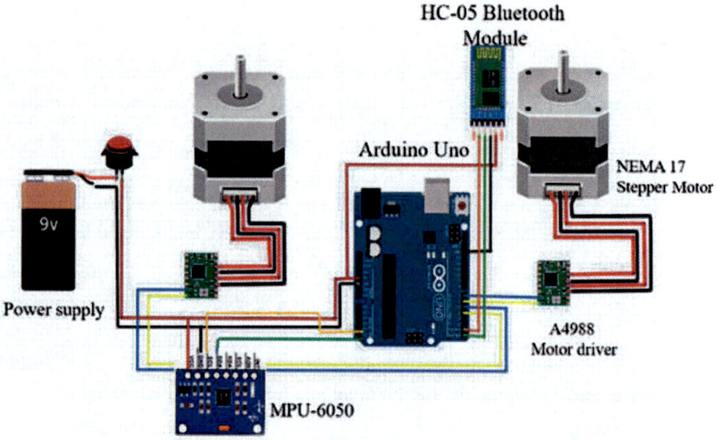


**Fig. 2** Interface diagram of the Di-mobile wheel robot

**Table 1** Summary of components used for design of Di-mobile wheel robot

| Components | Module | Description |
|---|---|---|
| Microcontroller | Arduino Uno | Arduino Uno is used as a microcontroller board based on the ATmega328P where all the program is stored and executed |
| Stepper motor | NEMA 17 | Bipolar stepper motor is used for traction control and mobility |
| Stepper motor driver | A4988 | Stepper motor driver is used to control the motors |
| Bluetooth module | HC05 | Bluetooth module is used for wireless communication to command the bot for movement |
| Inertia measurement unit | MPU6050 | Module is used to obtain the values of gyro, accelerometer |

## 4 Methodology of Di-Wheel Robot

The robot is capable of carrying out the various tasks through controlling algorithm. The accelerometers sensor used to monitor the change in speed of the robot and provide information to controller. Another sensor gyroscope is used to know the rate of rotation. The angle of tilt in the robot can be determined using data from gyroscope which provides a balancing state to the robot. The PID controller is used to controlling the motion of robot [11]. The correction factor need to incorporate in to the algorithm with following equation.

$$\mathrm{Cr} = \mathrm{Kp} * (\mathrm{Dev}) + \mathrm{Ki} * \mathrm{Terr} + \mathrm{Kd} * (\mathrm{Dev}) - \mathrm{Dev(previous)} \tag{1}$$

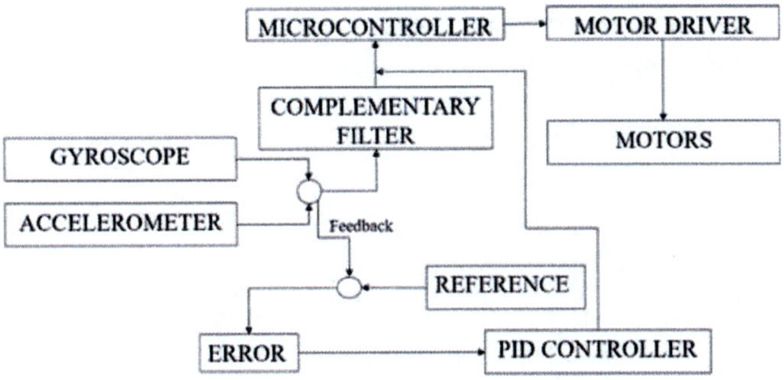


**Fig. 3** Overall controlled mechanism with respect to sensors and PID controller

$$\text{Total Error} = \text{Total Error} + 1 = \text{correction} \tag{2}$$

The operation of robot with respect to sensor and controlled feedback has been shown in Fig. 3.

## 5 Algorithm of Di-Mobile Wheel Robot

- Setup: Raw yaw and pitch values are taken from the gyro at rest and average offset value is calculated.
- Loop: MPU6050 settings: Accelerometer Raw data is pulled from the MPU5050 using as in function and current angle is recorded.
- When the accelerometer angle is zero, set variable to start PID and angle is stored in a variable A.
- Gyro raw data is pulled from the MPU6050 and pitch value is added to the variable A.
- Complementary Filter: To reduce the instability in gyro values, the accelerometer value is added to the variable A multiplying constants PID calculations.
- Error is calculated by calculating the difference between the desired angle (set point) and the actual angle.
- Then the I controller value is calculated. Calculate the PID output and limit the PI controllers.
- To control the direction of the robot, the set points and the turning speed are changed according to the received data from the sender, and set points are achieved.
- Then the right and left motor pulses are calculated according to the PID outputs.

| Angle | PID Output |
|---|---|
| SET POINT = Zero Degree | |
| -0.05 | -7.33 |
| 0.01 | 0 |
| 0.29 | 12.49 |
| 0.55 | 25.27 |
| 0.78 | 32.27 |
| 0.69 | 22.58 |
| 0.36 | 0 |
| 0.11 | -10.52 |
| 0.15 | 0 |
| 0.33 | 20.83 |
| 0.36 | 27.56 |
| 0.05 | 8.49 |
| -0.4 | -24.72 |
| -0.55 | -24.05 |

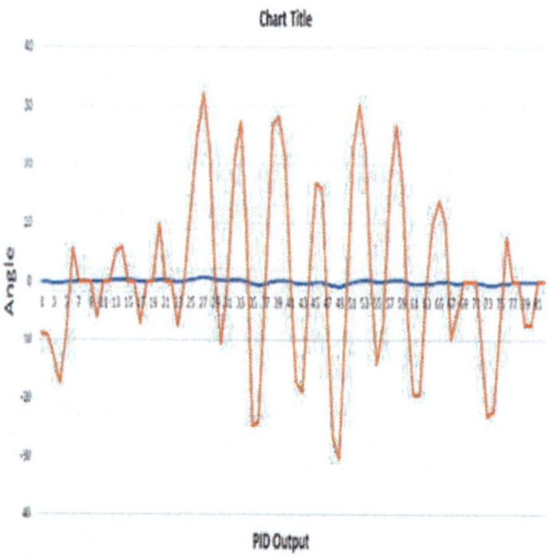


**Fig. 4** Change in angle with PID output

## 6 Results and Comparison

The Arduino IDE is used to compile and upload the program to the microcontroller. Arduino uses its serial monitor feature to debug the Arduino software sketches or to view data sent by a working sketch. The data from angle and PID output is listed. A set of data plot the graph between Angle and PID output, is shown in Fig. 4. From the graph it shows that as angle changes the output also changes to compensate the change in set-point which is at zero degree.

## 7 Conclusion

The Di-wheel robot provided a suitable control and balancing which is used for farming application. This paper presnted the design and implementation of the Di-Mobile wheel robot which can control and balance for farming application. The algorithem used PID control concept with Kalman filter. The design has two parts one is hardware setup including mechnical parts, controller and sensors. The second part is development of the algorithm to control the robot. The structure modeling and hardware interfacing has been done till now but subsequent development and simulation will be continued in future work. The Di-wheel robot can be used in small and medium size farms also the mobile robot can be integrated in an intelligent manufacturing system for material conveying.

## References


1. M. Li, K. Imou, K. Wakabayashi, S. Yokoyama, Review of research on agricultural vehicle autonomous guidance. Int. J. Agric. Biol. Eng. (IJABE) **2**(3), 1–16 (2009)
2. G.J. Nistala, Development of an inexpensive guidance system for agricultural purposes. Thesis, Louisiana State University, Baton Rouge, 2006
3. J. Koleosho, C.M. Saaj, System design and control of a di-wheel rover, in *Annual Conference Towards Autonomous Robotic Systems* (Springer, Cham, 2019)
4. A. Chouhan et al., Fabrication of di-wheel by computational and experimental approach. Mechatron. Appl. Int. J. (MECHATROJ) **2**(1) (2019)
5. B. Åstrand, A.J. Baerveldt, An agricultural mobile robot with vision-based perception for mechanical weed control. Auton. Robot. **13**(1), 21–35 (2002)
6. M. Li, K. Imou, K. Wakabayashi, S. Yokoyama, Review of research on agricultural vehicle autonomous guidance. Int. J. Agric. Biol. Eng. **2**(3), 1–16 (2009)
7. L. Emmi, M. Gonzalez-de-Soto, G. Pajares, P. Gonzalez-de-Santos, New trends in robotics for agriculture: integration and assessment of a real fleet of robots. Sci. World J. (2014)
8. P. Huang, Z. Zhang, X. Luo, Feedforward-plus-proportional–integral–derivative controller for agricultural robot turning in headland. Int. J. Adv. Rob. Syst. **17**(1), 1729881419897678 (2020)
9. S. Paul et al., Fuzzy tuned PID controller for envisioned agricultural manipulator. Int. J. Autom. Comput. 1–13 (2021)
10. K. Huang, M. Wang, S. Zhen, Udwadia-Kalaba equation based adaptive robust control for two-wheeled inverted pendulum system: underactuation and uncertainty. Int. J. Control Autom. Syst. **19**(7), 2420–2430 (2021)
11. S. Khatoon et al., Design kinematics and control for a differential drive mobile robot, in *Renewable Power for Sustainable Growth* (Springer, Singapore, 2021), pp. 189–196

# Proceedings of the International Conference on Smart and Sustainable Technologies (ICSST 2021)

This book features a collection of high-quality research papers presented at the International Conference on Smart and Sustainable Technologies: Rural and Tribal Development Using IoT and Cloud Computing held at GIET University, Gunupur, India, during December 16–18, 2021. Contributions are from smart and sustainable technologies that can support existing and future applications and services, which cover topics such as Internet of things, big data, cloud computing, smart communication systems, sustainable power systems, smart healthcare management, AI-based diagnosis system, biometrics, robotics for precision agriculture, and smart and sustainable technology for smarter living.



S. Patnaik et al. (eds.), *Smart and Sustainable Technologies: Rural and Tribal Development Using IoT and Cloud Computing*, Advances in Sustainability Science and Technology,
https://doi.org/10.1007/978-981-19-2277-0

# Author Index






S. Patnaik et al. (eds.), *Smart and Sustainable Technologies: Rural and Tribal Development Using IoT and Cloud Computing*, Advances in Sustainability Science and Technology,
https://doi.org/10.1007/978-981-19-2277-0